Chemical Analysis

Modern Instrumental Methods
and Techniques

Chemical Analysis

Modern Instrumental Methods and Techniques

English Edition

Francis Rouessac

Annick Rouessac

Translated by

Michel Bertrand and Karen Waldron

University of Montreal

JOHN WILEY & SONS, LTD

Chichester · New York · Weinheim · Brisbane · Singapore · Toronto

English language translation copyright © 2000 John Wiley & Sons, Ltd, The Atrium, Southern Gate, Chichester, West Sussex PO19 8SQ, England
Telephone (+44) 1243 779777

Email (for orders and customer service enquiries): cs-books@wiley.co.uk

Translated into English by Karen Waldron and Michel Bertrand.
Problems/Solutions translated by Steven Brooks.

First published in French © 1992 Masson
2nd edition © 1994 Masson
3rd edition © 1997 Masson
4th edition © 1998 Dunod

Reprinted September 2001, October 2002, January 2004, November 2004, August 2005

This work has been published with the help of the French Ministère de la Culture – Centre national du livre.

Other Wiley Editorial Offices

John Wiley & Sons Inc., 111 River Street, Hoboken, NJ 07030, USA

Jossey-Bass, 989 Market Street, San Francisco, CA 94103-1741, USA

Wiley-VCH Verlag GmbH, Boschstr. 12, D-69469 Weinheim, Germany

John Wiley & Sons Australia Ltd, 33 Park Road, Milton, Queensland 4064, Australia

John Wiley & Sons (Asia) Pte Ltd, 2 Clementi Loop #02-01, Jin Xing Distripark, Singapore 129809

John Wiley & Sons (Canada) Ltd, 22 Worcester Road, Etobicoke, Ontario M9W 1L1

Wiley also publishes its books in a variety of electronic formats. Some content that appears in print may not be available in electronic books.

Library of Congress Cataloging-in-Publication Data
Rouessac, Francis.
   [Analyse chimique, French]
   Chemical analysis : modern instrumental methods and techniques / Francis Rouessac,
Annick Rouessac.
      p. cm.
   Includes bibliographical references and index.
   ISBN 0-471-98137-0 (alk. paper) – ISBN 0-471-97261-4 (pbk. : alk. paper)
   1. Instrumental analysis.   I. Rouessac, Annick.   II. Title.

QD79.I5 R6813 2000
543–dc21   99-058872

British Library Cataloguing in Publication Data
A catalogue record for this book is available from the British Library

ISBN-10: 0-471-98137-0 (cloth)   ISBN-13:978-0471-98137-4 (cloth)
ISBN-10: 0-471-97261-4 (pbk)   ISBN-13:978-0471-97261-7 (pbk)

Typeset in 10/12pt Times by C.K.M. Typesetting, Salisbury, Wiltshire
Printed and bound in Great Britain by Biddles Ltd, King's Lynn, Norfolk
This book is printed on acid-free paper responsibly manufactured from sustainable forestry, in which at least two trees are planted for each one used for paper production.

Table of contents

Chapter 10   Infrared spectroscopy   161

Chapter 11   Ultraviolet and Visible absorption spectroscopy   189

Chapter 12   Fluorimetry   221

Chapter 13   X-ray fluorescence spectrometry   237

Chapter 14   Atomic absorption and flame emission spectroscopy 253

Chapter 18 Potentiometric methods 347

Chapter 19 Coulometric and voltammetric methods 359

Chapter 20 Some sample preparation methods 377

Foreword to the French edition

Several years ago, colleagues specialising in Analytical Chemistry explained to me that I was confusing their area of activity with chemical analysis. Analytical Chemistry was a science disembodied and remote from that of base applications, those of chemical analysis (note that one term is capitalised while the other is not). However, I did not believe them and I will never know whether the Analytical Chemistry course that I introduced at Strasbourg 15 years ago was either the one or the other. This book is either one or the other. Its subtitle is, however, more explicit: 'Modern Instrumental Methods and Techniques'. However, this title is not entirely informative. Chemical analysis has always been carried out with instruments that have never ceased being modernised. The analytical balance has for a long time been the most important one of these instruments: it is still essential (and it is hoped that a future edition of this book will describe what the balance has become – it would only require a few pages).

Today a variety of complex methods are used in laboratories. This book describes these methods in a simple, systematic, yet precise, way. Well-balanced introductions are provided for all chromatographic separation techniques and for methods of structure identification. Its many explanations make this book an indispensable teaching aid for aspiring analytical chemists.

A complimentary foreword must invoke the possibility of future editions. In future versions of this book, I hope that F. Rouessac and his wife will include a few additions: on the analytical balance, as I have mentioned above, but also on electrochemical methods, on sampling – so rarely discussed – and on the treatment of numerical data (archiving of primary data, statistical treatment, graphical representation, etc.). Or perhaps, in a volume 2?

23 April, 1992.

Foreword to the English edition

In Dutch, 'scheikunde' (German: 'Scheidekunst' meaning the art of separation) is synonymous with 'Chemistry'. Chemistry would not exist whatsoever without separation, without analysis! Analysis does not exist without 'Instrumentation Methods and Techniques', in other words without the sub-title of the present book!

Instrumentation was, for a long time, rather crude by today's standards: the furnace, the alembic, the separatory funnel, the filter, the balance ... crude and cheap. Today, no modern analytical laboratory is equipped without millions of pounds spent on investments in optical, mass and NMR spectrometers, in high performance chromatographs and in electro-analytical equipment.

How heavy this responsibility is, resting on the shoulders of the analytical chemist! He is the one who, in the first place, is responsible for the forced closing of a dioxin-delinquent waste incineration plant, for the approval of a new non-persistent pesticide, for the demotion of an athlete from his olympic title for having used illegal drugs, for the identification of a criminal by the traces of gunpowder on his hands, for the quantification of environmental contaminants, for the detection of diabetes, or the detection of poisoning, for the establishment and the enforcement of standards used in world trade .... The analyst, with his power to say 'yes' or 'no', is one of the most influential of our contemporaries!

There are many books on analytical chemistry, but there were very few, and rather old ones at that, in French, until Francis and Annick Rouessac published the first French edition of the present book, eight years ago: they had a niche to fill. Four successive editions have confirmed that they had filled it well: their book was simple, highly informative, and it was kept up-to-date. Through its successive improvements, it has become ripe for translation. I am sure the present English version, for which there is no equivalent that I am aware of, will now be useful worldwide to students, as well as to professionals. Fare well, Rouessac and Rouessac!

14 February 2000

Guy Ourisson
President of the French Academy of Sciences

Preface to the French edition

The objective of this book is to provide an overview of the basic knowledge about the most widely and currently used methods in chemical analysis, whether they are qualitative, quantitative or structural. A wide area of applications is covered, including the chemical industry, foodstuffs, environment, pollution and biomedicine.

The techniques reviewed in this book are classified into *separation methods*, *spectral methods* and *other methods*. In each of these, the basic principle of the method is presented and, later, the corresponding instrumental techniques are reviewed. Schematics of several instruments, provided by the principal manufacturers, are also included. Methods of decreasing importance or those rarely used have been omitted to keep the manuscript to a reasonable size.

The text was written as clearly as possible for the benefit of a large variety of students at the Institut Universitaire de Technologie (IUT) (in the departments of chemistry, physical measurements and applied biology), for specialised technicians, for university students at degree level, as well as for graduate-level science students who wish to have fundamental concepts reviewed at a basic level. This book could also be used by people involved in continuous training and industrial applications, where problems of chemical analysis are constantly being faced. The invasion of analysis techniques in the professional sector coupled with the growing complexity of current instruments justifies the training upgrade of personnel who are insufficiently prepared.

This book was conceived also to answer the readership needs in the area of Analytical Chemistry, whether it be for study within this discipline or as a tool used in other experimental sciences and diverse areas. The background knowledge required to profit from this book is essentially that possessed by students in their first years of university. Thus, the authors have limited themselves to fundamental principles and have considered that students already have basic training in mathematics and in the approach to studying physical phenomena. Throughout this book, in-depth studies of phenomena have been avoided in order not to put off the majority of readers at whom this book is targeted. Those interested will, if necessary, be able to consult specialised works without any major difficulties, after having acquired from this book a relatively complete overview of the most currently used methods.

The content of this book also reflects the profound changes in laboratory work habits that result, among other things, in the evolution of material, in the use of computers and in the growing demand for chemical analysis.

To review some 20 techniques in less than 450 pages may appear to be a challenge, each technique could merit a lengthy discussion because of its use in many applications and in a variety of areas. This is part of the reason why the authors have chosen to present only the tools instead of a full description of what these tools can accomplish. The occasional discussion of applications is for illustrative purposes.

This book originates from a course given to the students at the IUT in Le Mans for many years. Some friends and colleagues agreed to read the text of the first edition (in 1992) and provide comments and suggestions. This, the fourth edition, has been completed with two additional chapters directed at electro-analytical methods and a chapter on basic statistics used in analytical chemistry. The translation of this text from French to English is thus directed at an even larger readership.

We would like to thank our sons: Vincent for helping with the illustrations and for his expertise in computers and Xavier for useful technical and practical information.

We would also like to acknowledge the French and foreign societies that have provided us with the correct information for this book. Their assistance has been very precious to us, particularly since the topics treated here are highly related to technology and progress in instrumentation.

We would also like to express our deepest thanks to Guy Ourisson who has honoured us by writing the foreword to this book.

Le Mans, January 2000.

F. Rouessac
A. Rouessac

Preface to the English edition

The book entitled 'Chemical Analysis: Modern Methods and Instrumental Techniques', written originally in French by Professor F. Rouessac and his wife A. Rouessac, has been revised several times and is now in its 4th edition. It is an ongoing project that provides updated versions and increases the usefulness of the manuscript.

The purpose of the work has been to provide basic information on methods of chemical analysis and new instrumental techniques that have been developed and improved in recent years. Its objective is to provide the analyst with a reference manual while providing students with a teaching tool that covers the basics of most instrumental techniques presently used in chemical analysis. It incorporates basic principles, describes commonly used instruments and discusses the main application for most of the analytical techniques.

The book classifies methods of analysis according to three categories: separation techniques, spectroscopy techniques and other methods. It was written for undergraduate students in chemistry but with the view that it may be of interest for students in other disciplines (physics, biology, etc.) where chemical methods of analysis and instrumental techniques are used. Thus, it provides sufficient information to understand the techniques and their application and allows students to find additional information in more advanced works that discuss specialised instrumental techniques in more detail.

Professor Rouessac gathered the material presented in this book during his teaching career at the University of Mans and he has made an effort to integrate theory and practice in a remarkable way. The chapters contain detailed descriptions of instruments and techniques with a few applied examples that are useful to appreciate the scope of the techniques as well as their strengths and limitations in the applied world. The philosophy behind the manuscript is to show that although analytical chemistry and chemical analysis are sometimes considered as different topics, they are inherently intertwined.

Over the years, we have seen a tremendous evolution in chemical analysis. Because of developments in electronics and computer sciences, many new approaches have been developed based on physical measurements and these approaches are now widely used. Nowadays, there is a legion of instrumental techniques that are more sensitive, more selective and can be applied to analytical problems in many areas of science where the structure determination and quantitation of chemical species is needed. For example, physical methods of chemical analysis are being used overwhelmingly in the biological sciences. Moreover, the combination of two or more instrumental analysis techniques had led to the introduction of hyphenated methods that are extremely powerful and require the basic knowledge of the underlying principles. This manuscript provides the essential knowledge for the understanding of these techniques and opens the door to their areas of applications. It also treats some older techniques that maintain their important place in industrial processes.

It has been a pleasure for us to translate Professor Rouessac's work. Although we have been able to translate the technical material relatively precisely, there is a flavor of expression used by the authors in their native language that cannot be transposed, as is usually the case with translations. In spite of this limitation, we believe that the content of this book will be extremely useful to readers that are seeking knowledge and information on chemical analysis and analytical instrumentation.

Michel J. Bertrand, Professor
Karen C. Waldron, Assistant Professor
Department of Chemistry
Université de Montréal
September 1999

Acknowledgements

We are very pleased that the English edition of 'Analyse Chimique' has been published. The authors wish to thank Professor Karen Waldron of the University of Montreal for the very hard work she did in translating the manuscript.

We also gratefully acknowledge Professor Guy Ourisson, President of the French Académie des Sciences who agreed to write a new foreword and also Dr Steve Brooks of Institut Pasteur de Lille who translated the exercises and solutions.

Compared with the French edition we have simply introduced a few revisions and incorporated improved drawings. Some suggested selected web sites are also provided in order to illustrate topics and techniques.

Our special thanks also to the staff of John Wiley & Sons, Ltd (Chichester) for producing a book of pleasing appearance. We greatly appreciate the help of Publishing and Production editors Andy Slade and Dawn Booth who render possible the material realisation of the book. We also thank the reviewers and the proofreader Jane Hammett for her valuable suggestions.

The following is a list of companies that have graciously agreed to provide us with information and documents, a certain number of which have been reproduced in this book:

Agilent Technologies, AII, American Stress Technologies, Anotec, Amptek, Arelco, Asoma, ATI, ATS, Bio-Rad, Bosch, Bruker, Camag, Carbone-Lorraine, Chrompak, Ciba, CTTM, Daiiso Company, Desaga, Dionex, DuPont, EG&G-Ortec, Erba Science, ETP Scientific, Eurolabo, Finnigan, Fisons-Instruments, Galileo, Grasby-Electronics, Hamamatsu, Hamilton, Hewlett-Packard, Hitachi, Imaging Sensing Technology, Jeol, Jenway, Jobin-Yvon, Jordan Valley, Kevex, Labsystems, Leeman Labs, Leybolds, Merck, Metorex, Metrohm, Mettler-Toledo, Microsensor Technology, Nicolet, Ocean Optics, Oriel, Ortec, Oxford Instruments, Perkin-Elmer, PESciex, Pharmacia-Biotech, Philips, Photovac, Polymer Lab, PSS, Rheodyne, Scientec, Scientific Glass Company, Servomex, Shimadzu, Siemens, SMIS, Supelco, Tekmar, Teledyne, Thermo-Optek, Thermo-Quest, Thermo Jarrell Ash, Tosohaas, Varian, VG Instruments, Waters, Wilmad Glass.

The authors, Annick and Francis Rouessac

Introduction

Analytical chemistry is a science close to physical chemistry, which is a branch of *pure chemistry*. The objective of analytical chemistry is essentially to develop and apply new methodology and instrumentation with the goal of providing information on the *nature* and *composition* of matter. Analytical chemistry also allows the determination of a compound's *structure*, either partially or totally, in samples of differing complexity. Finally, part of the role of analytical chemistry is to provide an interpretation of the results obtained. The term *chemistry* is a reminder that analytical chemistry involves the analysis of *chemical elements* and the *defined compounds* derived from these.

From a more applied point of view, analytical chemistry is the basis of *chemical analysis*, which corresponds to the study of the methods and their diverse techniques applied to solving problems of analysis.

The vast discipline of analytical chemistry has implications in all experimental sciences. Its study requires knowledge of many different areas. As a multidisciplinary science, also sometimes referred to as transferable, analytical chemistry calls upon many phenomena, which may be remote from chemistry in the usual sense, in order to provide results. Thus, modern chemical analysis is based on physico-chemical measurements obtained through the use of a variety of instruments, which have greatly benefited from the appearance of microcomputers.

Gradually a tremendous arsenal of processes has been developed, allowing the analyst to respond to an increasing number of diverse demands. Furthermore, the study of modern chemical analysis techniques is far removed from traditional descriptive chemistry. Many analyses are conducted in non-specialised environments, either on site or at simple workbenches. The determination of compounds is currently quite remote from the use of chemical reactions, which are often avoided for many reasons. Former wet chemistry methods, at the origin of the term *analytical chemistry*, have become less important because they lack sensitivity, are lengthy and their precision can too easily be altered by the use of insufficiently pure reagents. Nonetheless, wet chemistry methods are still interesting to study.

Chemical analysis has become indispensable in many areas other than traditional chemistry or para-chemistry. Chemical analysis is now used in the medical sciences (and for diagnostics), in biochemistry, food chemistry, environmental sciences (pollution) and in numerous industrial areas. Analytical chemistry is no longer confined to chemistry. For example, development of regulations in Europe regarding the free circulation of products and the agency in charge of environmental protection both require a great number of analyses; evidence that analytical chemistry will assuredly have a privileged position in the future.

More than ever, analytical chemistry is a lively discipline with applied research, specialised journals and many international events (Pittsburgh Conference, Salon du Laboratoire, Analytica, etc.) attended by internationally recognised manufacturers.

■ One way of determining the *importance* of a technique is to examine the economic statistics related to the sale of corresponding instruments. The diffusion of a technique increases the probability that an analyst will encounter it during his professional career. For example, the statistics above show that chromatographic techniques alone represent more than half of the sales of instruments for *molecular* analysis (as opposed to *elemental* analysis, which is half of this). However, economic indicators are not the only factor taken into account when assigning the importance of a method. For certain analyses, even a rarely used method can be the most important if it represents the only available means of solving a problem.

Molecular Analysis

For the year 2000,
representing 9 billion Dollars

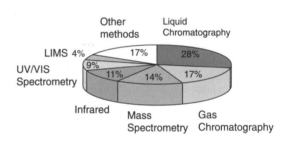

Chemical analysis is the proof of many innovations: the evolution of technologies has led to the development of high-performance instruments allowing new possibilities, notably hyphenated methods and non-destructive methods. *Non-destructive tests* can be conducted on very small samples that do not necessitate extensive sample preparation. Users can finally acquire instruments that meet the quality and precision requirements necessary to obtain certification. This latter requirement is an important if not a sufficient condition to officially recognise the quality of results produced by a laboratory. Certification procedures are enforced by a number of testing bodies all over the world.

Being an analyst requires scientific competence, austerity and honesty. The analyst must not only understand the technique but also understand the chemical system under examination in order to avoid errors.

■ A given compound can often be measured by several methods. The choice of which method to use is not an easy one because it involves the knowledge of all parameters and a number of questions have to be answered:

— Are qualified personnel available to conduct the analysis?

— Must the analysis be repetitive?
— What is the precision needed?
— Does it require a partial or complete analysis of the sample?
— Is the analyte a major component (1 to 100%), minor component (0.01 to 1%) or trace level component (less than 0.01%) of the sample?
— What is the cost of analysis?
— Must the sample be recovered after measurement?
— How long will the analysis take?
— What will the reliability of the results be for the method chosen?
— What are the consequences of a possible error in measurement?

When confronted with an analysis, the problem must be tackled methodically. The science called *chemometrics* is aimed at helping to find the best method required for solving an analytical problem as a function of imposed constraints and according to three different directions: methodology, data treatment and interpretation of results.

Taking into account the nature of the analyte to be determined, the starting point consists of *choosing a method* of analysis: the spectroscopic method, electrochemical method, separation method, etc.

The second step concerns that of *choosing a technique*. For example, if chromatography has been chosen as the method, will gas-phase or liquid-phase chromatography be better?

The third decision to make concerns the *choice of sample procedure* and the sample preparation method that will be required to obtain a good result.

As far as experimental *protocol* is concerned, this will correspond to the mode of operation chosen. The protocol is essentially the analysis recipe and is generally a process controlled by predefined standards. This procedure involves standardisation at each stage of analysis from sample preparation to experimental measurement.

Finally, the results are presented and the raw data from analysis is archived, if possible, as non-rewritable computer files. All of these important aspects of an analysis can be found in official texts under the heading Good Laboratory Practice or GLP.

Good laboratory practice

In instrumental analysis, results produced by an instrument are too readily accepted, when provided in numerical form, because we have a tendency to confuse the precision of a number with that of a measurement. Unfortunately, in chemical analysis, as is the case elsewhere, an error can have serious consequences. For this reason, the heads of analysis laboratories have been essentially forced to redefine the norms that define laboratory quality.

On 9 March, 1990, a series of general recommendations on chemical analysis appeared for the first time in the *Official Journal of the French Republic* (No. 90–206, p. 2891). These recommendations, named *Good Laboratory Practice* (GLP), apply not only to chemical analysis but also to a number of other areas.

GLP typically includes not only norms on the organisation of the laboratory but also on several analytical conditions such as planning, procedure, control, recording and dispersion of the results.

This decree takes into consideration the entire analysis environment necessary for quality assurance. Thus, it describes how the material organisation of a laboratory should be made, what the responsibility of personnel should be at every level of analysis as well as general rules regarding the content of a study and associated controls.

A series of directives specify that raw data, obtained most often in computerised form on modern instruments, have to be kept for 10 years. The same applies to the plans of the analytical study, reference standards, samples and the final report.

The archiving of raw data becomes a necessity when the end results depend on data treatment calculations carried out by software whose operations on the data are often unknown to the user. For this reason, more and more instruments are designed with built-in replication of the raw data as binary files that cannot be modified, in order to comply with regulations.

This decree is an international consensus to recognise studies carried out in different countries. Any laboratory wanting to conform to GLP can ask for an inspection. If it passes the inspection, the laboratory is declared GLP compliant, which allows it to operate within the scope of the European community regulations. However, GLP compliance does not give a laboratory a label of quality that is valid for all future operations.

In a way, these norms correspond to the ideal organisation of a laboratory. Certain rules are difficult to respect, particularly those of recording the flow of information acquired by instruments, which would require enormous amounts of computer memory. As a result, no laboratory presently has the capacity to respond 100% to the norms of archiving all raw data acquired by an instrument. Nevertheless, GPL represents a charter of laboratory operations, which has the advantage that it defines problems, provides answers and gives responsibilities to every person implicated in each stage of analysis.

SEPARATION METHODS

CHAPTER 1

General aspects of chromatography

Chromatography, the process by which the components of a mixture can be separated, has become one of the prime analytical methods for the identification and quantification of compounds from a liquid or homogeneous gas phase. The principle is based on the concentration equilibrium of the compounds of interest between two phases; one of which is called stationary because it is immobilised in a column, the other of which is called mobile because it is the transport mechanism through the system. The differential migration of compounds through the column leads to their separation.

Of all the analytical techniques, chromatography is the one with the widest scope of applicability. The sales of chromatographic instrumentation is a sector that represents at least half of all the sales of analytical equipment and material in the world.

1.1 General concepts of analytical chromatography

Chromatography is a physico-chemical process that belongs to fractionation methods in the same vein as distillation, crystallisation or fractionated extraction. It is believed that the *separation method* in its modern form originated at the turn of the century from the work of Tswett to whom we attribute the terms *chromatography* and *chromatogram*.

Preparative chromatography, which is used on a large scale for the separation and purification of components of a mixture, has progressed into a stand-alone *analytical* technique with impressive performance, and is now used for microseparation.

The use of analytical chromatography was limited until 1940, when it was discovered that an optical device, placed at the end of a column, could identify the migration time of the components without the need for their collection. Later, the development of new, sensitive detectors made possible the miniaturisation of the column thus leading to the analysis of very small quantities (of a few nanograms). The detector signal, which is registered in continuum, leads to a chromatogram that indicates the variation of the composition of the eluting phase with time.

When the column is used with reliable accessories, the chromatogram is fairly reproducible and the parameters of separation very well controlled. Thus it is possible to obtain, during successive analyses of the same sample conducted within a few hours, chromatograms that can be superimposed to within a second.

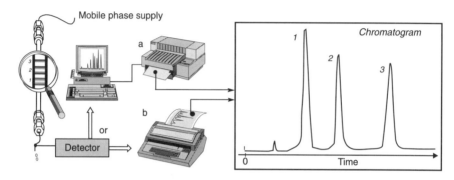

Figure 1.1—*Chromatograph and chromatogram.* The chromatogram obtained from the variation of an electrical signal as a function of time is often reconstructed from values that are digitised and stored to a microcomputer and then reproduced with the desired format (**a**, printer). Until recently, a chromatogram was obtained on-line by measuring the variation of the voltage and reproduced on a recording device (chart recorder or integrator, **b**). The reconstructed graph in the figure shows the separation of a mixture of at least three components in solution, injected from a solvent (the first peak, not numbered, belongs to the solvent).

1.2   Classification of chromatographic techniques

Chromatographic techniques can be classified into three categories depending on the *physical nature* of the phases, on the *process* used, or on the *physico-chemical phenomenon*, which is at the basis of the Nernst distribution coefficient K, also defined as:

$$K = \frac{C_S}{C_M} = \frac{\text{concentration of solute in the stationary phase}}{\text{concentration of solute in the mobile phase}} \qquad (1.1)$$

Certain phases combine distribution coefficients of very different natures. The K values are much higher (e.g. up to 1000) if the mobile phase is a gas and the stationary phase is a condensed phase. The classifications that follow give priority to the nature of the phases present.

1.2.1   Liquid–solid chromatography (LSC)

This is the earliest type of chromatography. The mobile phase is a liquid and the stationary phase is a solid. This category, which is widely used, can be subdivided depending on the retention phenomenon.

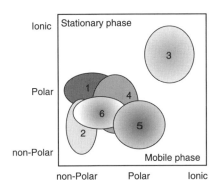

Ionic
Polar
non-Polar

non-Polar   Polar   Ionic

1 - Liquid–solid adsorption chromatography

2 - Gas–liquid partition chromatography (GC)

3 - Ion chromatography

4 - Liquid/liquid partition chromatography

5 - Size exclusion chromatography

6 - Supercritical chromatography

Figure 1.2—*Classification of some chromatographic techniques as a function of the polarity of the stationary phase*. The existence of a wide range of techniques permits the most suitable one to be selected for a specific analytical problem.

— **Adsorption chromatography.** This chromatographic technique is best known because of its use in the last century as a preparative method of separation. Stationary phases have made a lot of progress since Tswett, who used calcium carbonate or sugar. The separation of organic compounds on a thin layer of silica gel or alumina with solvent as a mobile phase are examples of this type of chromatography. Solutes bond to the stationary phase because of physisorption or chemisorption interactions. The physico-chemical parameter involved is the *coefficient of adsorption*.

— **Ion chromatography.** The mobile phase in this type of chromatography is a buffered solution and the stationary phase consists of spherical particles of a polymer, micrometres in diameter. The surface of the particles is modified chemically in order to generate ionic sites. These phases allow the exchange of their mobile counter ion, with ions of the same charge present in the sample. This separation relies on the *coefficient of ionic distribution*.

— **Molecular exclusion chromatography.** The stationary phase in molecular exclusion chromatography is a material containing pores, the dimensions of which are chosen to separate the solutes present in the sample based on their molecular size. This can be perceived as a molecular sieve allowing selective permeation. This technique is known as gel filtration or gel permeation, depending on the nature of the mobile phase, which is either aqueous or organic. The distribution coefficient in this technique is called the *coefficient of diffusion*.

1.2.2 Liquid–liquid chromatography (LLC)

In this chromatographic technique, the stationary phase is a liquid *immobilised* in the column. It is, therefore, important to distinguish between the inert support which only has a mechanical role and the stationary phase immobilised on the support. The impregnation of a porous material, the simplest way to immobilise a liquid, is abandoned when the mobile phase is a liquid because there is an elevated risk of losing the stationary phase, which is called bleeding. However, this method can still be used when the mobile phase is a gas (cf. 1.2.3).

In order to immobilise the stationary phase, it is preferable to fix it to a mechanical support using covalent bonds. In spite of the chemical link, the stationary phase still acts as a liquid and the separation process is based on the partition of the analyte between the two phases at their interface. The parameter involved in the separation mechanism is called the *partition coefficient*. This mechanism is comparable to a liquid–liquid extraction between an aqueous and organic phase in a separatory funnel.

1.2.3   Gas–liquid chromatography (GLC)

In this technique, commonly called *Gas phase chromatography* (GPC), the mobile phase is a gas and the stationary phase is a liquid. The liquid can be immobilised by impregnation or bonded to a support, which, in the case of capillary columns, is the capillary inner surface (the partition coefficient K is also involved).

■ In 1941 Martin and Synge suggested replacing the liquid phase by a gas to improve separation. It is at this point that analytical chromatography really began to take off.

1.2.4   Gas–solid chromatography (GSC)

In this technique, the stationary phase is a porous solid (such as graphite or silica gel) and the mobile phase is a gas. This type of gas chromatography demonstrates very high performance in the analysis of gas mixtures or components that have a very low boiling point.

1.2.5   Supercritical fluid chromatography (SFC)

In this technique, the mobile phase is a fluid in its supercritical state, such as carbon dioxide at about $50\,^{\circ}C$ and at more than 150 bars (15 MPa). The stationary phase can be a liquid or a solid. This approach combines the advantages of the techniques explained in sections 1.2.2 and 1.2.3.

1.3   The chromatogram

Chromatographic analysis is based on a chromatogram. This chromatogram is a two-dimensional diagram (Fig. 1.3) traced on chart paper or a screen that reveals, as a function of time, a parameter that depends on the instantaneous concentration of the solute as it exits the column. Time (or alternatively elution volume) appears on the abscissa and the detector signal appears as the ordinate.

The separation occurring in the column leads to a series of peaks that are more or less resolved from one another as they rise from the *baseline*, which is the trace obtained in the absence of analyte. If the detector signal varies linearly with the concentration of analyte, the same variation will occur for the area under the peak in the chromatogram.

A constituent is characterised by its retention time, t_R, defined by the time taken between the moment of injection into the chromatograph and the peak maximum recorded on the chromatogram. In an ideal case, the retention time t_R is independent of the quantity injected. A compound not retained will elute out of the column at time t_M, called the *void time* or the *dead time*[1] (sometimes designated by t_0). The separation is complete when as many peaks are seen returning to the baseline as there are components in the mixture. In quantitative analysis, it suffices to separate only the components that need to be measured.

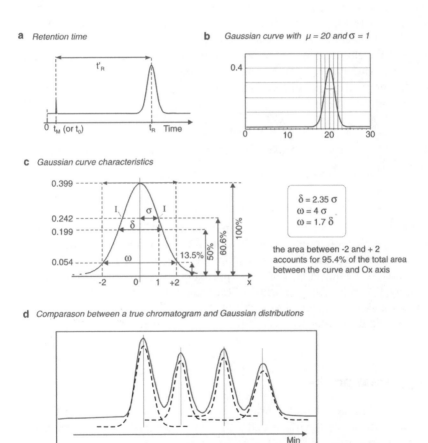

Figure 1.3—*Chromatographic peaks.* a) Retention time; b) Distribution of the peak; c) Significance of the three basic parameters and features of a Gaussian distribution; d) Example of a real chromatogram that shows the elution of components leading to peaks that resemble Gaussian distributions.

---

[1] Symbols used follow IUPAC recommendations – *Pure & Appl. Chem*, 65(4), 819 (1993).

■ The identification of a compound only by its retention time is somewhat arbitrary. A better method consists of using two different techniques that are complementary. Coupling a chromatograph with a second instrument on-line such as a mass spectrometer or an infrared spectrophotometer realises these objectives. These coupled methods, often called hyphenated techniques, allow us to obtain two different types of information that are independent. Therefore, it is possible to determine with more certainty the nature and concentration of the components in a mixture in nanograms or smaller units of measurement.

1.4   The ideal chromatogram and Gaussian peaks

The features of an ideal chromatogram are the same as those obtained from a normal distribution of random errors (Gaussian distribution equation (1.2), cf. 21.3). In keeping with the classical notations, μ would correspond to the retention time of the eluting peak, σ to the standard deviation of the peak (σ^2 represents the variance) and y represents the signal, as a function of time, from the detector located at the end of the column (see Fig. 1.3).

$$y = \frac{1}{\sigma\sqrt{2\pi}} \cdot \exp[-(x - \mu)^2/2\sigma^2] \tag{1.2}$$

Ideal chromatographic peaks are usually described by the probability density function (1.3).

$$y = \frac{1}{\sqrt{2\pi}} \cdot \exp[-x^2/2] \tag{1.3}$$

This is an even function that has a maximum of 0.399 for $x = 0$. The function also has two inflection points $x = \pm 1$ with a corresponding ordinate value of 0.242, which represents 60.6% of the maximum value of the function. The width of the function at the inflection points is approximately 2σ ($\sigma = 1$). In modern chromatography, $w_{1/2}$ represents the width of the peak at the half-height ($w_{1/2} = 2.35\sigma$) and σ^2 the *variance* of the peak. The width of the peak at the base, w, is measured at 13.5% of the height. At this position, if the curve is Gaussian, $w = 4\sigma$ by definition.

1.5   Real peaks

In a real chromatogram the peaks often have profiles that are non-Gaussian. There are several reasons for this. Besides the accepted approximations, such as invariance of the distribution coefficient K with concentration, there are irregularities of concentration in the injection zone at the head of the column. Furthermore, the speed of the mobile phase is zero at the walls and maximum at the centre of the column. The asymmetry observed in the peak shape is measured by a parameter called the skewing factor, which is calculated at 10% of the peak height (Fig. 1.4):

$$F_t = b/a \tag{1.4}$$

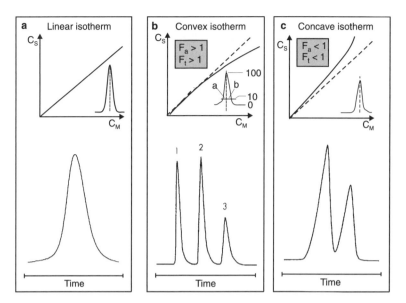

Figure 1.4—*Distribution isotherms.* a) Ideal situation corresponding to the invariance of the concentration isotherm; b) Situation in which the stationary phase is saturated, hence the front of the peak rises faster than the end of the peak descends (in this case, the skewing factor is greater than one); c) Inverse situation: the analyte is strongly retained by the stationary phase. The retention time is elongated and the rise time of the front of the peak is slower than the descent at the end of the peak, which appears normal. For each type of column, the manufacturers usually indicate the capacity of the column in ng/solute loaded before peak deformation is seen. Situations a, b and c are illustrated with real chromatograms in HPLC.

1.6 The theoretical plate model

Many theories have been suggested to explain the mechanism of migration and separation of analytes in the column. The oldest one, called the *theoretical plate model*, corresponds to an approach now considered obsolete but which nevertheless leads to relations and definitions that are universal in their use and are still employed today.

In this model, each analyte is considered to be moving progressively through the column in a sequence of distinct steps, although the process of chromatography is a dynamic and continuous phenomenon. The sequence of these steps describes the migration of the fluids through the column in a cartoon-like manner where each image successively projected gives the illusion of movement. Each step corresponds to a new equilibrium of the entire column. In liquid–solid chromatography, for example, the elementary process is described as a cycle of adsorption/desorption. This classical approach uses a polynomial expansion to calculate the movement of each analyte mass between the two phases present.

The total migration time t_R of the solute can be separated into two terms: t_M represents the time that the analyte is in the mobile phase and travels at the same speed as this phase, and t_S represents the time that the analyte spends in the stationary phase and is immobile. Within two successive transfers from one phase to the other, it is understood that the concentrations can re-equilibrate. At each new equilibrium, the analyte has progressed through the column by a small distance called a *theoretical plate*.

These successive equilibria are the basis for the static model for which the column length L is partitioned into N theoretical plates numbered from 1 to N, all with the same height. For each of these plates, the concentration of analyte in the mobile phase is in equilibrium with the main concentration of analyte in the stationary phase. The *height equivalent to a theoretical plate* (HETP or H) is thus given by equation (1.5):

$$H = \frac{L}{N} \qquad\qquad (1.5)$$

■ In chromatography there are at least three equilibria: analyte/mobile phase, analyte/stationary phase and mobile phase/stationary phase. The origin of the term *theoretical plate* in chromatography comes from the adaptation of an older plate theory for distillation described by Martin and Synge (Nobel Prize for Chemistry, 1952). This term which is universally used for historical reasons, has no physical significance. It may have been preferable to call it a Tswett!

At a given time I, plate J contains a total mass of analyte m_T. This total mass is composed of m_M, the quantity of analyte carried from plate $J - 1$ by the mobile phase in equilibrium at time $I - 1$, and m_S the quantity of analyte already present in the stationary phase of plate J at time $I - 1$.

$$m_T(I, J) = m_M(I - 1, J - 1) + m_S(I - 1, J)$$

If it is assumed for each theoretical plate that $m_S = Km_M$ and that $m_T = m_M + m_S$, then m_T can be calculated using a recursive formula (as can m_M and m_S). Because the analyte is in concentration equilibrium between the two phases for each theoretical plate, the total mass of analyte in the volume of the mobile phase V_M is constant up to the time when the analyte has reached the end of the column. As for the chromatogram, it corresponds to the mass carried by the mobile phase at the $N + 1th$ plate (Fig. 1.5) during the successive equilibria. One of the limitations of this theory is that it does not take into account the dispersion in the column caused by the diffusion of the compounds.

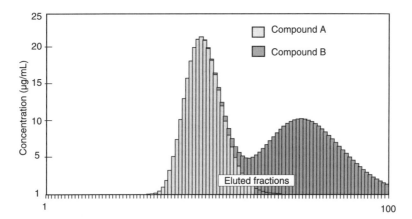

Figure 1.5—*Theoretical plate model.* Computer simulation of the elution of two compounds, A and B, chromatographed on a column with 30 theoretical plates ($K_A = 0.5$; $K_B = 1.6$; $M_A = 300\,\mu g$; $M_B = 300\,\mu g$) showing the composition of the mixture at the outlet of the column after the first 100 equilibria. As is evident from the graph, this model leads to a non-symmetrical peak. However, when the number of equilibria is very large, and because of diffusion, the peak looks more and more like a Gaussian distribution.

■ **Chromatography and thermodynamics.** Thermodynamic relationships can be applied to the distribution equilibria defined in chromatography. $K(= C_S/C_M)$, the equilibrium constant defining the concentration C of analyte in the mobile phase (M) and stationary phase (S) can be determined from chromatographic experiments. If the temperature of the experiment is known, it is possible to determine the variation of the standard free energy ΔG^0 for this transformation:

$$C_M \Leftrightarrow C_S \qquad \Delta G^0 = -RT \ln K$$

In gas chromatography, if the constant K is determined at two different temperatures, it is possible to obtain the variations in enthalpy, ΔH^0, and entropy, ΔS^0, if it is assumed that the enthalpy and entropy have not changed.

$$\Delta G^0 = \Delta H^0 - T\Delta S^0$$

The values of these three parameters are all negative, as expected for a spontaneous transformation. It is also logical that entropy is decreased when the analyte moves from the mobile phase into the stationary phase where it is essentially fixed. Because of this, Van't Hoff's equation can be used in a fairly rigorous way to predict the effect of the temperature on the retention times of the analytes.

$$\frac{\mathrm{d}\ln K}{\mathrm{d}T} = \frac{\Delta H}{RT^2}$$

These examples show that for any study of chromatography, classical thermodynamics can be used.

1.7 Column efficiency

1.7.1 Theoretical efficiency (number of theoretical plates)

As the analyte migrates through the column, it occupies an increasing area (Fig. 1.6). This linear dispersion σ_L, measured by the variance σ_L^2, increases with the distance of migration. When this distance is equal to L, the column length, the variance will be:

$$\sigma_L^2 = H \times L \tag{1.6}$$

By comparison with the model of theoretical plates, this more recent approach also leads to the value for the height equivalent to one theoretical plate, H. Stated in a simpler way, any chromatogram that shows an elution peak with a temporal variance σ^2, permits the determination of H or, directly, the number of the theoretical plates $N = L/H$, which is called the *theoretical efficiency* for the compound under investigation.

$$N = \frac{L^2}{\sigma_L^2} \qquad \text{or} \qquad N = \frac{t_R^2}{\sigma^2} \tag{1.7}$$

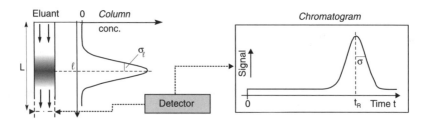

Figure 1.6—*Dispersion of a solute in a column and its translation into a chromatogram*. The curve on the left-hand side corresponds to an isochronic image of the concentration of an eluted compound at a particular instant. The chromatogram on the right-hand side corresponds to the variation of the concentration at the outlet of the column as a function of time. t_R and σ have the same ratio as L and σ_L.

On the chromatogram, σ represents the width of the peak at 60.6% of its height and t_R the retention time. t_R and σ have to be measured using the same units (time, distance, or eluted volume if the flow is constant). If σ is expressed in units of volume (using the flow), then 4σ corresponds to the volume of the peak that is 95% of the injected compound. Because of the properties of the Gaussian distribution, equation (1.8) is preferred. Equation (1.9) is less used because of the distortion of most peaks at the base.

$$N = 5.54 \frac{t_R^2}{w_{1/2}^2} \tag{1.8}$$

$$N = 16 \frac{t_R^2}{w^2} \tag{1.9}$$

The parameter N is relative because it depends on the analyte chosen and on the operating conditions of the chromatograph. It is a reference value that is not sufficient to indicate whether the column will in fact allow a given separation.

1.7.2   Effective plate number

If the performance of different columns has to be compared for a given compound, more realistic values are obtained by replacing the total retention times t_R, which appear in equations (1.7) to (1.9), by the *adjusted retention times* t'_R, which do not take into account the void time t_M spent by the compound in the mobile phase. The mathematical relationships thus become:

$$N_{eff} = \frac{t_R'^2}{\sigma^2} \tag{1.10}$$

$$N_{eff} = 5.54 \frac{t_R'^2}{w_{1/2}^2} \tag{1.11}$$

$$N_{eff} = 16 \frac{t_R'^2}{w^2} \tag{1.12}$$

These corrected parameters are only useful if the void time is large compared to the retention time of the compound. This is the case in gas chromatography particularly when the performance of capillary columns is compared to that of packed columns.

■ For an empty column without a stationary phase or for a column containing a material without any chromatographic effect, then $t_R = t_M$. This would not lead to a separation but at most to a filtration. The relationship using the corrected parameters would lead to a zero efficiency (contrary to the relationships shown in equations (1.7) to (1.9)). In gas chromatography, the effective height of a theoretical plate (H_{eff}) is sometimes obtained using the true efficiency. This is done using equation (1.13):

$$H_{eff} = \frac{L}{N_{eff}} \tag{1.13}$$

1.8   Retention parameters

In chromatography, three volumes are usually considered.

1.8.1   Volume of the mobile phase in the column (dead volume)

The volume of the mobile phase in the column (called the dead volume) V_M corresponds to the accessible interstitial volume. It can be determined from the chromatogram provided a solute not retained by the stationary phase is used. V_M can be expressed as a function of t_M and flow rate F:

$$V_M = t_M \cdot F \qquad (1.14)$$

1.8.2   Volume of the stationary phase

This volume, designated by V_S, does not appear on the chromatogram. It can be determined by subtracting the volume of the mobile phase from the total volume inside the column.

1.8.3   Retention volume

The retention volume V_R of each analyte represents the volume of mobile phase necessary to cause its migration throughout the column. On the chromatogram, the retention volume corresponds to the volume of mobile phase that has passed through the column from the time of injection to the peak maximum. If the flow rate is constant:

$$V_R = t_R \cdot F \qquad (1.15)$$

■ The injector and the detector have dead volumes that affect the total retention volume. In gas phase chromatography, because the mobile phase is compressible, the flow rate measured at the end of the column has to be corrected by the compressibility factor J, which accounts for increased pressure at the head of the column (c.f. 2.2).

1.8.4   Retention factor k (historically called capacity factor k′)

When a compound is injected onto a column, its total mass m_T is divided in two quantities: m_M, the mass in the mobile phase and m_S, the mass in the stationary phase. The values of these quantities are dependent on m_T and K (c.f. 1.2) but their ratio, the *retention factor*, is constant:

$$k = \frac{m_S}{m_M} = K \frac{V_S}{V_M} \qquad (1.16)$$

The factor k, which is independent of the flow rate and length of the column, can vary with experimental conditions. k is the most important parameter in chromatography for determining the behaviour of columns. The value of k should not be too high, otherwise the time of analysis is unduly elongated.

■ For *capillary columns*, it is possible to use the equation below derived from Golay (cf. 2.5.2) to relate the minimum theoretical H value to the retention factor, where r represents the radius of the column. The *coefficient of efficiency* of the column is equal to 100 times the ratio of the value given by the equation to that obtained from the van Deemter plot.

$$H_{\text{min. theor.}} = r\sqrt{\frac{1 + 6k + 11k^2}{3(1+k)^2}} \tag{1.17}$$

1.8.5 Experimental determination of the retention factor k

The mobile phase progressively transports the analyte towards the end of the column. It is assumed that the ratio of the retention volume V_R to the dead volume V_M is identical to the ratio which exists between the total mass of the compound and the mass dissolved in the dead volume. Consequently:

$$\frac{V_R}{V_M} = \frac{m_T}{m_M} = \frac{m_M + m_S}{m_M} = 1 + k$$

hence

$$V_R = V_M(1 + k) \tag{1.18}$$

Alternatively, using equations (1.14) and (1.15):

$$t_R = t_M(1 + k) \tag{1.19}$$

Therefore, the value of k is accessible from the chromatogram and can be obtained using the following equation:

$$k = \frac{t_R'}{t_M} \tag{1.20}$$

■ The retention volume V_R of each solute is related to the distribution coefficient K of the solute between the two phases. Equation (1.18) can be rewritten as:

$$V_R = V_M\left(1 + K\frac{V_S}{V_M}\right)$$

which leads to equation (1.21), as demonstrated in the theoretical plate model:

$$V_R = V_M + KV_S \tag{1.21}$$

1.9 Separation factor between two solutes

The separation factor, designated by α, allows the comparison of two adjacent solutes 1 and 2 present in the same chromatogram (Fig. 1.7).

$$\alpha = \frac{t_{R_2}'}{t_{R_1}'} \tag{1.22}$$

Using equation (1.20), it can be concluded that the separation factor is equal to the ratio of the retention factors for two compounds on the same column.

$$\alpha = \frac{k_2}{k_1} \tag{1.23}$$

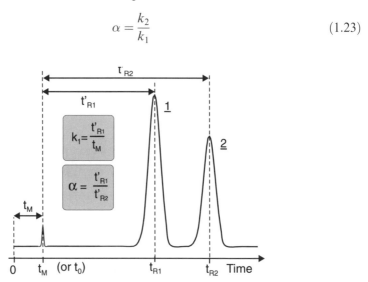

Figure 1.7—*Separation factor (or selectivity factor) between two adjacent components.* α alone does not determine if the separation is, in fact, possible. Here, the separation factor is in the order of 1.3. For two non-adjacent peaks, the *relative retention factor* can be obtained and it is designated by *r*. By definition, neither α nor *r* can be less than one.

1.10 Resolution factor between two peaks

To quantify the separation between two peaks, the resolution factor *R* is used and can be obtained from the chromatogram (see Fig. 1.8).

$$R = 2\frac{t_{R_2} - t_{R_1}}{w_1 + w_2} \tag{1.24}$$

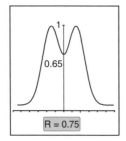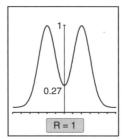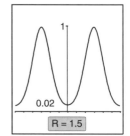

Figure 1.8—*Resolution factor.* Simulation of chromatographic peaks using two identical Gaussian curves side by side, and the visual aspect of a separation corresponding to the *R* values indicated on the diagrams. At *R* = 1.5, it is said that the peaks are baseline resolved; the valley between the peaks does not exceed 2%.

■ Other relationships derived from the preceding equations can be used to express resolution. These relationships are used to express resolution as a function of one parameter or another or, alternatively, to accommodate simplifications. Therefore, equations (1.25) and (1.26) are often used (equation (1.26) is obtained from (1.24) when $w_1 = w_2$). Equation (1.26) shows the influence on resolution of the efficiency, the capacity factor and the selectivity.

$$R = 1.18 \frac{t_{R_2} - t_{R_1}}{w_{1/2,1} + w_{1/2,2}} \tag{1.25}$$

$$R = \frac{1}{4} \sqrt{N_2} \cdot \frac{\alpha - 1}{\alpha} \cdot \frac{k_2}{1 + k_2} \tag{1.26}$$

1.11 The van Deemter equation in gas chromatography

In the previous relationships used to express the characteristics of a separation, the speed of the mobile phase in the column does not appear. Obviously, the speed has to affect the progression of the solutes, hence their dispersion within the column, and must have an effect on the quality of the analysis. These kinetic considerations are collected in a famous equation proposed by van Deemter. First used in gas chromatography, this equation has been expanded to liquid chromatography and relates H (HETP) to the mean linear velocity of the mobile phase in the column, $\bar{u}$ (see Fig. 1.9). The simplified form of this equation is given below:

$$H = A + \frac{B}{\bar{u}} + C \cdot \bar{u} \tag{1.27}$$

The three experimental parameters A, B and C are related to column parameters and also to experimental conditions. If H is expressed in cm, A will be expressed in cm, B in cm^2/s and C in s (where velocity is measured in cm/s). The function is a hyperbolic function that goes through a minimum (H_{min}) when:

$$\bar{u} = \sqrt{\frac{B}{C}} \tag{1.28}$$

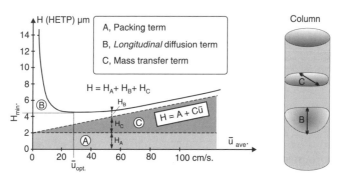

Figure 1.9—*Van Deemter plot for gas phase chromatography showing domains for A, B and C.* A similar equation exists in which H is plotted as a function of column temperature for each parameter: $H = A + B/T + CT$.

■ In practice, the values of coefficients *A*, *B* and *C* can be determined by conducting several experiments for the same solute at different flow rates. The method of multiple linear regression is then used to find the hyperbolic function that best matches the experimental values.

Equation (1.27), which has been expanded to different types of liquid chromatography (Knox equation), shows that there is an *optimum flow rate* for each separation and that this does indeed correspond to the minimum on the curve represented by equation (1.27). The loss in efficiency that occurs when the velocity is increased represents what occurs when trying to rush the chromatographic separation by increasing the flow rate of the mobile phase. However, intuition can hardly predict the loss in efficiency that occurs when the flow is too slow. To explain this phenomenon, the origins of the terms *A*, *B* and *C* have to be considered. Each of these parameters has a domain of influence that can be seen in Fig. 1.9. Essentially, this curve does not depend on the nature of the solute.

1.11.1   A, packing term (Eddy diffusion)

The term *A* is related to the flow profile of the mobile phase as it traverses the stationary phase. The size of the stationary phase particles, their dimensional distribution, and the uniformity of the packing are responsible for a preferential path and add mainly to the improper exchange of solute between the two phases. This phenomenon is the result of Eddy diffusion or turbulent diffusion, considered to be non-important in liquid chromatography or absent by definition in capillary columns, and WCOT (wall coated open tubular) in gas phase chromatography (Golay's equation without term *A*, cf. 2.5).

1.11.2   B, diffusion coefficient in the mobile phase

The term *B* in equation (1.27) is related to the longitudinal molecular diffusion in the column. It is especially important when the mobile phase is a gas. This term is a consequence of entropy, telling us that a system will tend towards the maximum degrees of freedom as demonstrated by a drop of ink that diffuses after falling into a glass of water. Hence, if the flow rate is too slow, compounds being separated will mix faster than they will migrate. This is why one must never interrupt the separation process, even momentarily.

1.11.3   C, mass transfer coefficient

The coefficient *C*, related to the resistance to mass transfer between the two phases, becomes important when the flow rate is too high for equilibrium to be obtained. Local turbulence within the mobile phase and concentration gradients slow down the equilibrium process ($Cs \Leftrightarrow Cm$). The diffusion of solute between the phases is not instantaneous, hence the solute will be in a non-equilibrium process.

1.12 Optimisation of a chromatographic analysis

Because analytical chromatography is used inherently in quantitative analysis, it becomes crucial to precisely measure the areas of the peak. Therefore, the substances to be determined must be well separated. In order to achieve this, the analysis has to be optimised using all the resources of the instrumentation and, when possible, software that can simulate the results of temperature modifications, phases and other physical parameters. This optimisation process requires that the chromatographic process is well understood.

■ In gas phase chromatography, separations can be so complex that it is difficult to assess whether the temperature should be decreased or increased. The choice of the column, its length, its diameter, the choice of the stationary phase and of the phase ratio are all parameters that can have an impact on the separation. Furthermore, all of these parameters can affect each other.

In the optimisation process, *resolution* and *elution time* are the two most important dependent variables. The goal is to conduct a separation of the components of interest in the minimum time without neglecting the time it will take for the instrument to come back to its initial state and be ready for the next analysis. Chromatography corresponds to a slow type of analysis. If the resolution is excellent, optimisation can still be conducted in order to save time in the analysis. This can be done using a shorter column, knowing that resolution varies with the square root of the length of the column (cf. N in equation (1.26)).

Figure 1.10 shows the optimisation of a separation of a mixture of aromatic hydrocarbons. In this case, optimisation of the liquid chromatography separation has been carried out by successive modifications of the composition of the mobile phase. It can be seen that this optimisation results in a significant increase in the cycle time for analysis.

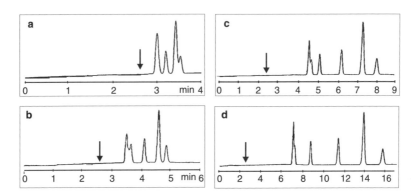

Figure 1.10—*Chromatograms of a separation.* The mobile phase in each trace is a binary mixture of water/acetonitrile: a) 50/50; b) 55/45; c) 60/40; d) 65/35. The arrow indicates the dead time t_M in minutes (from J.W. Dolan, *LC–GC Int.* 1994, **7**(6), 333).

If only certain compounds in the mixture are of interest, it is possible to use selective detection that will detect only those compounds. However, in other cases, it is important to separate as many compounds as possible within the mixture.

The chromatographer is always a prisoner of a triangle whose apexes correspond to resolution, speed and capacity – three parameters that are in conflict (see Fig. 1.11). An optimised analytical separation uses the potential of the most efficient parameter: selectivity. Thus, in this triangle, the optimised conditions are close to the apex corresponding to resolution.

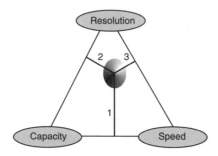

Figure 1.11—*The chromatographer's triangle.* The shaded zone in the upper half of the figure indicates the domain that corresponds to analytical chromatography. It is based on five parameters: K, N, k, α, and R.

Depending on the type of chromatography, optimisation can be fairly rapid. Optimisation in gas phase chromatography is easier than in liquid chromatography where the composition of the mobile phase plays a role. Computer software is available that has been specially designed to help determine the correct composition of the mobile phase.

When conventional methods of manual optimisation are no longer sufficient, it is possible to use deconvolution software or 'fuzzy logic' to deconvolute unresolved peaks into Gaussian components for which the area can be easily measured (see Fig. 1.12).

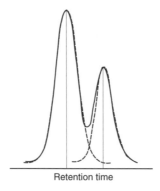

Retention time

Figure 1.12—*Deconvolution of unresolved peaks.* Under certain hypotheses (Gaussian peaks) it is possible to convolute the areas of two peaks leading to the observed signal.

Problems

1.1 A mixture placed in an Erlenmeyer flask comprises 6 ml of silica gel and 40 ml of a solvent containing, in solution, 100 mg of a non-volatile compound. After stirring, the mixture was left to stand before a 10 ml aliquot of the solution was extracted and evaporated to dryness. The residue weighed 12 mg.

Calculate the adsorption coefficient, $K = C_S/C_M$, of the compound in this experiment.

1.2 The retention factor (or capacity factor), k of a compound is defined as $k = m_S/m_M$, that is by the ratio of the masses of the compound in equilibrium in the two phases. Show, from the information given in the corresponding chromatogram, that the expression used – $k = (t_R - t_M)/t_M$ – is equivalent to this. Remember that for a given compound the relation between the retention time t_R, the time spent in the mobile phase t_M (hold-up or dead time) and the time spent in the stationary phase t_S, is as follows:

$$t_R = t_M + t_S$$

1.3 Calculate the separation factor (or selectivity factor), between two compounds, 1 and 2, whose retention volumes are 6 ml and 7 ml, respectively. The dead volume of the column used is 1 ml. Show that this factor is equal to the ratio of the distribution coefficients K_2/K_1 of these compounds ($t_{R(1)} < t_{R(2)}$).

1.4 For a given solute show that the time of analysis – which can be compared with the retention time of the compound held longest on the column – depends, amongst other things, upon the length of the column, the average linear velocity of the mobile phase and upon the volumes V_S and V_M which indicate respectively the volume of the stationary and mobile phases.

1.5 Equation (2) is sometimes employed to calculate N_{eff}. Show that this relation is equivalent to the more classical equation (1):

$$N_{eff} = 5.54 \frac{(t_R - t_M)^2}{w_{1/2}^2} \tag{1}$$

$$N_{eff} = N \frac{k^2}{(k+1)^2} \tag{2}$$

1.6 The resolution factor R for two solutes 1 and 2, whose elution peaks are adjacent, is sometimes expressed by equation (1):

$$R = \frac{t_{R(2)} - t_{R(1)}}{w_{1/2(1)} + w_{1/2(2)}} \tag{1}$$

$$R = \frac{1}{4}\sqrt{N_2} \frac{\alpha - 1}{\alpha} \frac{k_2}{1 + k_2} \tag{2}$$

1. Show that this relation is different from the basic one by finding the expression corresponding to the classic equation which uses W_b, the peak width at the baseline.
2. If it is revealed that the two adjacent peaks have the same width at the baseline ($W_1 = W_2$), then show that relation (2) is equivalent to relation (1) for the resolution.

1.7 1. Show that if the number of theoretical plates N is the same for two neighbouring compounds 1 and 2, then the classic expression yielding the resolution, equation (1) below, can be transformed.

$$R = \frac{\sqrt{N}}{2} \frac{k_2 - k_1}{k_1 + k_2 + 2} \tag{1}$$

$$R = \frac{1}{2} \sqrt{N} \frac{\alpha - 1}{\alpha + 1} \frac{\bar{k}}{1 + \bar{k}} \tag{2}$$

2. Show that if $\bar{k} = (k_1 + k_2)/2$ then expressions (1) and (2) are equivalent.

1.8 The effective plate number N_{eff} may be calculated as a function of the separation factor α for a given value of the resolution, R. Derive this relationship.

$$N_{\text{eff}} = 16R^2 \frac{\alpha^2}{(\alpha - 1)^2}.$$

1.9 Consider two compounds for which $t_M = 1$ min, $t_1 = 11.30$ min and $t_2 = 12$ min. The peak widths at half-height are 10 s and 12 s, respectively. Calculate the values of the respective resolutions using the relationships from the preceding exercise.

1.10 Certain gas phase chromatography apparatus allows a constant gas flow in the column when operating under programmed temperature conditions.
1. What is the importance of such a device?
2. Deduce from the simplified version of Van Deemter's equation for a full column the expression which calculates the optimum speed and which conveys a value for the HETP.

1.11 Explain how the resolution can be expressed in terms of the retention volumes in the following equation:

$$R = \frac{\sqrt{N}}{2} \frac{V_{R(2)} - V_{R(1)}}{V_{R(1)} + V_{R(2)}}$$

1.12 Which parameters contribute an effect to the widths of the peaks on a chromatogram? Is it true that all of the compounds present in a sample and which can be identified upon the chromatogram have spent the same amount of time in the mobile phase of the column? Can it be said that a reduction in the retention factor k of a compound enhances its elution from the column?

CHAPTER 2

Gas chromatography

Gas chromatography (GC) is a widely used technique. Its applications, dating back to the 1940s, were essentially in the analyses of light fractions from petroleum refineries. The use of GC has grown due to its sensitivity, versatility, wide range of stationary phases, and speed with which new analyses can be developed. The technique has also gained great interest due to its ease of automation. Because separation occurs in the gas phase, liquid and solid samples must first be vaporised. This represents the main constraint of this technique; compounds that are analysed by GC have to be thermo-stable and sufficiently volatile. Usually the stationary phases used in GC are liquids; rarely solids. Applications of this technique are used in the fields of petrochemistry, pharmaceuticals, the environment, flavour analyses, and so on.

2.1 Components of a gas chromatograph

The gas chromatograph is composed of several components. These components include the injector, the column and the detector (see Fig. 2.1). The mobile phase that transports the analytes through the column is a gas and is referred to as the *carrier gas*. The carrier gas flow, which is precisely controlled, allows great precision in the retention times.

The analysis starts when a small quantity of sample in liquid or gaseous state is injected. The dual role of the injector is to vaporise the analytes and to mix them uniformly in the mobile phase. Once the sample is vaporised in the mobile phase, it is swept into the column, which is usually a tube coiled into a very small section with a length that can vary from 1 to over 100 m. The column containing the stationary phase is situated in a variable temperature oven. At the end of the column, the mobile phase passes through a detector before it exits to the atmosphere. Some gas chromatograph models have their own power supply, permitting them to be used in the field (see Fig. 2.16).

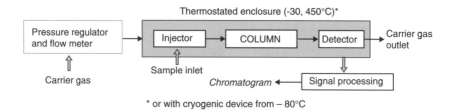

Thermostated enclosure (-30, 450°C)*

* or with cryogenic device from – 80°C

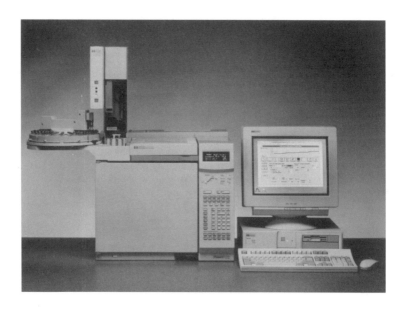

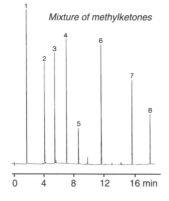

Mixture of methylketones

Phase : polysiloxane, film 0.25 μm
Column : 25 m, Int. Diam. : 0.22 mm
Temp. from 60 to 240°C
Inj. Split

1 Acetone
2 2-Heptanone
3 2-Octanone
4 2-Nonanone
5 2-Decanone
6 2-Tridecanone
7 2-Pentadecanone
8 2-Heptadecanone

Figure 2.1—*A typical gas chromatograph.* Schematic of the gas chromatograph and photograph of a Model 6890-GC with programmable carrier gas pressure. The instrument shown here is equipped with an auto-sampler (reproduced by permission of Hewlett-Packard). Example of a GC-separation.

2.2   Carrier gas and flow regulation

The mobile phase is a gas (helium, nitrogen or hydrogen) which can be purchased from industrial sources or generated on-site in the case of nitrogen and hydrogen since the flows are relatively small (1 to 25 ml/min depending on column type). The carrier gas should not contain traces of water or oxygen because these are very deleterious to the stationary phase. Typically, a filtering system containing a drying agent and a reducing agent is used between the gas source and the chromatograph.

In gas chromatography, the nature of the carrier gas does not significantly alter the partition coefficient K between the stationary and mobile phases. However, the viscosity of the carrier gas and its flow rate have an effect on the analyte dispersion in the column (cf. van Deemter equation) thus affecting the efficiency and sensitivity of detection (Fig. 2.2). The pressure at the head of the column (several tens to hundreds of kPa) is stabilised either mechanically or through the use of an electronic device ensuring that flow rate (linear velocity of the gas) in the column is at its optimal value. When a temperature program is used during analysis, the viscosity of the mobile phase is increased thus increasing the resistance to carrier gas flow. It is, therefore, desirable to correct the pressure to compensate for this effect.

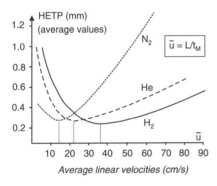 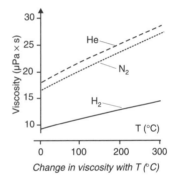

Figure 2.2—*Optimum linear velocity and viscosity of carrier gas.* The optimal mean linear velocities of the various carrier gases are dependent on the diameter of the column. The use of hydrogen as a carrier gas allows a faster separation than the use of helium while giving some flexibility in terms of the flow rate (which can be calculated or measured). This is why the temperature program mode is used. The significant increase in viscosity with temperature can be seen for gases. In addition, the sensitivity of detection depends on the type of carrier gas used.

■ If a chromatogram contains a peak for a compound that is not retained on the stationary phase, it is possible to calculate the mean linear velocity $\bar{u}$ of the carrier gas in the column. It is also possible to determine $\bar{u}_0$ at the *outlet* of the instrument at atmospheric pressure P_0 by putting a flow meter at the end of the column. The ratio between these two velocities is equal to the compression factor J, which is related to the pressure differential between the inlet and outlet, P/P_0 (P is the pressure at the head of the column):

$$J = \frac{\bar{u}}{\bar{u}_0} = \frac{3}{2} \times \frac{(P/P_0)^2 - 1}{(P/P_0)^3 - 1} \tag{2.1}$$

2.3   Sample introduction and the injection chamber

2.3.1   Sample introduction

The sample, usually in the form of a solution in the order of 0.5 µl, is introduced into the chromatograph with a microsyringe (Fig. 2.3). Several types of syringe exist because of the diversity of injectors and columns. For gaseous samples, loop injectors similar to those in liquid chromatography can be used (cf. 3.4). In order to automate the injection and improve reproducibility – simply changing the user can cause substantial deviations – manufacturers provide autosamplers in which the syringe and injection procedure are totally automated. These autosamplers, which can handle several samples, are very reliable. They operate in a cyclic fashion, taking the sample, injecting it rapidly (0.2 s) and rinsing the syringe. The latter is very important to avoid cross-contamination of successive samples that have similar composition.

Figure 2.3—*A typical 10 µl syringe frequently used in gas chromatography*. A guide is used to protect the fragile piston. In some models (0.1 to 1 µl), the plunger enters the needle in order to deliver all of the sample and avoid dead volume effects (reproduced by permission of Hamilton).

■ For the analyses of very volatile samples, there is a technique known as head space chromatography which can be used either in a static or dynamic mode (cf. Chapter 20).

2.3.2   Injectors

Besides its role as an inlet for the sample, the injector has to vaporise and mix the sample with the carrier gas before the sample enters the head of the column. The injection is of paramount value to the quality of the separation. The modes of injection and the characteristics of injectors vary depending on the type of column used in the analysis.

— **Direct vaporisation injection.** For packed columns and megabore columns of 530 µm, which typically use a flow rate of 10 ml/min, direct vaporisation is a simple way to introduce the sample. All models of this type of injector are a variation of a simple assembly which uses a metal tube with a glass sleeve or *insert*. The glass insert is swept by the carrier gas and heated to the vaporisation temperature for the analytes undergoing chromatography. One end of the injector contains a *septum* made of silicone rubber that allows the syringe needle to pass through it into the system. The other end of the injector is connected to the head of the column (Fig. 2.4). The entire sample is injected into the column in a few seconds.

— **Split/splitless injection.** When *capillary columns* are used with small flow rates, even the smallest of injection volumes can saturate the column. Injectors that can operate in two modes, with or without flow splitting, are used (called split/

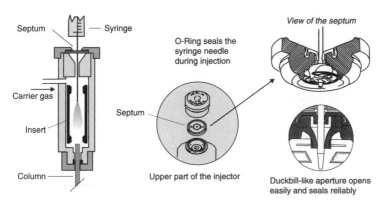

Figure 2.4—*Direct vaporisation injector used for packed columns.* Typical septum injector with a microseal option (the Merlin microseal) that can be used for thousands of injections (reproduced by permission of Hewlett-Packard).

splitless). In the split mode, the carrier gas arrives in the vaporisation chamber with a relatively large flow (see Fig. 2.5). A vent valve separates the carrier gas flow into two parts of which the smallest enters the column. A device is used to regulate the vent rate (generally between 50 and 100 ml/min). The split ratio varies between 1 : 20 and 1 : 500.

When working with capillary columns, the splitless mode is used for very dilute samples. In this mode, the injection is made very slowly, leaving valve no. 2 in the closed position (Fig. 2.5) for approximately 0.5 to 1 min. This allows vaporisation of the compounds and solvent in the first decimetre of the column by a complex mechanism of dissolution in the stationary phase, which is saturated with solvent. Compound discrimination is very weak using this method. The proper use of this injection mode, which demands some experience, requires a temperature program that starts with a colder temperature so that the solvent can precede the analytes in the column. This mode is typically used for trace analyses. The opening of valve no. 2 eliminates, from the injector, compounds which are less volatile and that can interfere with the analyses.

■ In quantitative analyses, the use of the split/splitless injector can cause concentration errors because of the discrimination of compounds that are of different volatility. The composition of the fraction that enters the column can be very different from that which is eliminated. This mode of operation should be excluded when an external standard is used (cf. 4.9). However, this problem can be corrected, up to a point, by use of a proper glass sleeve inserted into the injector.

— **Cold on-column injection.** In this approach, the sample is injected directly *on-column* and its vaporisation occurs after the injection. This requires a very special microsyringe. The needle (steel or silicone) has a diameter of approximately 0.15 mm and it penetrates the column, or pre-column, kept at 4 °C before raising it to its normal operating temperature. This approach, which is very difficult to master without the aid of an autosampler, is useful for thermolabile compounds (applications in

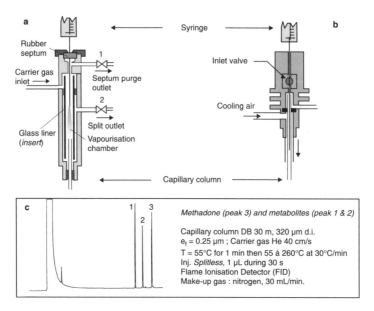

Figure 2.5—*Injectors.* a) A split/splitless injector (the split is regulated by valve 2). The exit labelled 1 is called the septum purge. b) A cold on-column injector. A typical feature of a chromatogram obtained in the splitless mode is the interference of the solvent with the analytes. This can be avoided using a selective detector.

biochemistry) and is known not to discriminate against compounds of different volatilities (Fig. 2.5).

2.4   Oven

The gas chromatograph has an oven with sufficient volume to hold the column easily and which can be heated to the desired temperature (between 40 and 450 °C, stabilised to within 0.1 °C). The atmosphere inside the oven, which usually has a very small thermal inertia, is constantly agitated by forced ventilation. Using a cryogenic valve through which nitrogen or carbon dioxide can be introduced, the oven can be regulated at low temperatures.

2.5   Columns

Three types of columns can be used in gas chromatography. From the oldest to the most recent, these are: *packed* columns, *capillary* columns and *wide bore* or '*530*' columns (which have a 530 μm inner diameter) (shown in Fig. 2.6). For packed columns, the stationary phase is deposited onto a porous support. For the latter two, the stationary phase is deposited onto or bound to the inner surface of the column. The performance of these columns differs.

2.5.1   Packed columns

These columns, less commonly used today, are made of stainless steel or glass. They have diameters of 1/8 or 1/4 in (3.18 or 6.35 mm) and range in length from 1 to 3 m. The internal surface of the tube is treated to avoid catalytic interactions with the sample. These columns use a carrier gas flow rate of typically 10 to 40 ml/min. Although they are still used in approximately 10% of cases for routine GC work, packed columns are not well adapted to trace analyses.

Packed columns contain an inert and stable porous support on which the stationary phase can be impregnated or bound (varying between 3 to 25%). The solid support is made of spheres of approximately 0.2 mm in diameter, obtained from diatomites, silicate fossils (such as kieselguhr, tripoli) whose skeleton is chemically comparable to amorphous silica. These materials, which have a specific surface area ranging from 2 to 8 m^2/g, have been commercialised by several companies such as Johns Manville, under the name of Chromosorb$^®$, and are used universally. Other synthetic materials have been developed such as Spherosil$^®$, made of small silica beads. All of these supports have a chemical reactivity comparable to silica gel because of the presence of silanol groups.

2.5.2   Capillary columns

Capillary columns are usually prepared from high purity fused silica obtained by the combustion of SiH$_4$ (or SiCl$_4$) in an oxygen-rich atmosphere. The internal diameter varies from 0.1 to 0.35 mm and the length from 15 to 100 m. Capillary columns are usually coated on the outside with polyimide or a thin aluminium film. Polyimide mechanically and chemically protects the column ($T_{max} = 370\,°C$). The columns are coiled around a lightweight, metallic support. The internal surface of the silica is usually treated or silanized, depending on the technique used to bond the stationary phase.

For *wall-coated open tubular* (WCOT) columns, the stationary phase covers the inside surface of the column. The film thickness of the stationary phase can vary from 0.05 to 5 μm. It can be simply deposited on the surface, can originate from the reticulation of a polymer on the silica surface or can be bound to the silica through covalent bonds. The surface of the silica is treated before the stationary phase is deposited to avoid problems of wetability, desorption and stability over time. This treatment can involve attack by HCl at 350 °C or the deposition of a fine coat of alumina particles. Afterward, the stationary phase is either deposited or prepared *in situ* by polymerisation at the inner surface of the column. Covalent bonding via Si–O–Si–C allows organic compounds to be bound to the silica surface. In the latter case, the columns are particularly stable and can be rinsed periodically allowing them to recover their initial performance. The efficiency of these columns can reach 150 000 theoretical plates.

■ In order to deposit a film of known thickness, a silanised column is filled with a solution of known concentration of stationary phase (e.g. 0.2% in ether) so that the desired thickness is obtained after evaporation. This layer can then be reticulated by peroxide or gamma radiation. The process is similar to the fixation of a dye on a

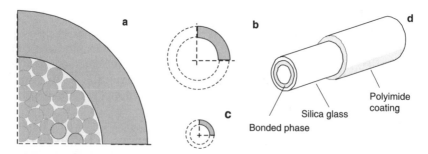

Figure 2.6—*Representation on the same scale of three types of GC column.* a) Stainless steel column, 1/8 in. inner diameter; b) A '530' column of 0.53 mm inner diameter; c) Capillary column of 0.25 mm inner diameter; d) Details of a capillary column. On this scale, the thickness of the stationary phase can hardly be seen. The standard length is approximately 30 m. However, for analyses of high molecular weight materials or for screening purposes, the length is 15 m.

fabric. For example, the dye incorporates a reactive site that allows it to be fixed to the alcohol functional group of cellulose in the cotton fibres.

■ Representing the average linear velocity by $\bar{u}$, the kinetic equation for chromatography for capillary columns, as introduced by Golay, is written as:

$$H = \frac{B}{\bar{u}} + (C_G + C_L) \cdot \bar{u} \qquad (2.2)$$

This relationship, which can be very accurate when the film thickness is negligible compared to the diameter of the column, does not contain an A term, contrary to van Deemter's equation (cf. 1.11). The coefficients C_G and C_L are related to the coefficients of diffusion of the solute in the gaseous and liquid phases. Equation (1.17) is obtained from this equation.

2.5.3   The '530 μm' or wide bore column

Made from a silica tube of 0.53 mm internal diameter with length varying from 5 to 50 m, these columns maintain the features of capillary columns.[1] These columns appeared fairly recently (1983) following developments in the area of flexible silica tubing. Depending on the supplier, they are also called Megabore®, Macrobore® or Ultrabore®. The flow rates used in these columns can be as high as 15 ml/min, close to that used in packed columns. Thus it is possible to replace a packed column by a 530 μm column while retaining the same injector and detector. However, the resolution with these columns is lower than with capillary columns (resolution is higher with columns of smaller inner diameter). The advantage of wide bore columns over packed columns is their lack of bleeding (loss of stationary phase with time).

■ The intermediate performance of these columns is related to the phase ratio ($\beta = V_M/V_S$). Taking d_c as the column diameter and d_f as the film thickness, the following approximation can be obtained:

---

[1] There are several models of packed '530 μm' columns.

$$\beta = \frac{V_M}{V_S} = \frac{d_c}{4d_f} \tag{2.3}$$

If the compounds to be separated are volatile, a column with a small phase ratio should be chosen. A 320 μm column with a stationary phase of 1 μm film thickness leads to a β ratio of 80 while a 250 μm column with a film thickness of 0.2 μm gives a ratio of 310. Since $K = k\beta$, k for a given compound and a given stationary phase will increase if β decreases. Values of β leading to extremely high partition coefficients K (e.g. 1000) are due to the fact that the mobile phase is a gas.

2.6 Stationary phases (liquid type)

Over 100 stationary phases of various types have been described in the literature for packed columns, which are slowly being abandoned. However, for bonded phase capillary columns the choice of stationary phase is limited because the generation of the film at the surface of the column requires a different principle than impregnation. Generally, two families of compounds are used to modify the polarity: *polysiloxanes* and *polyethylene (silicones) glycols.* Very special phases such as cyclodextrins can be used for enantiomeric separations. Stationary phases can be used between a minimum temperature under which equilibrium is too slow to occur and a maximum temperature above which degradation of the polymer occurs. The maximum temperature depends on the film thickness and the nature of the polymer.

2.6.1 Polysiloxanes

Polysiloxanes (also known as silicone oils or gums) have a repetitive backbone that consists of two hydrocarbon chains per silicon atom (see Fig. 2.7). These phases are the most widely used for capillary columns because of their wide temperature range ($-50 < T < 325\,°C$, with $R_1 = R_2 = CH_3$). Approximately 20 types of polysiloxane phases have been commercialised worldwide.

In Fig. 2.7, R_1 and R_2 are either simple alkyl or aryl chains (methyl or phenyl) or incorporate functional groups (e.g. cyanopropyl, trifluoropropyl). Combined in different proportions, R_1 and R_2 modify the polarity and the characteristics of the columns. One of the processes used to obtain a bonded polydimethylsiloxane phase is to allow a solution of tetradimethylsiloxane to flow through the column, then heat to 400 °C after evaporation of solvent and closure of the the extremities (Fig. 2.7).

■ Some columns support temperatures up to 450 °C (e.g. DEXSIL 400® or PETROCOL®). A particular application for these columns is the simulated distillation of petroleum fractions. This is done instead of the conventional distillation which may take up to hundreds of hours to conduct (Fig. 2.8).

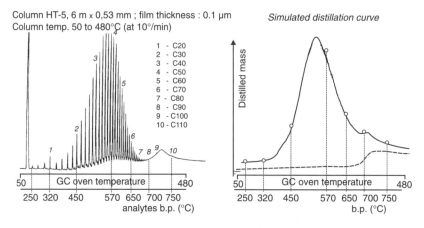

Bonded polysiloxanes (examples) :

ex. R_1 and R_2 = Ph m = 95% and n = 5%

Method of formation of a bonded phase

$T > 400°C$

Capillary wall
($-CH_4$)

Capillary wall

Polyethyleneglycols

"carbowax"

Figure 2.7—*Structure of polysiloxanes (silicones) and polyethylene glycols.* An inventory of all the compositions of these phases that can be used for impregnation or bonding would be lengthy. The surface of the silica column can be treated with tetradimethylsiloxane in order to obtain the bound phase, which is polymerised and then cross-linked.

Column HT-5, 6 m x 0,53 mm ; film thickness : 0.1 μm
Column temp. 50 to 480°C (at 10°/min)

1	- C20
2	- C30
3	- C40
4	- C50
5	- C60
6	- C70
7	- C80
8	- C90
9	- C100
10	- C110

GC oven temperature

50   480
250 320 450   570 650 700 750
analytes b.p. (°C)

Simulated distillation curve

Distilled mass

GC oven temperature

50   480
250 320 450   570 650 700 750
b.p. (°C)

Figure 2.8—*Simulated distillation of lubricating oil (Polywax).* Using a column that can operate at high temperatures, a correlation is made between retention times and boiling temperatures for a series of oligomers. The sample to be distilled is then run under the same chromatographic conditions. Software using the chromatogram reproduces a distribution curve identical to that which would be obtained from the mixture if it were distilled, a much longer process (document SGE 712–0546 and –0547).

■ A well-known phase which is used as a reference because it is the only one that is well defined is *squalane*. Its polarity is zero by the McReynolds scale (2.13). This saturated hydrocarbon ($C_{30}H_{62}$) is derived from squalene, a natural terpene extracted from shark liver or skin. On this stationary phase, which can be used between 20 and 120 °C (deposition or impregnation), compounds are eluted according to their boiling temperature (retention time is inversely proportional to analyte vapour pressure). Several bonded phases of low polarity based on poly-alkylsiloxanes can replace squalane.

2.6.2 Polyethylene glycols

The most widely known example of this family of compounds is Carbowax® (see Fig. 2.7). These polar polymers ($M = 1500$–$20\,000$ for Carbowax 20M®) can be used for deposition, impregnation or as bonded phases ($60 < T < 260$ °C, depending on column diameter and film thickness).

2.7 Stationary phases (solid type)

These phases are composed of adsorbing materials: molecular sieves, alumina, porous glass and gels (such as Chromosorb® 100, Porapak® and PoraPLOT®), and graphitised carbon black. They are mainly used to separate gases or volatile compounds. Capillary columns made by deposition of these materials in the form of very fine particulates are called PLOT (porous layer open tubular) columns.

■ Historically, silica gel, which is thermally stable and insensitive to oxygen, was one of the first compounds used as a stationary phase for gas chromatography columns. Today, solid phases are made with more elaborate materials. The efficiency of graphite-based columns is very high (see Fig. 2.9 below).

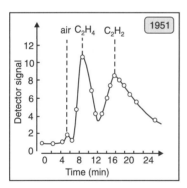
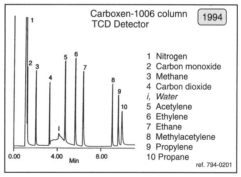

Figure 2.9—*Gas analysis*. On the left is one of the first chromatograms ever obtained, point by point, representing a mixture of air, ethylene and acetylene separated on silica gel (E. Cremer and F. Prior, *Z. Elektrochem*. 1951, **55**, 66). On the right is a gas analysis obtained on a PLOT column (reproduced by permission of Supelco).

2.8 Common detectors

Some detectors are universal; that is they are sensitive to almost every compound that elutes from the column. However, most detectors are sensitive to a particular type of compound. These are called selective detectors. A selective detector is one that can detect only certain compounds, yielding a very simple chromatogram. The ideal situation for determination of an analyte is to have a detector that can detect only this type of analyte (cf. 2.10.2.).

Detectors can be classified into two groups depending on whether they provide only retention time information or retention time plus structural information of an analyte. The latter are described in section 2.9. Nonetheless, all detectors give a response that is dependent on the concentration of analyte in the carrier gas.

2.8.1　Thermal conductivity detector (TCD)

This universal detector has been in use since the beginning of gas chromatography and has been essential to the separation technique. It has moderate sensitivity compared to other detectors and can be miniaturised for use with capillary columns (with a dynamic range of 10^6). Its basic operating principle relies on the thermal conductivity of gaseous mixtures. The thermal conductivity affects the resistance of the thermistor as a function of the temperature. The detector incorporates two identical thermistors, resembling minuscule filaments, which are placed inside a metallic block held at a temperature above that of the column (see Fig. 2.10). One of the filaments is flushed by the carrier gas re-routed prior to the injector while the other is flushed by the carrier gas exiting the column. In the steady state, a temperature equilibrium exists, which depends on the resistance and which in turn is a function of the thermal conductivity of the gas and of the electrical current flowing through the filament. When a solute elutes from the column there is a change in the composition of the mobile phase and thus in the thermal conductivity. This results in a deviation from thermal equilibrium, causing a variation in the resistance of one of the filaments. This variation is proportional to the concentration of analyte, provided its concentration in the mobile phase is low.

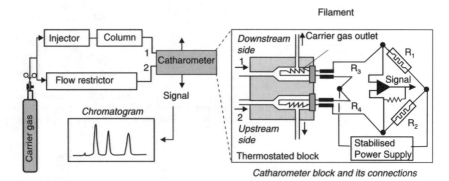

Figure 2.10—*Thermal conductivity detector*. To the left is a schematic showing the path of the carrier gas. To the right is a schematic of the TCD and its operating principle, based on an electrical Wheatstone bridge (equilibrium exists when $R_1/R_2 = R_3/R_4$).

2.8.2 Flame ionisation detector (FID)

This detector, considered to be universal for the analysis of organic compounds, appears ideal for gas chromatography. The gas flow exiting the column passes through a small burner fed by hydrogen and air.

This detector essentially destroys the sample. Combustion of the organic compounds flowing through the flame creates charged particles that are responsible for generating a small current between two electrodes (voltage differential of 100–300 V). The burner, held at ground potential, acts as one of the electrodes. A second annular electrode, called the collector, is kept at a positive voltage and collects the current that is generated (10^{-12} A). The signal is amplified by an electrometer that generates a measurable voltage (Fig. 2.11).

For organic compounds, the intensity of the signal is considered to be proportional to the *mass flow* of carbon. Thus, the area under the peak will reflect the amount of compound present (dm/dt integrated between the beginning and end of the peak will give the total mass m). The presence of heteroatoms, like halogens, will modify this rule. The sensitivity of this detector to organic compounds is high and is expressed in Coulombs/grams of carbon. The detection limit is in the order of 2 to 3 pg/s and the dynamic range can reach 10^8 (8 decades).

■ The FID detector, for which the signal depends on the *instantaneous mass flow*, is essentially free from variations due to flow rate that can cause errors in detectors whose response is a function of the *instantaneous molar concentration* (Fig. 2.11).

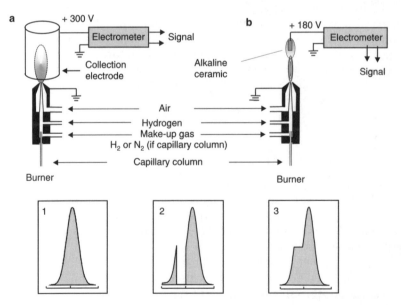

Figure 2.11—*(a) FID detector; (b) NPD detector; and (c) effect of flow rate on detector signal and difference between the mass flow detector and concentration dependent detector.* 1, normal situation (constant flow); 2, mass flow detection (i.e. FID) with an interruption in the flow rate (the area remains constant); 3, TCD detection with an interruption in the flow rate (the area does not represent the mass of the compound flowing through the detector).

■ In order to evaluate the concentration of volatile organic compounds (VOCs), there exist small, portable instruments that contain, essentially, a flame ionisation detector that allows the measurement of the *carbon content* of the atmosphere without chromatographic separation.

2.8.3   Nitrogen phosphorus detector (NPD)

This thermoionic detector is very sensitive to compounds that contain nitrogen or phosphorous. It operates in a different mode from the FID detector (shown in Fig. 2.11). The NPD detector incorporates, between the flame and collector, a piece of ceramic doped with an alkaline salt (Rb or Cs). Because of the catalytic effect of the alkaline salt, compounds containing nitrogen or phosphorous produce more ions than other molecules. Nitrogen present in air does not, however, yield any signal. There are several types of NPD detector and, depending on the type, compounds are ionised in different ways. The flame used in these detectors is much cooler than that used in an FID and an electrical current is used to heat the ceramic, producing an alkaline plasma necessary for the operation of this detector. The sensitivity of this detector is 0.1 pg/s for nitrogen-containing analytes.

2.8.4   Electron capture detector (ECD)

In this detector, a flow of nitrogen, ionised by electrons generated from a β-emitting material of low energy (a few mCi of ^{63}Ni), passes between two electrodes that are maintained at a voltage differential of several hundred volts (see Fig. 2.12). At equilibrium, a base current I_0 is generated. If an organic substance M containing chlorine or fluorine passes through the cell, it will capture some of the electrons leading to a decrease in the stationary current and hence to a reduction in the signal. The measured signal follows an exponential form of the type $I = I_0 e^{-kC}$.

$$N_2 \xrightarrow{\beta^-} N_2^+ + e^-$$

$$M + e^- \longrightarrow M^-$$

$$M^- + N_2^+ \longrightarrow M + N_2$$

The ECD is well suited as a selective detector for compounds with high electron affinity and limited linear response (dynamic range 10^4 with nitrogen). ECD detection is mainly used for analyses of chlorine-containing pesticides. Because of the presence of a radioactive source in this detector, it is subject to special regulations (e.g. inspections; location and maintenance visits).

2.8.5   Flame photometry detector (FPD)

The flame photometry detector is specific for compounds containing sulphur or phosphorous. Compounds eluting from the column are burned in a flame hot enough to excite these elements and induce photonic emission, which is detected by a photomultiplier (see Fig. 2.12). Optical filters are used in the detection system to monitor wavelengths that are characteristic of these substances (526 nm for phos-phorous and 319 nm for sulphur). For sulphur-containing compounds, the detection limit is in the order of 1 ng/s.

■ Some detectors, such as the FID, are constructed to provide maximum perfor-mance with a carrier gas flow rate in the order of 30 ml/min. When using capillary columns, this flow rate is obtained at the exit of the column by using a *makeup gas* either identical to or different from the carrier gas.

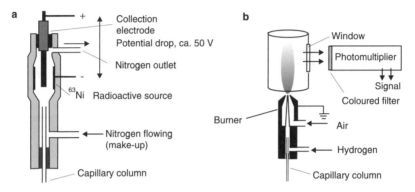

Figure 2.12—*Detectors.* a) Electron capture (ECD); b) Flame photometry (FPD).

2.9 Detectors yielding structural information

The detectors described thus far do not give any information as to the nature of the compounds that are eluting from the column. They are, at best, selective. With these detectors, compound identification has to proceed with the use of internal calibra-tion based on retention times. When the chromatogram is very complex, some confusion can occur. Because of these limitations, other detectors have been devel-oped that can provide structural information based on spectroscopic data. In this case, one can use retention times and a specific characteristic for each compound to identify the components of a sample. These detectors lead to stand-alone analysis techniques for which the results depend only on the proper separation of the com-pounds eluting from the column.

2.9.1 Atomic emission detector

It is possible to extend the principle of photometric emission (FPD) by replacing the flame with a microwave plasma that has a temperature high enough to induce any element to radiate light. This is equivalent to atomic emission where each solute is atomised and gives rise to specific emission bands.

2.9.2 Other detectors

A *mass spectrometric detector* (MSD) can be placed at the end of the column thus producing fragmentation spectra of each of the eluting compounds. With the use of

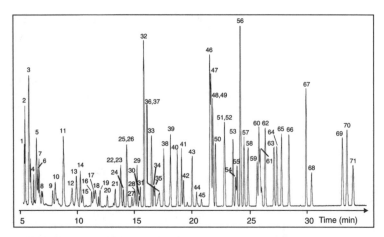

Figure 2.13—*Detection by mass spectrometry*. TIC chromatogram obtained with a mass spectrometer as a detection system. The instrument is capable of obtaining hundreds of spectra per minute. The above chromatogram corresponds to the total ion current at each instant of the elution profile. It is possible to identify each of the components using its mass spectrum. In many instances, compounds can be identified with the use of a library of mass spectra. (Chromatogram of a mixture of 71 volatile organic compounds (VOCs), reproduced by permission of Tekmar and Restek, USA.)

the total ion current (TIC) or selective ion monitoring (SIM), a chromatogram is obtained that represents the compound eluting from the column (Fig. 2.13). Although in some cases this technique is less sensitive than some of the conventional detectors, it has become essential for many types of analyses such as those of environmental importance. Similarly, an *infrared detector* can be put at the end of the column to provide the IR spectrum of each of the eluted compounds.

These detectors represent techniques that are widely used for trace analysis. Both modes of detection mentioned above can be used with capillary columns.

APPENDIX

Retention indices and constants related to stationary phases

A great number of stationary phases are listed in catalogues and it is sometimes difficult to choose the best column for a particular analysis. The chemical nature of the phases and their polarities do not always allow one to predict which column will be optimal for a given separation. Therefore, a technique called the *retention index* system has been developed with the use of reference compounds whose retention factors differ with different stationary phases. Using retention indices obtained on columns of different stationary phases, it is possible to characterise a compound and facilitate its identification.

2.10   Kovats relationship

To determine the constants that are related to a given stationary phase, a series of n-alkanes are injected on the column in the isothermal mode. Under these conditions, the logarithm of the corrected retention times $t'_{R(n)}$ $(t'_{R(n)} = t_{R(n)} - t_M)$ increases linearly with the number n of carbon atoms present in the alkane (Fig. 2.14). A graph of carbon number n versus log $t'_{R(n)}$ is usually a straight line:

$$\log t'_{R(n)} = a \cdot n + b \tag{2.4}$$

In this semi-empirical mathematical relationship, $t'_{R(n)}$ represents the retention time of the alkane containing n atoms of carbon, and a and b are numerical coefficients. The slope of the graph depends on the type of stationary phase used and the operating conditions of the chromatograph.

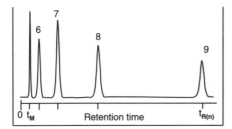 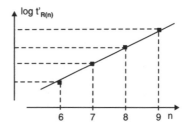

Figure 2.14—*Kovats relationship.* Isothermal chromatogram of a serie of n-alkanes (C_6–C_9) and corresponding plot of the Kovats relationship.

■ The chromatogram that gives the Kovats relationship for a given stationary phase can also be used to evaluate the performance of a given column. For this, the *separation number*, also known as the *trennzahl* number (TZ), is calculated using equation (2.5). The two retention times occurring in the equation are related to two alkanes differing by one carbon number (n and $n + 1$) or to two compounds of a different type. The separation number indicates how many compounds can be baseline separated and inserted between the two reference compounds. For the chromatogram shown in Fig. 2.14, *TZ* is in the order of 11.

$$TZ = \frac{t_{R_2} - t_{R_1}}{w_{1/2,1} + w_{1/2,2}} - 1 \tag{2.5}$$

$$TZ = \frac{R}{1.18} - 1 \tag{2.6}$$

2.11   Kovats retention indices

Without changing the tuning of the instrument, a compound X is now injected onto the column. The new chromatogram obtained allows the calculation of the *Kovats retention index* of X on the specific column used. This index is obtained by multiplying the equivalent alkane carbon number that has the same retention time as X by 100.

Two methods can be used to obtain the equivalent carbon number n_X of compound X:

— The graphical method, which uses the Kovats relationship previously obtained (Fig. 2.14).

— A method based on the corrected retention times of two alkanes that bracket compound X in the chromatogram (see Fig. 2.15) and equation (2.7):

$$I_X = 100 \cdot n + 100 \frac{\log t'_{R(X)} - \log t'_{R(n)}}{\log t'_{R(n+1)} - \log t'_{R(n)}} \tag{2.7}$$

In contrast to the Kovats relationship, retention indices depend only on the stationary phase and not on the column dimensions or the flow rate of the carrier gas. In practice, compound X is injected with the two bracketing alkanes to ensure that the experimental conditions are uniform.

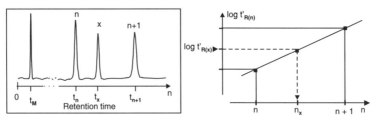

Figure 2.15—*Graphical measurement of Kovats index* $(I = 100\,n_X)$ *on a column in the isothermal mode.* The equivalent carbon number n_X is obtained using the logarithm of the corrected retention time $t'_{R(X)}$. When using a temperature program, a linear relationship can be obtained using a corrected formula. However, this is achieved with a lower precision.

2.12   Stationary phase constants

2.12.1   McReynolds constants

The basis for stationary phase characterisation is the comparison of the Kovats indices of five reference compounds belonging to different chemical classes for a given stationary phase to that of squalane, which is used as a standard phase for calculations. The five indices on a squalane column, the only truly reproducible phase because it consists of a pure compound, have been obtained (see Table 2.1).

The five McReynolds constants for a given stationary phase are obtained by calculating the differences between Kovats indices obtained on squalane (I_{squalane}) and those obtained on the stationary phase being studied (I_{Phase}):

$$\text{McReynolds constant} = \Delta I = (I_{\text{Phase}} - I_{\text{Squalane}}) \tag{2.8}$$

The sum of the five calculated values using equation (2.8) is used to define the total polarity of the phase under study.

■ Assuming that the following Kovats indices are found for a column containing polypropyleneglycol: benzene = 775 and pyridine = 918, the McReynolds constants using the above calculation will lead to: benzene = 122, pyridine = 219 (cf. Table 2.1).

These constants, which are related to the structure of the molecules, allow an evaluation of the forces of interaction between the stationary phase and the solute for different classes of compounds. An index with an elevated value indicates that the stationary phase has a strong affinity for compounds that contain particular organic functions. This leads to a greater selectivity for this type of compound. For example, in order to separate an aromatic hydrocarbon contained in a mixture of ketones, a stationary phase for which benzenes have a very different constant than butanone will be selected. These differences in indices appear in most manufacturers' catalogues of chromatographic components (Table 2.1). McReynolds constants have more or less replaced Rohrschneider constants, which are based on the same principle but use different reference compounds.

Table 2.1—McReynolds constants (ΔI) for several stationary phases

Stationary phase	benzene	1-butanol	2-pentanone	nitropropane	pyridine
Squalane	0	0	0	0	0
SPB-Octyl	3	14	11	12	11
SE-30 (OV-1)	16	55	44	65	42
Carbowax 20M	322	536	368	572	510
OV-210	146	238	358	468	310
Kovats indices for the five reference compounds above (X' Y' Z' U' S') on Squalane.					
$I_{(Squalane)}$	653	590	627	652	699

■ The calculation of retention indices implies that measurements are made under isothermal conditions. When using temperature programming, good results can be obtained by substituting retention times in equation (2.7) by the corresponding logarithms.

2.12.2 Table of retention indices of organic compounds

The retention index of a compound obtained on a given stationary phase under given experimental conditions constitutes worthwhile information. However, if several indices of the same compound obtained on different stationary phases are available, better identification of this compound can be made. Because of the excellent reproducibility of retention times on modern chromatographs, this method is reliable for known control compounds. While obtaining retention indices does not constitute absolute identification of a compound, this method can be quite useful to identify unknowns if the proper retention indices tables are available on the most common stationary phases (Squalane, Apiezon, SE30, Carbowax 20M). However, the use of retention indices is now of lesser interest because of capillary columns that involve new stationary phases. This limits the information that can be obtained from the retention indices. Hyphenated techniques are currently more popular. They represent excellent methods for compound identification but depend on instruments that are more complex and more expensive.

Problems

2.1 Consider a GC capillary column where the length, diameter and thickness of the film of the stationary phase could all be modified (one factor at a time and without adjustment of the apparatus' physical characteristics, such as temperature and pressure, yet maintaining a flow such that the linear velocity of the gas remains the same).

The three parameters to be modified are displayed in the first column of the table below. Complete the different cases by indicating, using the symbols provided, the observations anticipated: + symbolises an increase, 0 shows a weak variation in either direction and − for a decrease.

	Speed of separation	Column capacity	Retention factor, k	Selectivity factor α	Effective plate number, N
increase in column length					
increase in column diameter					
increase in film thickness					

2.2 The best-known method for estimating the dead time, t_M, consists of measuring the retention time of a compound not retained upon the column. Described here is another method for calculating the dead time, which recalls the relation used in establishing factors of retention. Knowing that for a homologous series of organic compounds and where the temperature of the column is constant, it can be written:

$$\log(t_R - t_M) = a \cdot n + b$$

(in which t_R represents the total retention time of a compound having n atoms of carbon, while a and b are constants which depend upon the type of solute and the stationary phase chosen).

1. Recall the chromatographic parameters for which it is essential to know t_M.
2. Give examples of compounds that might be used to determine t_M in GPC.
3. Calculate t_M from the following experiment, employing the method above: A mixture of linear alkanes, possessing six, seven and eight atoms of carbon, is injected into the chromatograph. The total retention times for these compounds were respectively, 271 s, 311 s, and 399 s, under a constant temperature of 80 °C. (Length of column 25 m, $ID = 0.2$ mm, $d_f = 0.2$ µm and the stationary phase is made up of polysiloxanes).
4. If the retention index for pyridine on squalane is 695, what is the McReynolds constant of this compound on the column studied, if it is known that under the conditions of the experiment, the retention time is 346 s?

2.3 Find an expression to calculate the thickness of film d_f of a capillary column from K, k and ID, (the column's interior diameter).

A sample of hexane is injected onto a column having an internal diameter of 200 μm for which K is equal to 250. The retention time for the hexane is 200 s, while an unretained compound has a retention time of only 40 s. Calculate the thickness of the stationary phase d_f from this information.

2.4 Show that for a capillary column, the average flow can be calculated from the following formula:

$$D_{ml/min} = u_{cm/s} \times 0.47 d^2_{imm}$$

(where u represents the average linear velocity in a column of internal diameter d_i).

2.5 The table below contains values of the retention factor k for four refinery gases, studied at three different temperatures upon the same capillary column (length $L = 30$ cm, internal diameter $= 250$ μm), whose stationary phase is of type SE-30. The chromatograph is supplied with a cryogenic accessory.

		Temperature of the column (°C)		
Compound	b.p. (°C)	−35	25	40
ethene	−104	0.249	0.102	0.0833
ethane	−89	0.408	0.148	0.117
propene	−47	1.899	0.432	0.324
propane	−42	2.123	0.481	0.352

1. Can the polarity or non-polarity of the phase SE-30 be deduced from the elution order of the compounds?
2. Calculate the selectivity factor α for the couple propene–propane at the three temperatures indicated.
3. Why for the same compound does k decrease in response to an increase in temperature?
4. What is the number of theoretical plates of the column for the propane at 40 °C, if it is known that at this temperature the resolution factor for the couple propene–propane $= 2$?
5. What would be the minimum theoretical value of HETP for the propane at 40 °C?
6. Supposing that the HETP resulting from the van Deemter curve is the same as that resulting from the calculation which might be performed when considering Question 4, calculate the coating efficiency.
7. If for a given compound, k is linked to the absolute temperature of the column by the expression: $\ln k = (a/T) + b$, find, for the case of ethane, numerical values for both a and b.

2.6 In a GPC experiment a mixture of n-alkanes (up to n carbon atoms, where n represents a variable number) and butanol ($CH_3CH_2CH_2CH_2OH$) were injected onto a column maintained at a constant temperature and whose stationary phase was of silicone-type material. The equation of the Kovats' straight line derived from the chromatogram is: $\log t'_R = 0.39n - 0.29$ (where t'_R the adjusted retention time is in seconds). The adjusted retention time of butanol is 168 s. If it is known that the retention index for butanol on a column of squalane is 590 s then deduce its corresponding McReynolds constant upon this column.

2.7 A chromatogram reveals a resolution factor of 1.5 between two neighbouring peaks, 1 and 2. If $k_2 = 5$ and $\alpha = 1.05$, and knowing that the retention time of the second compound is 5 min, then:

1. Calculate the dead time of this chromatography.
2. What is the width of the second peak at half-height, if the scale of the chromatogram is such that 1 min corresponds to 1 cm?
   Use the following relation:

$$R = \frac{1}{4} \sqrt{N_2} \, \frac{\alpha - 1}{\alpha} \, \frac{k_2}{1 + k_2}$$

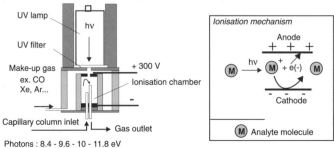

UV lamp
hν
UV filter
Make-up gas
ex. CO
Xe, Ar...
+ 300 V
Ionisation chamber
Capillary column inlet
Gas outlet
Photons : 8.4 - 9.6 - 10 - 11.8 eV

Ionisation mechanism
Anode
+ + +
hν
M → M$^+$ + e(-)   M
- - -
Cathode
M Analyte molecule

Figure 2.16—*A miniature chromatograph.* Instrument using a capillary column and a photoionisation detector. The instrument, weighing 4 kg including the carrier gas (CO_2), is mainly used for the analysis of volatile organic compounds (VOCs) in air pollution. The photoionisation detector, which is of limited use because of its variable sensitivity, is well suited for the analysis of hydrocarbons. The high powered UV source emits photons that have energies between 10 and 11 eV, ionising the compounds that exit the column, with the exception of the carrier gas. The ionic current generated is amplified using an electrometer and is proportional to the concentration of analytes (reproduced by permission of Photovac).

CHAPTER 3

High performance liquid chromatography

The field of applications of high performance liquid chromatography (HPLC) overlaps that of gas phase chromatography to a great extent. In addition, it allows the analysis ofcompounds that are thermolabile, very polar or of high molecular weight. Although column efficiency is less than gas chromatography, HPLC is successful because the resolution can be increased by modifying the composition of the mobile phase. It is thus possible to modify interactions that occur between analyte and mobile phase, analyte and stationary phase, mobile phase and stationary phase. Whilst many chromatographic techniques employ a liquid mobile phase, HPLC is the archetype of this domain because it uses polyvalent stationary phases. These stationary phases lead to hybrid types of chromatography that range from traditional HPLC to ion chromatography and even to exclusion chromatography. Interfacing HPLC to mass spectrometry, another powerful analytical tool, is now a routine technique.

3.1 Origin of HPLC

High performance liquid chromatography, HPLC, is an analytical technique in general use. It evolved from preparative column chromatography and its performance (efficiency, resolution) has been greatly enhanced by the use of elaborate stationary phases composed of spherical particles with diameters of between 2 and 5 μm. However, because the particles are small, the head pressure needed to force the mobile phase through the column packing must be greatly increased compared to that used in preparative liquid chromatography. Because of this, the letter *P* in HPLC has occasionally corresponded to the word *pressure*.

■ The forced migration of a liquid phase through a stationary phase is encountered in many chromatographic techniques. One of the aspects particular to HPLC is that of the partition mechanisms between analyte, mobile phase and stationary phase. These are based on coefficients of adsorption or partition.

3.2   General scheme of an HPLC system

HPLC systems are composed of several components with defined functions that represent independent yet interrelated entities, such as those one might find in a hi-fi audio system. The components can be, and often are, inserted in a common frame to form an integrated unit (Fig. 3.1).

The components within the HPLC system are interconnected by short segments of transfer tubing with a very small internal diameter (0.1 mm). Traditionally constructed from stainless steel, transfer tubes can also be made of PEEK® (polyetheretherketone), a coloured, flexible polymer that is less expensive to fabricate than stainless steel and is resistant to common solvents under elevated pressures (up to 350 bar).

■ A low flow rate of mobile phase minimises the dispersion of compounds, which occurs inevitably throughout the instrument. Detection limits are also improved. The laminar flow within the column follows Poiseuille's law, the velocity of the mobile phase being at its maximum in the centre of the tube and zero at the wall.

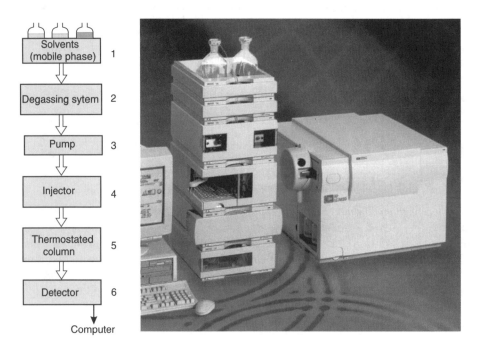

Figure 3.1—*Schematic of an HPLC system and example of a modular commercial HPLC.* A modular system allows users to tailor the instrument to their needs and budget. The HP 1100 chromatograph equipped with an autosampler and a mass detector that allows continuous operation is shown (reproduced by permission of Agilent Technologies). Regulation of the temperature of the column greatly improves the reproducibility of the separation.

3.3 Pumps and gradient elution

3.3.1 Pumps for the mobile phase

All HPLC systems include one or several pumps that are used to flow the mobile phase through the packing, which is fairly compact and responsible for a pressure drop across the column that can be as high as 20 000 kPa. This elevated pressure, created before the injector, depends on the flow rate, the viscosity of the mobile phase and the size of the particles that form the stationary phase. The pumps are *metered* in order to maintain a stable flow and avoid pulsations even when the composition of the mobile phase varies.

> ■ Ambient gases (N_2, O_2 etc.) are always present in the solvents that form the mobile phase. Separations can be affected by modifications in the compressibility of the mobile phase and eventual formation of bubbles. In particular, oxygen can interfere with electrochemical detection and shortens the lifetime of a column. It is thus desirable to degas the solvent using ultrasound, rapid bubbling of helium through the solvent or, alternatively, filtration through a membrane.
> Because corrosion effects increase with pressure, the components that come into contact with the mobile phase need to be inert. Thus, the piston and check valves of the pumps are typically made of sapphire, Teflon or special alloys.

The majority of HPLC instruments use pumps based on alternating pistons. In order to avoid variations in the flow rate that are caused by single piston pumps (solvent in/solvent out), manufacturers have developed pumps that use two pistons. These pumps can be classified into two categories: *series* and *parallel*.

In the serial arrangement (Fig 3.2), the first piston pushes a volume of solvent twice that of the second piston, due either to the fact that its diameter is greater or that it travels a greater distance. The volume of solvent contained in piston B is released into the column while piston A is filling with solvent. Then as piston A pushes the mobile phase, part of it is used to fill piston B, which regulates the nominal flow. These arrangements demand the use of check valves at the input and output of solvent. To regulate the flow rate, the piston speed is controlled by a motor.

> ■ In order to obtain very low flow rates without pulsation (e.g. 1 µl/min), pumps based on the principle of high volume motorised syringes are used. The piston, activated by a pneumatic amplifier, moves at a constant linear velocity. These pumps are still in widespread use.

When the mobile phase has a fixed composition (*isocratic mode*) a single pump is sufficient. However, if the composition of the mobile phase has to vary with time, as in a *gradient elution*, one of the methods described below can be chosen. The instrument must compensate for differences in solvent compressibility in order to attain the desired composition at a given pressure.

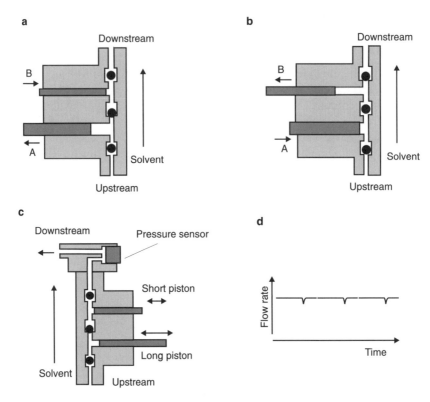

Figure 3.2—*Schematic of the core of a pump using two heads in series.* a) Solvent intake by the bigger piston, A, and delivery by the smaller one, B; b) Inverse situation to regulate the flow; c) Arrangement using two pistons of the same diameter, one of which travels through a distance (and at a speed) that is twice the other; d) Graph showing flow variation as a function of movement of the pistons with time.

3.3.2   Low and high pressure gradients

The present system incorporates electronically activated valves to feed the solvents at a programmed composition into a single pump (Fig. 3.3a). This is accomplished using a *low pressure* mixing chamber located before the pump.

Alternatively, in the *high pressure* arrangement, it is essential to have as many pumps as there are solvents. The final mixing is done in a tee located after the pump but before the column (Fig. 3.3b).

■ In all cases, to achieve a proper regulation of the flow, a pulsation damper is used. It operates on the principle of a mechanical ballast or can be electronically con-trolled. The simplest mechanical ballasting device is a coil with a low cross-section, several metres in length, placed between the pump and the injector. Under the influence of a wave of solvent, the tube uncoils slightly which increases its internal volume and dampens the variation in pressure.

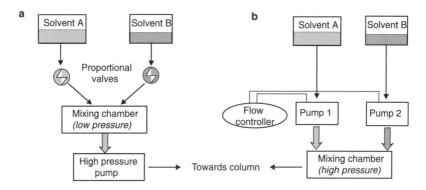

Figure 3.3—*Solvent and pump for (a) low pressure and (b) high pressure gradients.*

3.4 Injectors

In HPLC, the injection must be made as fast as possible to avoid perturbation of the dynamic regime that is already established in the column and in the detector. The main difficulty involves injecting, at the head of the column where the pressure is often higher than 10 000 kPa, a precise volume of sample without stopping the flow of the mobile phase. This is done using a precisely machined manual or motorised high pressure valve located just before the column (Fig. 3.4). The valve, which is mounted in the path of the mobile phase, provides several flow paths. In the *load* position, the valve connects the pump to the column (Fig. 3.5). Using a syringe, the sample contained in a solution is injected into a *loop* that has a small defined volume. Loops range in volume and are either integrated in the rotor of the valve or are connected to the outside of the valve. In the *inject* position, the sample, which is in the loop at atmospheric pressure, is directly inserted into the flow of the mobile phase. In manual injectors, this is done using a handle that allows the valve to rotate 60° thus connecting the sample loop to the mobile phase loop. Highly reproducible injections are achieved when the loop is completely filled with the sample.

Figure 3.4—*HPLC injection valve and assorted loops.* Rear view of the valve showing six entries/exits and a series of loops with different volumes (reproduced by permission of Rheodyne Inc.).

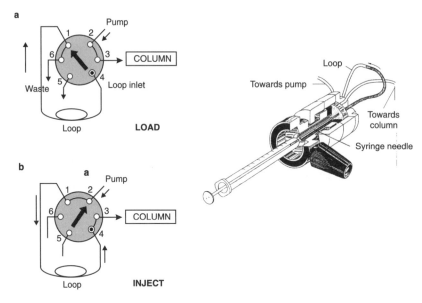

Figure 3.5—*The two positions of a loop injector.* a) Load sample; b) Inject sample. Schematic of the model 7125 valve from Rheodyne Inc. Injection valves can be manual, pneumatic or electrical (reproduced with permission of Rheodyne Inc.).

3.5 Columns

An HPLC column is a straight, stainless steel calibrated tube (sometimes coated on the inside with an inert material such as glass or PEEK®) between typically 3 and 25 cm in length. The internal diameter of the column can vary from 0.5 to 5 mm (Fig. 3.6). The stationary phase is held in the column by one porous disc at each extremity. Dead volumes within the column are kept as small as possible. The flow of the mobile phase into the column must not exceed a few ml/min. Some manufacturers offer microcolumns that have an internal diameter in the order of 0.3 mm and are 5 cm in length. Typical mobile phase flow rates for these narrow columns can be as low as a few μl/min. Because complete elution of compounds with these micro-columns requires only a few drops of mobile phase, the stationary phase and pumps have to be adapted accordingly. These narrow columns not only substantially reduce the amount of mobile phase which has to be used but also improve resolution by diminishing the diffusion inside the column. They are well suited for HPLC-MS (cf. Chapter 16).

■ To increase the lifetime of the column, a precolumn or *guard column* is often used. The guard column is usually filled with the same stationary phase as the analytical column is between 0.4 to 1 cm long. The guard column is inexpensive and can be changed periodically. It is also recommended to pass samples through a filter with a porosity of less than 0.5 μm (Fig. 3.6) before analysis to increase the column lifetime.

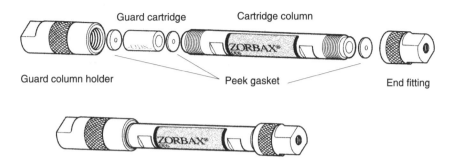

Guard cartridge   Cartridge column

Guard column holder   Peek gasket   End fitting

Figure 3.6—*HPLC column and guard column.* Zorbax® column assembly and its exploded view. The removable precolumn (Guard cartridge) which is not expensive, prevents clogging of the analytical column while extending its lifetime and preserving its performance (reproduced by permission of RTI).

3.6 Stationary phases

A successful separation depends on the right choice of stationary phase, which is characterised by several surface parameters.

3.6.1 Silica gel

With the exception of a few organic polymers, *silica gel* represents the basic material used to pack HPLC columns. The silica gel used for chromatography is quite different from crystalline silica (SiO_2), which is used for its preparation. It is prepared under conditions of controlled hydrolysis by polymerisation of tetraethoxysilane in the form of an emulsion giving rise to microspheres of uniform diameter in the order of 2 to 5 µm (Fig. 3.7). A *sol-gel* is formed in the process and these very small particles must grow in a regular manner in order to obtain the diameter of a few micrometres. The material has to be free of metallic ions. The silica gel particles obtained must be of uniform diameter to avoid the presence of preferential pathways in the packed bed in the column.

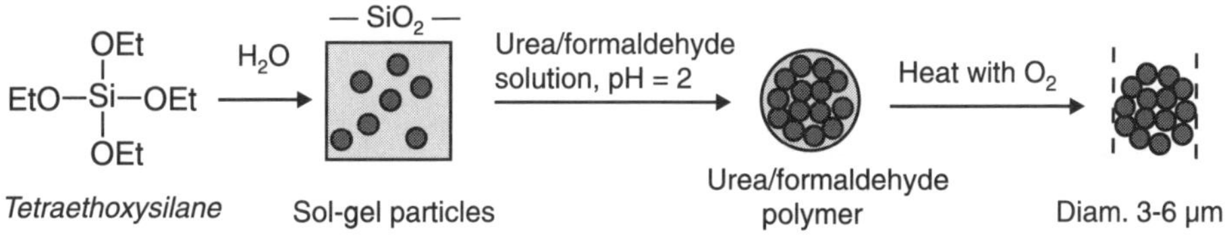

$$
\begin{array}{c}
\text{OEt} \\
| \\
\text{EtO—Si—OEt} \\
| \\
\text{OEt}
\end{array}
\xrightarrow{\text{H}_2\text{O}}
\begin{array}{c}
— SiO_2 — \\
\text{Sol-gel particles}
\end{array}
\xrightarrow[\text{solution, pH = 2}]{\text{Urea/formaldehyde}}
\begin{array}{c}
\text{Urea/formaldehyde} \\
\text{polymer}
\end{array}
\xrightarrow{\text{Heat with O}_2}
\text{Diam. 3-6 µm}
$$

Tetraethoxysilane

Figure 3.7—*Preparation of spherical particles of silica gel via a sol-gel.* The dispersive medium, called *sol*, is made of particles that are only a few nanometres in diameter. The small spheres are then agglutinated together with the aid of an organic binding agent (urea/methanol) to yield spheres with the requisite size (3–6 µm). Pyrolysis is then used in the final treatment.

■ Hydrothermal synthesis of the non-uniform silica gel used for preparative chromatography proceeds in a different fashion. Sodium silicate (Na_2SiO_3),[1] obtained by alkaline fusion of very pure sand, is acidified to yield orthosilicic acid ($Si(OH)_4$). This unstable acid initially dimerises then condenses further to yield a gel with a hydroxylated surface. Under conditions of controlled polymerisation, a *hydrogel* is obtained which is further calcinated to yield a very dense silica gel (xerogel). Some of the processes involved here are of the same type as those used to produce microspheres for analytical chromatography.

$$SiO_2 + NaOH \longrightarrow Na_2SiO_3 + H_2O$$

$$Na_2SiO_3 \xrightarrow{H_3O^+} \left[Si(OH)_4\right] \xrightarrow{-H_2O} \left[(HO)_3SiOSi(OH)_3\right] \xrightarrow{-n\,H_2O} (SiO_2)_n, (H_2O)_x$$

Orthosilicic acid   Orthodisilicic acid   "Silica Gel"

Amorphous silica gel corresponds to a three-dimensional molecular network that has a variable number of silanol groups at its surface and is formed during the final stage of the preparation. The gel does not retain the organised structure of the crystalline silica used for its preparation; however, it does have a tetrahedral geometry around the silicon atom. Cross-linking of polysilicic acid chains occurs through oxygen atoms. In effect, the gel is an inorganic polymer that resembles a reticulated organic polymer and whose structure is somewhat difficult to represent (Fig. 3.8). Solid state NMR of ^{29}Si (cf. Chapter 9) can be used to measure the concentration of silanol groups.

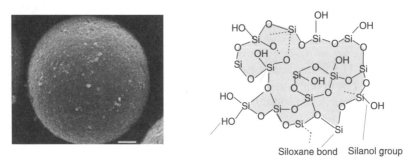

Siloxane bond   Silanol group

Figure 3.8—*Silica gel for chromatography*. Image of the spherical grains of silica gel and a representation of the silica gel surface (reproduced by permission of Daiso Ltd).

The quality of a silica gel depends on many factors such as its *internal structure*, the *size* of its particles, its *open porosity* (dimension and distribution of pores), its *specific area*, its *resistance to crushing* and its *polarity*.

---

[1] Aqueous sodium silicate was formerly known by pharmacists as pebble liquor.

■ Common silica gels used in HPLC are composed of spherical particles (diameter 2 or 5 µm) that can resist crushing at pressures of 1000 bar and typically contain 5 silanol groups per µm^2. The specific area is in the order of 350 m^2/g with pores of approximately 10 nm (total volume of pores 0.7 ml/g). These phases are somewhat different than the chalk and powdered sugar used by Tswett! The treatment that the silica undergoes converts it into a type of 'magical sand'.

ABSORPTION   ADSORPTION

Figure 3.9—*Adsorption and partition phenomena.* Unlike absorption, adsorption is a surface phenomenon (reproduced by permission of M. Laguës, *L'Actualité Chimique* 1990, (1) p. 17).

Silica gel is a polar material. The presence of silanol groups is responsible for the acidic catalytic effect of this material (the pK_a of Si–OH is comparable to that of phenol). The mode of action of silica gel is based on adsorption (Fig. 3.9), a phenomenon that leads to the accumulation of a compound at the interface between the stationary and mobile phases. In the simplest case, a monolayer is formed (known as a Langmuir isotherm) but there is also some attraction and interaction between molecules that are already adsorbed and those still in solution. This contributes to the asymmetry of the elution profile. Although it demonstrates good resolution and a high adsorption capacity, bare silica gel is seldom used for analytical purposes. For most applications, it must be deactivated by partial rehydration (in 3–8% water).

3.6.2 Bonded silica

The structure of silica gel tends to change with time and this creates problems of irreproducibility in the separations. To remedy this situation and reduce the gel's polarity, the reactivity of silanol groups can be used to covalently bind organic molecules. Bonded stationary phases behave like liquids. However, the separation mechanism now depends on the *partition coefficient* instead of *adsorption* (Fig. 3.9). Bonded phases, whose polarity can be easily adjusted, constitute the basis of *reversed phase partition chromatography*, which is used in the majority of analyses by HPLC.

The reaction of alkylchlorosilanes with silica gel in the presence of an alkaline agent (Fig. 3.10) represents a typical transformation procedure. This reaction is used to obtain phases such as RP-8 (dimethyloctylsilane groups), RP–18 or ODS (dimethyloctadecylsilane groups). This procedure converts approximately half of the silanol groups into bonded groups. The reaction is more complete with chlorotrimethylsilane (ClSiMe$_3$) or hexamethyldisilisane (Me$_3$SiNHSiMe$_3$). The sites that are not transformed because of their inaccessibility to the reagent are of little consequence since they should also be inaccessible to the analyte. The use of di- or trichlorosilanes in the presence of water vapour leads to a hydrophobic layer of cross-linked polymer.

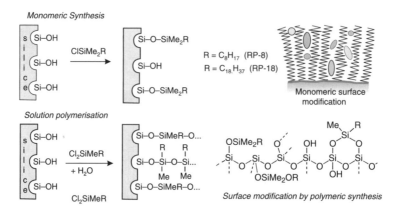

Figure 3.10—*Formation of bonded organosilanes at the interface of silica gel.* Representation of organic monomers and polymers at the surface of silica gel. The arrangement Si–O–Si–C is more stable than Si–O–C. This reaction leads to a carbon content of 4 or 5%. Other reactions can also be used (hydrosilylation in particular). When a monolayer of hydrocarbons is bonded to the surface of silica, they orient in a particular manner at the interface due to their lipophilic and hydrophilic character.

Besides bonded phases that incorporate linear alkyl chains with 8 or 18 carbon atoms and that can be utilised between pH 2 and pH 13, there are polar stationary phases with linear chains incorporating aminopropyl, cyanopropyl, benzyl or mixed moieties. For example, aminoalkyl chains can be used for the separation of sugars.

A variety of non-bonded phases is also available. These can be graphite based or make use of polymers such as styrene/divinylbenzene or hydroxymethylstyrene. Recently, the chemistry of these interfaces has been developed in order to generate phases that are hybrids between conventional HPLC and ionic chromatography.

3.7   Mobile phases

The degree of interaction between the mobile phase and the stationary phase, whether the latter be *normal* or *reversed* phase, affects the retention times of the analytes. In principle, the polarity of the stationary phase can lead to two situations:

— if the stationary phase is *polar*, then a *less polar* mobile phase is used and the technique is called normal phase chromatography.

— if the stationary phase is *nonpolar*, then a *polar* mobile phase is used (most commonly methanol or acetonitrile with water) and the technique is called reversed phase chromatography. Modification of the polarity of the mobile phase for either technique will affect the retention factor k of the analytes (see Table 3.1).

■ Organic solvents cannot be used to conduct the separation of fragile water soluble compounds such as proteins, because of denaturation. A reversed phase column is equilibrated with a concentrated buffer. Under these conditions, the compounds retained by the stationary phase can be eluted in decreasing order of their hydrophilic character by reducing the concentration of salt in the mobile phase.

Bonded silica phases are usually of relatively low polarity. Reversed bonded phases, which can be visualised as silica gel particles coated with paraffin, generate elution orders opposite to those usually encountered with normal phases. The use of a polar eluant will thus favour the migration of a polar compound over a nonpolar one. Under these conditions, hydrocarbons are strongly retained. Elution gradients are obtained by decreasing the polarity of the mobile phase as the separation proceeds. For example, this can be done by increasing with time the concentration of acetonitrile in a water-acetonitrile mobile phase. The separation of easily ionised compounds can be improved by adding ionic substances such as alkylsulphonates or ammonium salts to the mobile phase. This induces the formation of dipole-dipole interactions with ionic analytes rendering them less polar and relatively stable. Their retention on the reversed phase column is increased, permitting a better separation.

■ There are four types of interactions between analyte and solvent molecules:

— *dipolar* interactions that occur when both the solvent and the analyte have permanent dipole moments

— *dispersion* that occurs when neighbouring molecules interact

— *hydrogen bonding* that occurs when the combination of analyte and solvent corresponds to that of a proton donor and proton acceptor

— *dielectric* interactions that favour the dissolution of ionic species in a polar solvent.

Table 3.1—Eluotropic force and viscosity of solvents used as mobile phases. The eluant strength of the mobile phase can be adjusted by mixing different solvents together.

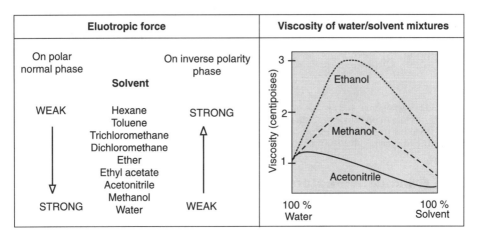

3.8   Chiral chromatography

If a pure compound containing a centre of asymmetry is chromatographed on a chiral stationary phase, two peaks corresponding to the R and S enantiomers will be observed and the area will be proportional to the abundance of each of the forms. The optical purity, defined in terms of enantiomeric excess (e.e.), can be obtained using the equation below where A_R and A_S represent the areas of the peaks for each enantiomer:

$$\text{OPTICAL PURITY (e.e.\%)} = 100 \frac{|A_R - A_S|}{A_R + A_S}$$

Many types of chiral stationary phase are available. Pirkle columns contain a silica support with bonded aminopropyl groups used to bind a derivative of D-phenyl-glycine. These phases are relatively unstable and the selectivity coefficient is close to one. More recently, chiral separations have been performed on optically active resins or cyclodextrins (oligosaccharides) bonded to silica gel through a small hydrocarbon chain linker (Fig. 3.11). These cyclodextrins possess an internal cavity that is hydro-phobic while the external part is hydrophilic. These molecules allow the selective inclusion of a great variety of compounds that can form diastereoisomers at the surface of the chiral phase leading to reversible complexes.

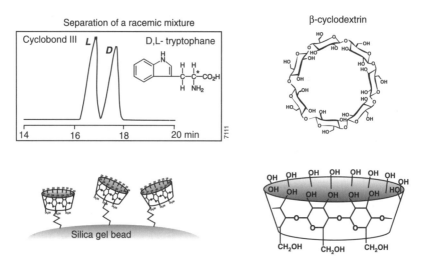

Figure 3.11—*Separation on a cyclodextrin-bound stationary phase*. Chromatogram of a racemic mixture; chemical formula of β-cyclodextrin (diameter, 1.5 nm; cavity, 0.8 nm; height, 0.8 nm); partial representa-tion of cyclodextrin bonded to a silica gel bead through an alkyl chain linker arm; side view of a cyclodextrin molecule with a hydrophobic cavity.

3.9   Principal detectors

The object of liquid chromatography is rarely to determine the global composition of a sample. Rather, it is used to precisely measure the concentration of a species that is present. Thus, it is important to choose the appropriate detector. In quantitative analysis, the use of a universal detector is not necessary. On the contrary, it can complicate the chromatogram. In many cases, relative peak areas in the chromatogram bear little or no relation to the actual molar composition of the mixture.

The detector, irrespective of its nature, is required to have a number of fundamental properties. It should give a response that is proportional to the instantaneous mass flow, be sensitive, have a small inertia, be stable with time and yield very low background noise.

The most widely used detectors for chromatography are based on the optical properties of the analytes: absorption, fluorescence and refractive index.

3.9.1 Spectrophotometric detectors

Absorbance detectors measure, at the end of the column, the analyte absorption at one or many wavelengths in the UV or Visible spectrum (cf. Chapter 11). In order to detect the compounds that are eluting from the column, it is essential that the mobile phase be transparent or possess a very small absorption.

— **Monochromatic detection.** A schematic of a monochromatic absorbance detector is given in Fig. 3.12. It is composed of a mercury or deuterium light source, a monochromator used to isolate a narrow bandwidth (10 nm) or spectral line (i.e. 254 nm for Hg), a flow cell with a volume of a few μl (optical path 0.1 to 1 cm) and a means of optical detection. This system is an example of a selective detector: the intensity of absorption depends on the analyte molar absorption coefficient (see Fig. 3.13). It is thus possible to calculate the concentration of the analytes by measuring directly the peak areas without taking into account the specific absorption coefficients. For compounds that do not possess a significant absorption spectrum, it is possible to perform derivatisation of the analytes prior to detection.

■ The current expansion of biotechnology has increased the demand for analyses of amino acids (protein hydrolysates): a photometric detection system can be used provided the compounds are reacted with ninhydrin, for example, before they pass through the detection cell (cf. Chapter 8).

Figure 3.12—*Schematic of the principle of photometric detection at a single wavelength.*

— **Polychromatic detection.** The most sophisticated detectors (e.g. DAD diode-array detectors) either allow the wavelength to be changed during the course of the analysis or simultaneously record the absorbance at several wavelengths. These detectors can be used not only to record the chromatogram but also to provide spectral information that can be used to identify the compounds. This is called *specific detection* (cf. 11.14).

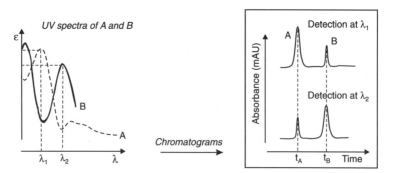

Figure 3.13—*Chromatogram of a sample containing two compounds A and B with different UV spectra.* The profile of the chromatogram will vary depending on the wavelength that is used for detection. For quantitative analysis, the response factor of each of the components that is being analysed must be determined prior to the analysis (cf. Quantitative analysis in chromatography, Chapter 4).

It is also possible to monitor a broad range of wavelengths without interrupting the flow in the column. This results in the full UV spectrum of the eluting mobile phase and provides a good means of identification of the analytes (Fig. 3.14). Such diode-array detectors (DAD) can be used in the gradient elution mode. Actually, they are frequently used in automated routine analysis.

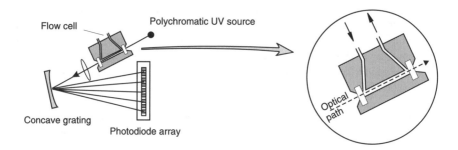

Figure 3.14—*Principle of the diode array detector.* The flow cell is irradiated with a polychromatic UV/Vis light source. The light transmitted by the sample is dispersed by reflection on a grating and the reflected intensities are monitored by an array of photodiodes. Several hundred photodiodes can be used, each one monitoring the mean absorption of a narrow band of wavelengths (i.e. 1 nm).

■ The successive spectra of the eluting compounds, monitored at the end of the column, are stored in the memory of the instrument (e.g. 1 spectrum/s) and can be treated later with the appropriate software. Such software can yield spectacular spectrochromatograms (Fig. 3.15). The ability to record thousands of spectra during a single analysis increases the potential of these detector systems. A topography of the separation can therefore be conducted, $A = f(\lambda, t)$ (isoabsorption diagram).

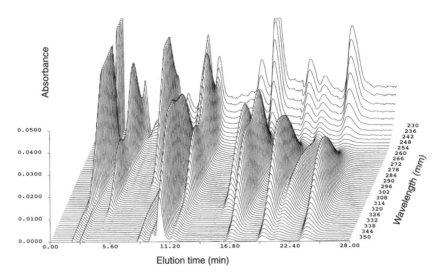

Figure 3.15—*Three-dimensional representation of a separation obtained with a rapid data recording system* (reproduced by permission of TSP Instruments).

3.9.2 Spectrofluorimetric detectors

Some compounds are fluorescent, in that they have the ability to re-emit part of the light absorbed from the excitation source (Chapter 12). In practice, fluorescence is measured perpendicular (at 90°) to the exciting radiation source (Fig. 3.16). The intensity of fluorescence is proportional to the concentration of analyte, as long as this concentration is kept low. The application of this very sensitive and selective detector (Fig. 3.17) can be extended by using derivatisation, either pre- or post-column.

■ A device can be inserted between the column and detector to automate the reaction, or reactions, of a fluorescent reagent with the eluted analyte to render it fluorescent. This principle is used to measure trace carbamates (pesticides) in the environment. The compound is reacted with sodium hydroxide then *o*-phthalalde-hyde, which transforms the methylamine group into a fluorescent derivative (Fig. 3.16).

3.9.3 Refractive index detector

This type of detector relies on the Fresnel principle of light transmission through a transparent medium of a refractive index of *n*. Schematically, a beam of light (mono- or polychromatic) travels through a cell that has two compartments. One compartment is filled with the pure mobile phase while the other is filled with the mobile phase eluting from the column (Fig. 3.18). A compound eluting from the column causes a change in refractive index which results in an angular displacement (refraction) of the beam. In practice, the signal corresponds to a measure of

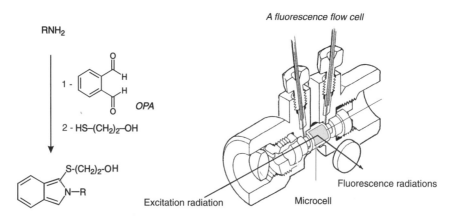

Figure 3.16—*Flow cell of a fluorimetric detector* (reproduced by permission of Hewlett-Packard Inc.). One example of a reagent that is used to form fluorescent derivatives with compounds containing primary amines. Reaction of OPA in the presence of monothioglycol.

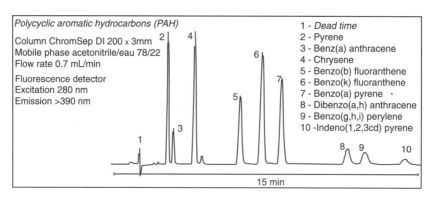

Figure 3.17—*Chromatogram of a mixture of polycyclic aromatic hydrocarbons.* Detection by fluorescence (reproduced by permission of Chrompack).

retroaction that must be provided to the optical element in order to compensate for the beam displacement.

The refractive index detector, considered to be almost universal, is often used in series with a UV detector in the isocratic mode to provide a supplementary chromatogram. This detector, which is not highly sensitive, has to be temperature controlled, as does the column (0.1 °C). The baseline of the chromatogram has to be set to an intermediate position because it can lead to either positive or negative signals (Fig. 3.18). The detector can only be used in the isocratic mode because in gradient elution the composition of the mobile phase changes with time, as does the refractive index. Compensation, which is easily obtained in the case of a mobile phase of constant composition, is difficult to carry out when the composition at the end of the column differs from that at the inlet.

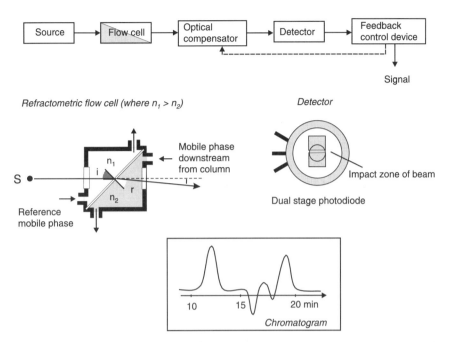

Figure 3.18—*Example of a differential refractive index detector. Optical path through the cell.* Control of the position of the refracted beam is obtained with a dual stage photodiode. Chromatogram of a mixture of sugars obtained with this type of detector.

3.9.4 Other detectors

The identification of a compound using only its retention time is vulnerable to error. It is essential that a standard compound is injected in order to verify the retention time. As is the case in gas chromatography, more sophisticated detectors can be used. These detectors provide complementary information and can be installed at the end of the column. These can be other types of spectrophotometers or a mass spectrometer and they are used simultaneously as classical detectors (to obtain the chromatogram) or for identification purposes of the analytes (cf. Chapter 16). For example, the coupling of HPLC to NMR, which has long been considered impossible, has now been realised through the miniaturisation of the probes and the increased sensitivity of the NMR instruments (cf. Chapter 9).

The present performance of high field NMR instruments allows the recording of a spectrum in quantities of micrograms. Under these conditions, it is possible to install a flow cell of only a few microlitres into the magnet of the instrument which allows the spectra of the analytes to be recorded. The experiment is conducted with a very small flow rate of the mobile phase (D_2O or CD_3CN) or in the *stop-flow* mode. In this mode, the mobile phase is momentarily stopped in order to record the spectrum. This technique, which requires very expensive materials, is of limited use. It is used mainly to isolate and identify very unstable compounds that cannot be isolated through classical means.

3.10   Applications

Because of the extremely wide field of application of this technique, it is used in almost all sectors, including the chemical the industry, the food industry, in the environment, pharmacy and biochemistry (Fig. 3.18). Three specific fields of application are:

— screening for antibiotic drugs in food of animal origin
— optical purity control of therapeutic molecules
— trace and residue analysis in environmental chemistry
— monitoring of the concentration of cytotoxic compounds (chemotherapy).

Problems

3.1   Current HPLC apparatus can use columns of internal diameter (written as ID in the text), of 300 µm, and for which the optimum flow rate advised is 4 µl/min.

1. Show by simple calculation that this flow rate will conduct the mobile phase at practically the same linear speed as in a column of the same type but possessing a standard diameter of 4.6 mm for which the advised flow rate is 1 ml/min.

   The chromatogram below corresponds to an example of separation obtained with a narrow column of 300 µm × 25 cm.

2. What is the dead volume of this column? (The arrow marked on the chromatogram indicates the dead time.)
3. What is the retention volume, expressed in ml, of the compound which has the greatest affinity for the stationary phase?
4. Calculate the efficiency of the column for this compound.
5. Calculate the retention factor of this compound on the column.
6. What is the likely nature of the stationary phase? State precisely the superficial structure of such a phase. (On the figure ACN represents acetonitrile.)

Two columns selected for an experiment are filled with the same stationary phase. One column has an internal diameter of 4.6 mm, the other 300 µm, while both have the same length and an equal rate of filling (V_S/V_M). It is decided to use them in succession with the same chromatograph and under identical conditions with the flow rates as advised above. An equal amount of the same compound is injected into each of them in turn.

7. When passing from one column to the other, would a difference in the retention volumes (or elution volume) of the compound be expected?
8. If the sensitivity of the detector has not been adjusted between the two experiments, will the intensity of the corresponding elution peaks be different?

Separation of PAHs (SRM 1647)

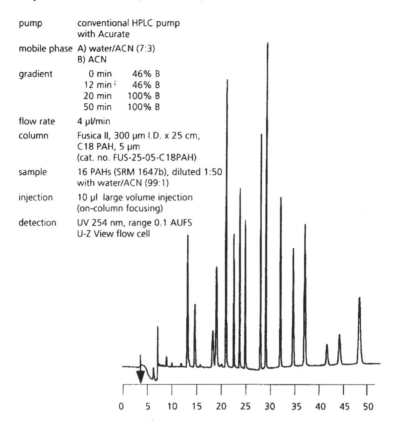

pump	conventional HPLC pump with Acurate
mobile phase	A) water/ACN (7:3) B) ACN
gradient	0 min 46% B
	12 min 46% B
	20 min 100% B
	50 min 100% B
flow rate	4 µl/min
column	Fusica II, 300 µm I.D. x 25 cm, C18 PAH, 5 µm (cat. no. FUS-25-05-C18PAH)
sample	16 PAHs (SRM 1647b), diluted 1:50 with water/ACN (99:1)
injection	10 µl large volume injection (on-column focusing)
detection	UV 254 nm, range 0.1 AUFS U-Z View flow cell

0 5 10 15 20 25 30 35 40 45 50

3.2 What is the order of elution of the following acids from an HPLC column whose stationary phase is of type C18 while the mobile phase is a formate buffer $C = 200\,mM$ of pH 9.

Mixture:
1. Linoleic acid $CH_3(CH_2)_4CH=CHCH_2CH=CH(CH)_7CO_2H$
2. Arachidic acid $CH_3(CH_2)_{18}CO_2H$
3. Oleic acid $CH_3(CH_2)_7CH=CH(CH_2)_7CO_2H$

3.3 Indicate for each of the chromatographic techniques below the term which best expresses the analyte interaction with the stationary phase.

1. Reverse phase   a. Molecular mass
2. Gel permeation   b. Hydrophilicity
3. Ionic chromatography   c. Hydrophobicity
4. Normal phase   d. Protonation/ionisation

3.4 Consider a study, by HPLC, of the separation of three nucleotides (AMP, ADP and ATP), with a column of type RP-18. The mobile phase is a binary mixture of $H_2O - KH_2PO_4$, 0.1 M (pH 6)/methanol (90/10). The compounds appear in the order ATP, ADP, AMP. If a solution of 4 mM tetrabutylammonium hydrogen sulphate is added to the mobile phase, the order of elution of these compounds is reversed.

Explain the reasons for this phenomenon.

3.5 The separation of two compounds A and B is studied by HPLC on a column of type RP-18. The mobile phase is a binary mixture of water and acetonitrile. A linear relationship exists between the logarithm of the retention factor and the % of acetonitrile within the binary mixture (H_2O/CH_3CN).

Two chromatograms were obtained, one in which the mixture of water and acetonitrile was 70/30 v/v, respectively and the other where the same two components were mixed 30/70 v/v. The graph below was constructed from the results. The equations of the straight lines are,

for compound A:

$$\log k_A = -6.075 \times 10^{-3} \, (\% \ CH_3CN) + 1.3283$$

and for compound B:

$$\log k_B = -0.0107 \, (\% \ CH_3CN) + 1.5235$$

1. Find the composition of the binary phase which leads to the selectivity factor $= 1$.
2. Suppose that for each compound the width at half height of the corresponding peak in the chromatogram is the same and the efficiency of the column is not modified following the composition of the mobile phase. For which combination of the binary mixtures mentioned above is the resolution between the two peaks the best? Discuss the practicalities of your choice.

CHAPTER 4

Ion chromatography

Ion chromatography (IC) is a separation technique related to HPLC. However, because it has so many aspects such as the principle of separation and detection methods, it requires special attention. The mobile phase is usually composed of an aqueous ionic medium and the stationary phase is a solid used to conduct ion exchange. Besides the detection modes based on absorbance and fluorescence, which are identical to those used in HPLC, ion chromatography also uses electrochemical methods based on the presence of ions in a solution. The applications of ion chromatography extend beyond the measurement of cations and anions that initially contributed to the success of the technique. One can measure organic or inorganic species as long as they are polar.

4.1 The principle of ion chromatography

The separation of ions or polar compounds present in the sample and transported by the mobile phase is a consequence of their interaction with ionic sites on the stationary phase. The retention of a compound in the stationary phase depends on its charge density. The higher the charge density, the longer the compound will be retained on the stationary phase. This exchange process is much slower than that found in other types of chromatography.

■ Ion chromatography instruments have the same components as those found in HPLC (see Fig. 4.1). They can exist as individual components or as an integrated instrument. The components of the system are made out of inert materials because the mobile phase is composed of acids or alkaline entities that can be highly corrosive. Instruments that operate in the isocratic mode are used more often than those allowing gradient elution.

The detection of ionic species present in the sample is difficult because these analytes are present in low concentration in a mobile phase that contains high quantities of ions. This is why the detection mode based on conductivity needs a special arrangement. It requires a suppression device to be inserted between the column and the detector, which is used to eliminate the ions in the mobile phase via acid-base type reactions.

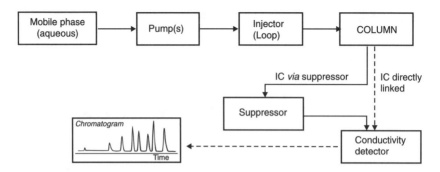

Figure 4.1—*Schematic of an ion chromatograph.* Configuration showing a suppression device in series with a conductivity detector.

4.2 Mobile phases

The mobile phase is an aqueous solution containing electrolytes and, if necessary, a small content of methanol used to dissolve samples that have a low degree of ionisation. In this medium, the solutes and the functional groups of the stationary phase are ionised. The elution mechanism is based on the displacement of ionic equilibria.

Two classical examples are given here, based on ion exchange resins:

— if ionic species of the type M^+ are to be separated, a column with a stationary phase capable of *exchanging cations* will be used (e.g. a polymer containing sulphonate groups). The mobile phase will be made acidic using a mineral or organic acid. The presence of M^+ species in the mobile phase will modify the equilibrium at the exchange sites in the column:

$$(\text{Eluant})M^+ + [\text{Resin-SO}_3]^-H^+ \rightleftharpoons (\text{Eluant})H^+ + [\text{Resin-SO}_3]^-M^+$$

— in a similar fashion, if anions of the type A^- are to be separated, a column capable of *exchanging anions* will be used and the mobile phase will be composed of a solution of hydrogenated carbonates (e.g. sodium). The column can be made of a polymer that contains quaternary ammonium groups. Under these conditions, a species A^- present in the mobile phase will give rise to the following equilibrium.

$$(\text{Eluant})A^- + [\text{Resin-NR}_3]^+(\text{CO}_3H)^- \rightleftharpoons (\text{Eluant})(\text{CO}_3H)^- + [\text{Resin-NR}_3]^+A^-$$

4.3 Stationary phases

Although stationary phases were initially made from resins designed to deionise water, the stationary phases presently used are as complex as those found in HPLC. These materials follow the same requirements of microporosity, granular-metric distribution, mechanical resistance and stability under extreme pH conditions.

4.3.1 Synthetic copolymers

Cross-linked copolymers of polystyrene and divinylbenzene are usually used. These polymers are hard enough to resist pressure in the column and are made of small spheres with a diameter of a few micrometres (Fig. 4.2).

The spheres are chemically modified at the surface in order to introduce functional groups that have acidic or basic properties. Thus, by sulfonation of the aromatic nuclei, a strongly acidic phase is obtained (cationic) on which the anion is fixed to the macromolecule and the cation can be reversibly exchanged with other ionic species present in the mobile phase. These materials, which are stable over a wide range of pHs, have an exchange capacity of a few mmol/g.

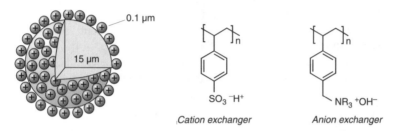

Figure 4.2—*Stationary phases in ion chromatography*. Schematic of a polystyrene sphere used as a cation exchanger. The polystyrene matrix is transformed into a resin that can exchange cations (e.g. DOWEX® 4) or into an anion exchange resin (e.g. DOWEX® MSA-1, or Permutite® if R = CH₃).

In order to obtain an anion exchange resin starting from the same copolymer, chloromethylation (binding of $-CH_2Cl$) can be used and then the tertiary or secondary amine can be reacted upon, depending on the basicity given to the stationary phase. After this transformation, the cation is fixed and the anion is exchangeable (Fig. 4.2).

■ Another approach is based on the copolymerisation of a mixture of two acrylic monomers. One is of the anionic type (or cationic) and the other one is poly-hydroxylated (Fig. 4.3). The latter is used to ensure the hydrophilic character necessary for the stationary phase. A limitation of these resins is their variable swelling, which depends on the composition of the mobile phase. They are normally used for medium pressure chromatography and certain biochemical applications.

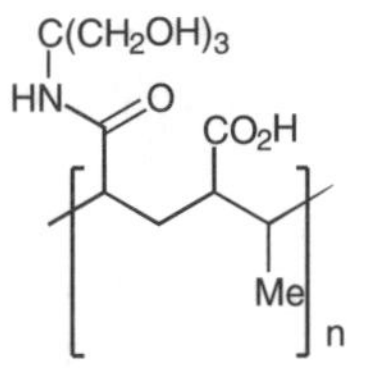

Figure 4.3—*Copolymerisation of two unsaturated monomers (acid and trihydroxyamide)*. Example of the structure obtained (CM-TRISACRYL M® of IBF France).

4.3.2 Modified silica

Porous silica particles can be used to support alkylphenyl chains substituted by sulfonated groups or quaternary ammonium groups using covalent bonding $(\text{silica} - \text{R}' - \text{NR}_3)^+\text{OH}^-$. This approach is similar to that used to obtain the bonded silica phases used in HPLC. Certain phases compound the particularities of each technique. In such cases, separation depends on ionic coefficients as well as the partition coefficient.

4.3.3 Film resins

A polymer called latex, prepared from a monomer that contains organic groups, is deposited as small spheres (0.1–0.3 μm in diameter) on the support to form a continuous film about 1–2 μm thick. The support is made of silica microspheres or spheres of polystyrene of about 25–50 μm diameter (Fig. 4.4).

For a polystyrene support, the latex is linked by polar bonds. The remaining double bonds are responsible for the hardening of the material by reticulation.

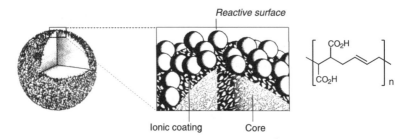

Figure 4.4—*The film resins.* Example of a resin made of hard nuclei on which has been deposited a copolymer made by the reaction of maleic acid on 1,3-butadiene (reproduced by permission of Dionex).

4.4 The principle of separation

The separation of ionic species can occur by many processes: ion exchange, ion suppression or ion pairing (see Fig. 4.9). Because the latter two processes occur in HPLC, only the first process will be described here in detail.

As an example, let us take an anionic stationary phase in which an E^- species is in equilibrium between the mobile phase and the stationary phase. This species, called a counterion, is present in high abundance in the mobile phase. Although the OH^- species would be a simple and logical choice for this counterion, hydrogenated carbonate forms are preferred (CO_3^- and HCO_3^- at 0.003 M). Carbonated species are much more efficient at displacing the ions to be separated. As an anionic species A^- is transported by the mobile phase, and a series of reversible equilibria are produced. These equilibria are dependent on the ionic equilibrium constant K (Fig. 4.5).

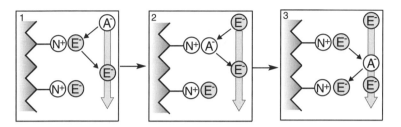

Figure 4.5—*Schematic showing successive exchanges of an analyte anion A^- and a counter anion E^- in the mobile phase in contact with an anion exchange stationary phase.* Initially, the counterion E^- fixed on the stationary phase is exchanged with A^- present in the mobile phase. Then the elution inverses the phenomenon and regenerates the stationary phase with the E^- ions.

If s designates a species A^- or E^- fixed on the stationary phase and m, a species in the mobile phase, an equilibrium will result that is characterised by an exchange constant α that describes the selectivity between the two ions E^- and A^- $(\alpha = K_A/K_E)$.

$$E_s^- + A_m^- \underset{2}{\overset{1}{\rightleftharpoons}} E_m^- + A_s^- \tag{4.1}$$

$$\alpha = \frac{[E^-]_m \cdot [A^-]_s}{[E^-]_s \cdot [A^-]_m} = \frac{1}{K_E} \cdot K_A = \frac{K_A}{K_E} \tag{4.2}$$

K_A and K_E are the distribution coefficients of the two species present in equation (4.2). Using a cation exchange resin, a similar situation can be described.

The exchange phenomenon that allows polar species to be retained on the resin is known as solid phase extraction. If the sample contains two ions X and Y and if $K_Y > K_X$, Y will be retained more than X on the column.

$$K_X = \frac{\text{moles of X per g of stationary phase}}{\text{moles of X per ml of mobile phase}} \tag{4.3}$$

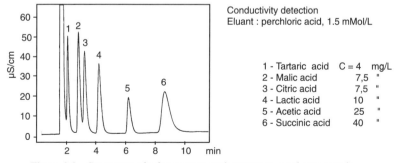

Figure 4.6—*Separation of a few organic acids present in a red wine sample.*

4.5 Conductivity detectors

Besides the spectrophotometric detectors seen in HPLC based on absorbance or fluorescence of UV/Vis radiation, another type of detector based on electrolyte conductivity can be used. This mode of detection measures conductance of the mobile phase, which is rich in ionic species (Fig. 4.6). The difficulty is to recognise in the total signal the part due to ions or ionic substances present in the sample at very low concentrations. In a mobile phase loaded with buffers with a high conductance, the contribution of ions due to the analyte is small. In order to do a direct measurement, the ionic loading of the mobile phase has to be as low as possible and the cell requires strict temperature control (0.01 °C) because of the high dependence of conductance on temperature. Furthermore, the eluting ions should have a small ionic conductivity and a large affinity for the stationary phase.

■ *Review of the conductivity of solutions* – the *conductance* G measured between two electrodes of area A and spacing d inserted into a conducting medium is the reciprocal of resistance R. $G = 1/R$ is expressed in *siemens* (S). For a given ion, the conductance of the solution will vary with the concentration of the electrolyte. This relationship is linear for very dilute solutions. The *specific conductance* (in $S\,mol^{-1}$) or *conductivity* k allows the measure to be independent of the detection cell:

$$k = G \cdot K_{cell} \tag{4.4}$$

K_{cell} designates the *cell constant* ($K_{cell} = d/A$). Its value cannot be obtained by direct measurement, but is determined using a standard solution for which the conductivity k is known. The *equivalent ionic conductance* ($S\,m^2\,mol^{-1}$) refers to the conductivity of an ion with a valence of z in a solution at 25 °C when the molar concentration C ($mol\,l^{-1}$) tends towards 0 (Table 4.1).

$$\Lambda_0 = 1000\, k/C \cdot z \tag{4.5}$$

The variation of the conductance of a solution (ΔG) depends on ΔK, the difference between the equivalent conductance of ion X and that of the elution ion E multiplied by the concentration C_X:

$$\Delta K = C_X(\Lambda_X - \Lambda_E) \tag{4.6}$$

Table 4.1—The equivalent ionic conductivities at infinite dilution in water at 25 °C

ion	H^+	Na^+	K^+	OH^-	HCO_3^-	Cl^-
Λ_0 $(cm^2 \cdot S)/mol$	350	50	74	198	44.5	76

The detection cell placed at the end of the column has a very small volume (*ca.* 2 µl). In order to increase the sensitivity of this detection method, a device neutralising the ions present in the electrolyte is placed between the column and the detector. This device, called the *suppressor*, was initially commercialised in 1975.

4.6 Ion suppressors

The suppressor is a device used when the mobile phase is highly ionic. It reduces the conductivity thus allowing the detection of species present in the sample. Ion exchange-type columns containing functional groups of opposite charge to those used in the separation column were initially used as suppressors. Their action can be described using the following example.

Suppose that the separation of a mixture containing the cations Na^+ and K^+ is to be made using a cation exchange column and a mobile phase composed of hydrochloric acid. In the acidic medium at the end of the column, the Na^+ and the K^+ ions are accompanied by Cl^- anions in order to conserve electroneutrality. After the separation column, the mobile phase flows through a second column that contains an *anionic exchange* resin of which the mobile ion is OH^-. The Cl^- anion will be fixed on the column thus displacing the OH^- ions that will react with H^+ species in solution to give water. After the suppressor, only Na^+OH^- and K^+OH^- species will remain, each of which has a higher conductivity than Na^+Cl^- and K^+Cl^-. H^+ and Cl^- species will have disappeared. This will amplify the detection of Na^+ and K^+ species (Fig. 4.7).

■ In summary, a chemical suppressor containing an anionic resin (e.g. $ArCH_2(NR)_3OH$) is associated to a cationic separation column ($ArSO_3H$) in order to neutralise the mobile phase. The limitation of this type of suppressor lies in its very large dead volume that diminishes separation efficiency by remixing ions before their detection. It must be periodically regenerated and can only be used in the isocratic mode.

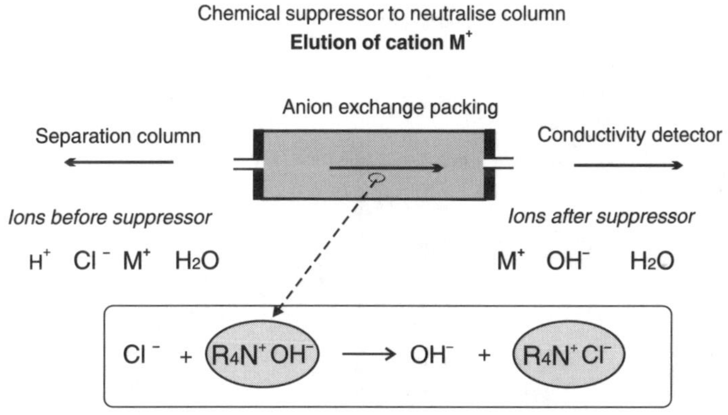

Chemical suppressor to neutralise column
Elution of cation M⁺

Anion exchange packing

Separation column   Conductivity detector

Ions before suppressor   Ions after suppressor

H^+ Cl^- M^+ H_2O M^+ OH^- H_2O

Cl^- + $\boxed{R_4N^+OH^-}$ ⟶ OH^- + $\boxed{R_4N^+Cl^-}$

Figure 4.7—*Chemical suppressor*. In this example describing the detection of a cation, the anionic suppressor purges the mobile phase of H^+ and Cl^- ions. This facilitates the detection of the M^+ cation.

Fibre or micromembrane suppressors of high ionic capacity have now taken over from chemical suppressors. With dead volumes in the order of 50 µl, they allow gradient elution with negligible drift in the baseline. Figure 4.8a shows the passage of an anion A^- in solution in a typical electrolyte used for anionic columns through a membrane suppressor.

Recently, autoregenerating suppressors using electrolytic reactions have been introduced. They are composed of either a special column containing a resin that can be regenerated by electrolysis or a membrane suppressor where the regenerating ions are produced *in situ* by electrolysis of water. The latter process is illustrated in Fig. 4.8b. It represents the passage of a cation in a hydrochloric acid solution through a suppressor, the membrane of which allows the passage of an anion.

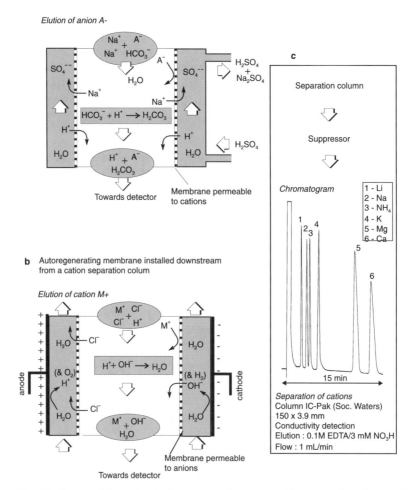

Figure 4.8—*Membrane and electrochemically regenerated suppressors.* Two types of membrane exist; those that allow the permeation of cations (H^+ and Na^+) and those that allow the permeation of anions (OH^- and X^-). a) The microporous cationic membrane model is adapted to the elution of an anion. Only cations can migrate through the membrane (corresponding to a polyanionic wall that repulses the anion in the solution); b) Anionic membrane suppressor placed after a cationic column and in which ions are regenerated by the electrolysis of water. Note in both cases the counter-current movement between the eluted phase and the solution of the suppressor; c) Separation of cations illustrating situation b).

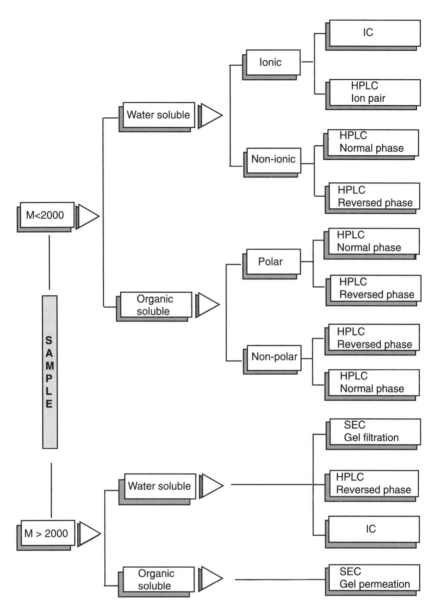

Figure 4.9—*Selection guide for chromatographic methods with various mobile phases (HPLC, IEC and SEC)*. The separation method is chosen as a function of the molar mass, solubility and the polarity of the compounds to be separated.

The separation of anions X^- can be used as another example. If the mobile phase contains diluted sodium hydroxide, a membrane allowing the diffusion of cations is used. The passage of hydronium ions towards the electrolyte will neutralise OH^- ions. At the anode, Na^+ ions will migrate and react with OH^- ions.

APPENDIX ———————————————————————

Quantitative analysis by chromatography

The widespread use of chromatography in quantitative analysis is mainly due to its reliability and to its use in standardised analyses. This type of analysis relies mainly on reproducibility of the separation and on the linear relationship that exists between the *injected mass* of the compound and the *area of the peak* in the chromatogram. The use of an integrating recorder or a microcomputer with the appropriate data treatment software allows automation of all the calculations associated with the analysis. Computer software can analyse the results and produce a computerised report. Trace and ultratrace analyses by chromatography are often the only recognised methods (*EPA Methods for Environmental Analyses*), although their costs are relatively high. The three most widely used methods are described below in their simplest formats.

———————————————

4.7 Principle and basic equation

In order to measure the concentration of a compound appearing as a peak in the chromatogram, two basic conditions have to be met. First, a reference compound of the species to be measured is used to establish the detector sensitivity to this compound. Second, a device giving the areas of the eluting peaks is also necessary. All quantitative methods in chromatography rely on these two principles.

For a given tuning of the instrument, it is assumed that there is a linear relationship between the area of each peak in the chromatogram and the quantity of the compound in the sample. This applies for a given concentration range, depending on the detector. It translates into the following equation.

$$m_i = K_i A_i \qquad (4.7)$$

m_i quantity of compound i injected on the column
K_i *absolute response factor* for compound i
A_i area of the eluting peak for compound i

To calculate the response factor K_i of a compound i, it is essential, according to equation (4.7), to know the injected quantity. However, it is difficult to know precisely the injected volume, which depends on the injector or injection loop or the precision of the syringe. Moreover, the absolute response factor K_i (not to be confused with the partition coefficient) depends on the tuning of the chromatograph. This factor is not an intrinsic property of the compound. This is why most chromatographic methods for quantitative analyses, whether they are pre-programmed into an integrating recorder or software, do not make use of the absolute response factor, K_i.

4.8 Recording integrators and data treatment software

In order to measure peak areas, recording integrators are used. Alternatively specialised software for data treatment can be loaded onto a microcomputer linked to the detector by an appropriate interface card.

The analogue signal produced by the detector is sampled by an ADC circuit with a frequency of a few hundred hertz. The sampling frequency must be sufficiently fast so that the area of the narrowest peak in chromatograms obtained with capillary columns is correctly measured. The values obtained (*digital units* that depend on the units used on both axes) are equivalent to elementary areas that are bracketed by the time interval. These areas depend on the measured signal and a reference signal. The reference signal is also used to determine the baseline in an artificial manner that allows each time rectangle to be closed (Fig. 4.10). Each of the rectangles can be compared to an element of a histogram. Although some variation exists between models, all of these instruments or software allow for baseline correction and treatment of negative signals and all incorporate a calculation method and an integration method (Fig. 4.11).

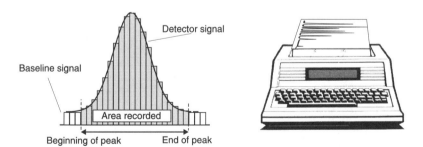

Figure 4.10—*Sampling of a chromatogram.* Because the signal from the detector is not digitised at the exit of the chromatogram, its analogue to digital conversion (ADC) is done at the input of the recording integrator. The chromatogram is stored in memory as numbers corresponding to a histogram that will be later used for data treatment. The area units correspond to the product of the units chosen for each axis, e.g. $\mu A \cdot min$.

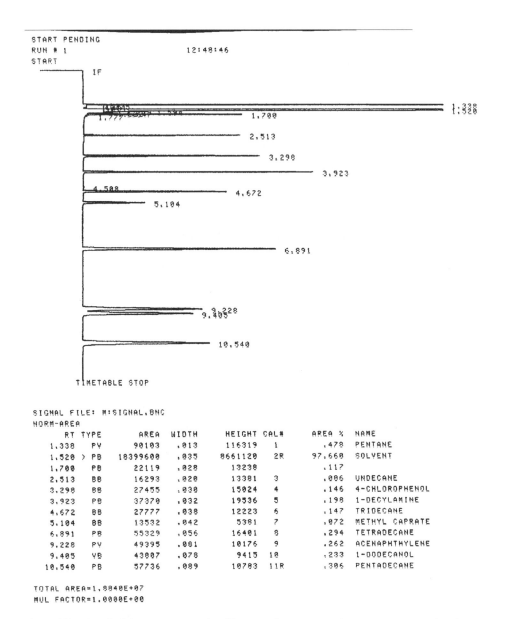

Figure 4.11—*Standard chromatogram produced by a recording integrator showing retention time (min), area of the peak (arbitrary units) and corresponding % of area.* The % area of the peaks must not be confused with the % quantity of the sample constituents since the response factors are not identical for all compounds (non-normalised report).

■ The manual triangulation method used to measure peak areas is no longer in use. However, it is useful to remember that for a Gaussian eluting peak, the product of its width at half height by its height corresponds to approximately 94% of the total area of the peak.

4.9   External standard method

This method, which is easy to use, is common to many quantitative analysis techniques. It allows the measurement of the concentration of one or more components that elute in a chromatogram containing, perhaps, many peaks.

This method, employing the *absolute response factor, K*, is used in the following way (shown in Fig. 4.12): a solution of known concentration C_{Ref} of the reference compound is prepared for which a known volume V is injected. The corresponding area A_{Ref} in the chromatogram is measured. Then, without changing the conditions of analysis, *the same volume V* of the sample containing an unknown concentration of analyte C_{Unk} is injected in the same solvent. The eluting peak that corresponds to the analyte has an area A_{Unk}. Because an identical volume of both samples has been injected, the ratio of the *areas* is proportional to the ratio of the *concentrations* (areas depend on the injected quantity; $m_i = C_i \cdot V$). Using the two chromatograms and equation (4.7), the following relation is obtained:

$$C_{Ref} = K \cdot A_{Ref} \quad \text{and} \quad C_{Unk} = K \cdot A_{Unk}$$

such that:

$$C_{Unk} = C_{Ref} \frac{A_{Unk}}{A_{Ref}} \tag{4.8}$$

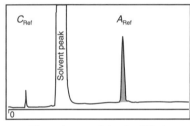 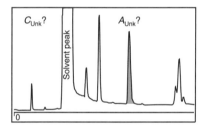

| Chromatogram of standard (C_{Ref}) | Chromatogram of sample solution (C_{Unk}) |

Figure 4.12—*Analysis using an external standard.* Sample chromatogram. This basic method can be improved by using several reference solutions varying in concentration to create a calibration curve. In certain cases, peak height can be used instead of peak area.

■ The single point calibration method assumes that the calibration line goes through the origin. In such a case, precision is increased if the concentration of the reference compound approaches that of the unknown in the sample. It is imperative that instrument tuning does not vary between injections. An excellent reproducibility in the injection volumes is also necessary.

Precision can be increased if several injections of the sample and the reference solutions are made, always using equal volumes. In a multilevel calibration several different amounts of the standard are prepared and analyzed. A regression method is used (e.g. linear least-square, or quadratic least square) and this leads to a more precise value for C_{Unk}. This quantitative method is the only one adapted to gaseous samples.

This simple method is used in industry for repetitive analyses. For such analyses, the chromatograph must be equipped with an autosampler, including a sample tray and an automatic injector. The single reference solution, periodically injected for control purposes, can be used to compensate for baseline drifts. It is not necessary to add an internal standard to each of the samples, as discussed below.

4.10   Internal standard method

For trace analysis, it is preferable to use a method that relies on the *relative response factor* for a compound against a reference compound. The areas of the compounds to be quantified are compared to the area of a reference compound, called an *internal standard*, present at a given concentration in each one of the samples (see Fig. 4.13). This approach can compensate for imprecision due to the injected volume and instrument instability between successive injections. It is superior to the preceding method where all of these factors influence the quantification.

In the following example, it is assumed that two compounds, 1 and 2, are to be quantified and that the internal standard is designated by IS.

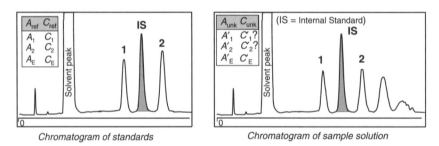

Figure 4.13—*Simple example illustrating the internal standard method.*

4.10.1   Calculation of relative response factors

A solution containing compound 1 at known concentration C_1, compound 2 at known concentration C_2 and the *internal standard* at known concentration C_{IS} is prepared. Assuming that A_1, A_2 and A_{IS} represent the areas of the elution peaks in the chromatogram due to the three compounds in the solution and the quantities of each compound are m_1, m_2 and m_{IS}, equation (4.7) can be used to derive the following:

$$m_1 = K_1 \cdot A_1$$

$$m_2 = K_2 \cdot A_2$$

$$m_{IS} = K_{IS} \cdot A_{IS}$$

whereby:

$$\frac{m_1}{m_{IS}} = \frac{K_1 \cdot A_1}{K_{IS} \cdot A_{IS}} \quad \text{and} \quad \frac{m_2}{m_{IS}} = \frac{K_2 \cdot A_2}{K_{IS} \cdot A_{IS}}$$

These ratios allow the *relative response factors* of 1 and 2 against IS to be derived, which are designated by $K_{1/IS}$ and $K_{2/IS}$:

$$K_{1/IS} = \frac{K_1}{K_{IS}} = \frac{m_1 \cdot A_{IS}}{m_{IS} \cdot A_1} \quad \text{and} \quad K_{2/IS} = \frac{K_2}{K_{IS}} = \frac{m_2 \cdot A_{IS}}{m_{IS} \cdot A_2}$$

Because the quantities m_i injected are proportional to the concentrations in solution C_i ($m_i = C_i \cdot V$), the above equations can be rewritten as follows:

$$K_{1/IS} = \frac{C_1 \cdot A_{IS}}{C_{IS} \cdot A_1} \quad \text{and} \quad K_{2/IS} = \frac{C_2 \cdot A_{IS}}{C_{IS} \cdot A_2}$$

4.10.2   Sample chromatogram – calculation of concentrations

The second step in the analysis is to obtain a chromatogram for a given volume of sample to which has been added a known quantity of compound IS, the internal standard. Assuming that A_1', A_2' and A_{IS}' are the areas in the new chromatogram, and that m_1', m_2' and m_{IS}' are the quantities of compounds 1, 2 and IS introduced into the column under the same experimental conditions, then:

$$\frac{m_1'}{m_{IS}'} = K_{1/IS} \frac{A_1'}{A_{IS}'} \quad \text{and} \quad \frac{m_2'}{m_{IS}'} = K_{2/IS} \frac{A_2'}{A_{IS}'}$$

Using the relative response factors obtained in the first experiment and with knowledge of the concentration of the internal standard C_{IS}', this leads to:

$$C_1' = C_{IS}' K_{1/IS} \frac{A_1'}{A_{IS}'} \quad \text{and} \quad C_2' = C_{IS}' K_{2/IS} \frac{A_2'}{A_{IS}'}$$

Expanding to n components, it is possible to calculate the concentration of component i in the sample using equation (4.9):

$$C_i' = C_{IS}' K_{i/IS} \frac{A_i'}{A_{IS}'} \tag{4.9}$$

and the % concentration of component i can be expressed using equation (4.10):

$$x_i\% = \frac{C_i'}{\text{mass of sample taken}} \times 100 \tag{4.10}$$

This method is more precise if several injections of the standard and the samples are carried out. The internal standard method is of general use and is reproducible but relies on the proper choice of compound for the internal standard, which should have the following characteristics:

- it must be pure and not present initially in the sample
- its elution peak must be well resolved from the other components of the sample

- its retention time must be close to that of the compounds that are to be quantified
- its concentration must be close to or above that of the compounds that are to be quantified in order to obtain a linear response
- it has to be inert with respect to the sample components.

4.11 Internal normalisation

The method of internal normalisation (normalised to 100%) is used for mixtures in which each compound has been identified by its elution peak. Each of the peaks must be well separated from the others in order to fully characterise the sample.

Assuming that the concentrations of three components (1, 2 and 3) present in a mixture are to be determined (Fig. 4.14), the analysis is again carried out in two steps.

4.11.1 Calculation of the relative response factors

A reference solution containing the three compounds at known concentrations C_1, C_2 and C_3 is prepared. The chromatogram obtained by injection of a volume V shows three peaks of areas A_1, A_2 and A_3. These areas will be related to the quantities m_1, m_2 and m_3 of the compounds contained in the injected solution according to equation (4.7).

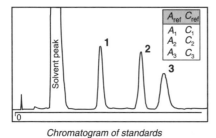
Chromatogram of standards
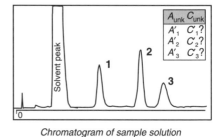
Chromatogram of sample solution

Figure 4.14—*Internal normalisation method.*

In this approach, one of the substances is considered for internal normalisation. For example, if compound 3 is used to determine the *relative response factors* $K_{1/3}$ and $K_{2/3}$ of compounds 1 and 2 with respect to 3, one obtains:

$$K_{1/3} = \frac{K_1}{K_3} = \frac{m_1 \cdot A_3}{m_3 \cdot A_1} \quad \text{and} \quad K_{2/3} = \frac{K_2}{K_3} = \frac{m_2 \cdot A_3}{m_3 \cdot A_2}$$

Because $m_i = C_i \cdot V$, the following expressions are obtained for $K_{1/3}$ and $K_{2/3}$:

$$K_{1/3} = \frac{C_1 \cdot A_3}{C_3 \cdot A_1} \quad \text{and} \quad K_{2/3} = \frac{C_2 \cdot A_3}{C_3 \cdot A_2}$$

4.11.2   Chromatography of the sample – calculation of concentrations

A mixture containing components 1, 2 and 3 is analysed by injection into the chromatograph. For the eluting peaks with areas A_1', A_2' and A_3', the % composition of the mixture represented as x_1, x_2 and x_3 can be obtained from an expression of the following form:

$$x_i\% = \frac{K_{i/3} \cdot A_i'}{K_{1/3} \cdot A_1' + K_{2/3} \cdot A_2' + A_3'} \times 100 \quad (\text{with } i = 1, 2, 3)$$

and the condition for normalisation is $x_1 + x_2 + x_3 = 100$.

■ If the procedure is extrapolated to n components, a general expression for reference to solute j can be obtained:

$$K_{i/j} = \frac{C_i \cdot A_j}{C_j \cdot A_i} \tag{4.11}$$

It is also possible to obtain $K_{i/j}$ by plotting a concentration response curve for each of the solutes. If A_i designates the area of the ith peak in the mixture containing n components and the internal reference is j, the content of component i can be expressed using the following equation:

$$x_i\% = \frac{K_{i/j} \cdot A_i'}{\displaystyle\sum_{i=1}^{n} K_{i/j} \cdot A'} \times 100 \tag{4.12}$$

Problems

4.1   0.604 g of an undiluted stationary phase, comprising SO_3H groups is introduced to an erlenmeyer. 100 ml of distilled water, is added and approximately 2 g of NaCl. The liberated acidity is measured by sodium hydroxide 0.105 M in the presence of helianthin (methyl orange), as an indicator to identify the equivalence point (towards pH 4). If it is known that 25.4 ml of the sodium hydroxide solution must be added to neutralise the acid liberated by the stationary phase, calculate its molar capacity in grams.

4.2   A mixture of proteins is separated on a column with a stationary phase of carboxymethylated cellulose. The internal diameter of the column is 0.75 cm and its length is 20 cm. The dead volume is 3 ml. The flow rate of the mobile phase is 1 ml/min. The pH of the mobile phase is adjusted to 4.8. Three peaks appear upon the chromatogram corresponding to the elution volumes V_1, V_2 and V_3 at 12 ml, 18 ml and 34 ml respectively.

   1. Does this arise from an anionic or cationic phase? Give reasons for your answer.

2. Why, when increasing the pH of the mobile phase, are the times of elution of the three compounds subject to modification? Predict whether these times are increased or decreased.

4.3 In measuring cyclosporin A (a treatment for skin and organ transplant rejection) by HPLC, according to a method derived from that of internal standard, the following procedure is employed.

Preparation of the samples: An extraction of 1 ml of blood plasma is made, to which is added 2 ml of a mixture of water, and acetonitrile (80/20), containing 250 ng of cyclosporin D as internal standard, which has a similar structure to cyclosporin A.

The 3 ml of new mixture is now passed on a disposable solid phase extraction column in order to separate the cyclosporins retained upon the sorbent. Following the rinsing and drying of the column, the cyclosporins are eluted with 1.5 ml of acetonitrile and are then concentrated to 200 μl following evaporation of solvent. A fraction of this final solution is injected into the chromatograph.

1. What is the concentration factor of the original plasma following this treatment?
2. Can the rate of recuperation following the extraction step on the solid phase be deduced?

Sample: The standards are created with a plasma originally possessing no cyclosporin. Six different solutions are prepared by adding the necessary quantities of cyclosporin A to each in order to create solutions of 50 ng/ml, 100 ng/ml, 200 ng/ml, 400 ng/ml, 800 ng/ml, and 1000 ng/ml. 1 ml from each of these solutions is subjected to the same extraction sequence following the addition of 250 ng of cyclosporin D to each, as above.

ng/mL en cyclo. A	50	100	200	400	800	1 000
Ratio of peaks heights cyclo.A/cyclo.D (R_h)	0.25	0.5	1.02	2.04	4.05	5.1

Column Supelco 75 × 4.6 mm. Silica gel 3 μm, phase RP-8.

3. Determine the concentration in units of ng/ml of cyclosporin A in the blood plasma giving rise to the chromatogram reproduced below. This question should be attempted in two ways:

a) By choosing a single point from a standard.
b) By using the gradient which it is possible to draw from the data in the table above (in both cases it will based upon the heights of the peaks).

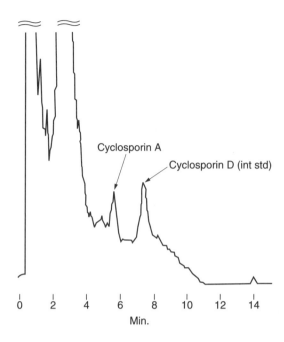

4.4 The method of internal normalisation was chosen to determine the mass composition of a sample comprising a mixture of four esters of butanoic acid. To
this end, a reference solution containing known % masses of these esters led to
the following relative values of the response coefficients of the butanoates of
methyl (ME), of ethyl (EE), and of propyl (PE), all three in ratio with butyl-
butanoate (BE).

$$K_{ME/BE} = 0.919 \qquad K_{EE/BE} = 0.913 \qquad K_{PE/BE} = 1.06$$

From the chromatogram of the sample under analysis, reproduced below,
and the information given in the table, find the mass composition of this
mixture (ignore the first peak at 0.68 min.)

Peak no.	t_R	compound	Area (mV. min)
1	0.68	–	0.1900
2	2.54	methyl ester (ME)	2.3401
2	3.47	ethyl ester (EE)	2.3590
3	5.57	propyl ester (PE)	4.0773
5	7.34	butyl ester (BE)	4.3207

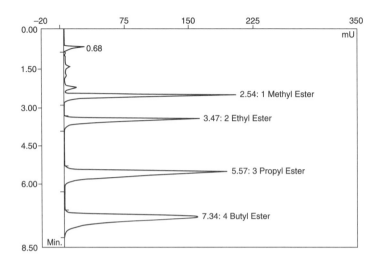

4.5 To measure serotonin (5-hydroxytryptamine), by the internal standard method, a 1 ml aliquot of the unknown solution is added to 1 ml of a solution containing 30 ng of N-methyl-serotonin. This mixture is then treated to remove all other compounds which could interfere with the experiment. The operation performed was an extraction in the solid phase to isolate the serotonin and its methyl derivative, diluted in a suitable medium.

1. Why is the compound forming the internal standard added before the extraction step?
2. Calculate the response factor of the serotonin compared to that of N-methyl-serotonin if it is known that the chromatogram yielded by the standards gave the following results.

Name	Area ($\mu V.s$)	Quantity injected (ng)
Serotonin	30 885 982	5
N-methyl-serotonin	30 956 727	5

3. From the chromatogram of the sample solution, find the concentration of serotonin in the original sample, if it is known that:
   — area serotonin   2 573 832 $\mu V/s$
   — area N-methyl-serotonin   1 719 818 $\mu V/s$

CHAPTER 5

Planar chromatography

Planar chromatography, also known as Thin Layer Chromatography (TLC), is a technique related to HPLC but with its own specificity. Although these two techniques are different experimentally, the principle of separation and the nature of the phases are the same. Due to the reproducibility of the films and concentration measurements, TLC is now a quantitative method of analysis that can be conducted on actual instruments. The development of automatic applicators and densitometers has lead to nano-TLC, a simple to use technique with a high capacity.

5.1 Planar chromatography: principles of application

Separation of sample components using planar chromatography (also known as Thin Layer Chromatography, TLC) is conducted on a rectangular plate made out of glass, plastic or aluminium. The plates, of a few centimetres in dimension, are coated with a thin film (100–200 µm) of stationary phase that is usually silica gel or modified silica gel. The cohesion of the phase and its binding to the support is ensured by an organic ligand or a mineral introduced when the film is deposited.

In order to conduct a separation, the sample is deposited about 1 cm from the bottom of the plate. The sample volume is small (a few nl to a few µl) and forms a small spot of about 1–3 mm in diameter. Sample deposition can be done manually or automatically, using a flat-end capillary (Fig. 5.1). The spot can also be a horizontal band of a few millimetres deposited by automatic spraying of the sample. This method has the advantage of having a high reproducibility of the quantity of sample deposited, essential for quantitative analysis.

The sample end of the plate is then put in a covered chamber. The chamber contains a small quantity of the mobile phase, which is in contact with the plate to a depth of a few millimetres. A typical model of the chamber is shown in Fig. 5.2. If the mobile phase contains water (reversed phase) it can be useful to add a salt (NaCl or LiCl) to limit diffusion phenomena and increase resolution.

The mobile phase migrates by capillary action through the fixed stationary phase, thus inducing differential migrations of the sample components. Migration time depends on many parameters. Once the solvent has travelled a sufficient distance – a few centimetres – the plate is withdrawn from the chamber and the mobile phase is evaporated.

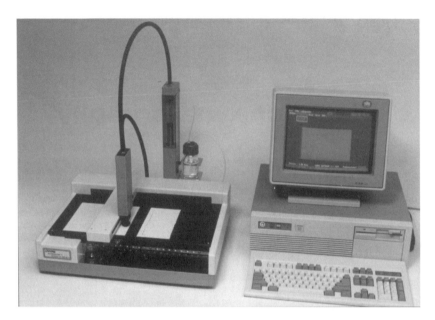

Figure 5.1—*Automatic deposition used in analytical TLC. Programmable ATS-3 applicator* (reproduced by permission of Desaga).

5.2   Post-chromatography: development of the plate

Although the detection of coloured compounds on the plate is straightforward, the detection of colourless compounds is difficult and the plate must be developed. Manufacturers sell plates on which a fluorescent salt of zinc is added to the stationary phase in order to facilitate the localisation of the spots. When the plate is irradiated with a mercury lamp, dark spots appear over a fluorescent background indicating the position of the compounds.

Another universal method of development consists of heating the plate after spraying it with sulphuric acid, which carbonises the compounds to make them visible. This approach, however, cannot be used in quantitative TLC. It is preferable to develop the plate by using a general (phosphomolybdic acid) or specific (ninhydin in alcohol for amino acids) reagent. Hundreds of reagents have been described that can introduce chromophores or fluorophores into the analyte molecules after separation.

■ The use of square plates allows two-dimensional chromatography to be performed using two different mobile phases. The second migration is performed after rotating the plate a quarter turn (Fig. 5.3). A typical application of this approach, although seldom used for quantitative analysis, is the separation of amino acids.

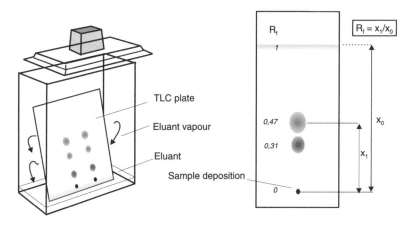

Figure 5.2—*Vertical TLC chamber and TLC plate.* Typical appearance of a TLC plate after development.

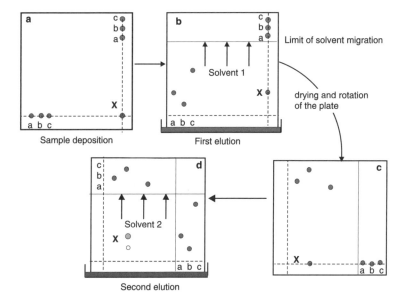

Figure 5.3—*Principle of two-dimensional TLC.* The use of distinctly different mobile phases is recommended for a better separation. a) Deposition of three standards and an unknown; b) Migration with the first solvent; c) drying and rotation of the TLC plate; d) migration with the second solvent. Conclusion: *x* is probably a, but the impurity is not b or c.

5.3　Stationary phases

The choice of a stationary phase involves the evaluation of several factors and physico-chemical parameters. The size of the particles, their specific surface area, the volume of the pores and the distribution of particle diameters are all factors that define the properties of the stationary phase. For nano-TLC, the particles are in the order of 4 µm and the pores are 6 nm.

The ratio between the silanols and siloxane groups determines the hydrophilic character of the phase. Like to HPLC, it is possible to use bonded silica phases in which various chains are bound to silanol groups at the surface through covalent bonds. Some phases incorporate simple alkyl chains (RP-2, RP-8, RP-18), while others include organic functional groups (nitriles, amines or alcohols) allowing these phases to be used with any type of mobile phase. For example, Fig. 5.4 shows the results obtained for the separation of a test sample containing three components of different polarity. The mixture has been chromatographed on six stationary phases using binary mobile phases, one polar and the other nonpolar.

Chemically modified cellulose supports are also used in TLC, either as fibres or a crystalline powder. The most widely known of these is DEAE-cellulose, a basic phase containing diethylaminoethyl groups. The hydrophilic character of other polar phases with ion exchange properties can be used for the separation of ampholytes.

■ The use of mineral ligands (types of plaster) that confer fragility to the stationary phase can be used when one wants to recover the compounds after their separation. This can be done by removing the zones of interest from the support and extracting them using a solvent.

5.4   Retention and separation parameters

In TLC each component is defined by its R_f value (retardation factor) that corresponds to its relative migration compared to the solvent:

$$R_f = \frac{\text{distance moved by the solute}}{\text{distance moved by the solvent front}} = \frac{x}{x_0} \tag{5.1}$$

The efficiency N of the plate for a compound with a migration distance of x and spot diameter of w is given by equation (5.2) and H (HETP) is given by equation (5.3):

$$N = 16\frac{x^2}{w^2} \tag{5.2}$$

and

$$H = \frac{x_0}{N} \tag{5.3}$$

In order to calculate the retention factor k of a compound or the selectivity coefficient between two compounds, the distance migrated along the plate is usually translated into the migration time for the chromatogram. Assuming that the *ratio* of the migration velocities u/u_0 is the same on the plate as on the column (this is only an approximation), then R_f can be related to k:

$$R_f = \frac{x}{x_0} = \frac{\bar{u}}{\bar{u}_0} = \frac{t_0}{t} = \frac{1}{k+1}$$

such that

$$k = \frac{1}{R_f} - 1 \tag{5.4}$$

Finally, by comparison to equation (1.24), the resolution can be given by:

$$R = 2\frac{x_2 - x_1}{w_1 + w_2} \tag{5.5}$$

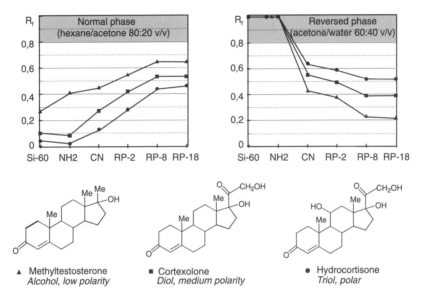

Figure 5.4—*Separation of three steroids on six stationary phases using two sets of solvents*. The evolution of the response factor R_f and reversal of migration order with changes in mobile phase should be noted (reproduced by permission of Merck).

5.5 Quantitative TLC

It is essential to be able to quantify the spots in order to use TLC as a quantitative method of analysis (Fig. 5.5). This is done by scanning the plate with a densitometer that can measure either absorption or fluorescence at one or many wavelengths. This instrument produces a pseudo-chromatogram that contains peaks whose areas can be measured. This corresponds to an isochronic image at the end of the separation.

■ The detection of radioactively labelled compounds (β^-) is done using special densitometers that are equipped with a video camera giving an image of the radio-activity on the plate. The old process of autoradiography, performed by putting a photoplate in contact with the TLC plate, had the disadvantage of low sensitivity (and a long exposure time of 48 hours). However, these new instruments, known as

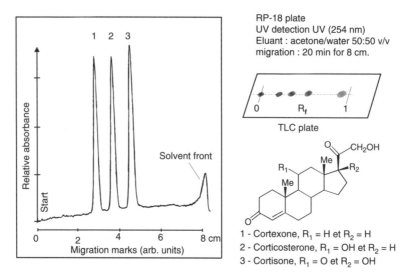

Figure 5.5—*A pseudo-chromatogram obtained by scanning a TLC plate.*

Charpak machines, which are also used in gel electrophoresis, have sufficient sensitivity to detect activities in the order of a few Becquerels per mm^2.

For many applications, TLC can now replace HPLC. Although TLC presently requires more manual manipulation than HPLC, new tools for spotting, gradient elution, plate development and recording yield adequate reproducibility. Results obtained by TLC are now comparable to those obtained by HPLC. Compared to HPLC, TLC can analyse four times more sample in a given period of time. Several analyses can be conducted simultaneously on a given plate under similar experimental conditions. A TLC plate, only used once, allows for more rapid sample preparation with less risk of contamination or loss of the compound to be analysed. It can be very useful for the analysis of biological samples. Moreover, medium polarity bonded stationary phases are available, some of which allow the separation of enantiomers. Thus TLC is far from being the screening technique which it is still perceived as by many chemists (see Fig. 5.7).

5.6   Aspects of TLC

Inasmuch as the vapour phase is in equilibrium with the mobile and the stationary phase, TLC corresponds to a three-phase system. As in HPLC, the migration process is composed of a series of adsorptions and desorptions but, in TLC, more complex phenomena exist for the following reasons:

— the support is only partially equilibrated with the liquid phase before the migration of the compounds

— the adsorption capacity of the surface is substantially diminished when parts of the adsorption sites are occupied. This creates an elongation of the spots. Thus the R_f of a compound in its pure state differs from the R_f of the same compound present in a mixture. Knowledge of the eluotropic properties of each solvent is thus very useful

— the speed of migration of the solvent front is not constant. It is a complex function in which the size of the stationary phase particles is an important factor. The migration velocity can be described by a quadratic function: $x^2 = kt$ (x represents the distance of the migration front, t the time and k is a constant). Thus the resolution between two spots depends greatly on the R_f values of the compounds. The resolution is usually a maximum for an R_f value of about 0.3 (see Fig. 5.6).

In summary, the efficiency N of a TLC plate is variable. The height equivalent to a theoretical plate has a minimum value, as in HPLC. However, it is not possible, unlike in HPLC, to vary the flow rate of the mobile phase in order to increase separation efficiency.

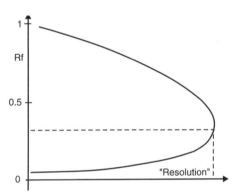

Figure 5.6—*Example of interdependency of R_f values on resolution in an HPLC column.* It is possible to transpose results obtained by TLC onto HPLC using the same phases. Resolution between two compounds on a TLC plate varies with the distance of migration on the plate. Resolution passes through a maximum for an R_f value in the order of 0.3.

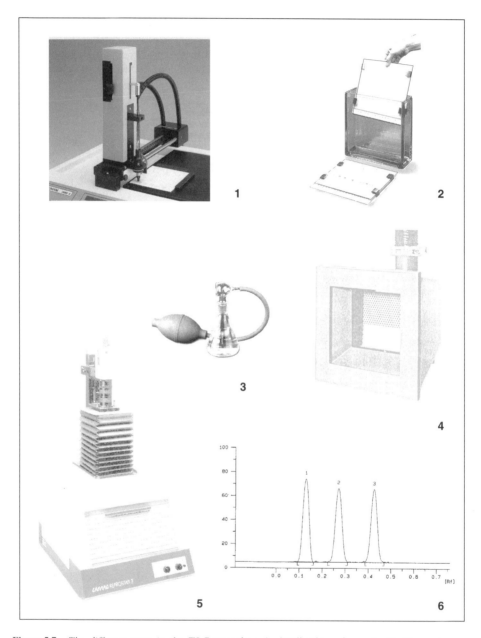

Figure 5.7—*The different steps in the TLC procedure.* 1. Application of sample. 2. Chromatogram development. 3. Atomiser operating from a rubber pump. 4. Spray cabinet. 5. Cabinet with a video camera for acquiring image of TLC (Adapted from Camag). 6. Integration of chromatogram.

Problems

5.1 A mixture of two compounds A and B migrates from the origin to leave two spots with the following characteristics (migration distance x and spot diameter w).

$$x_A = 27 \text{ mm} \qquad w_A = 2.0 \text{ mm}$$
$$x_B = 33 \text{ mm} \qquad w_B = 2.5 \text{ mm}$$

The mobile phase front was 60 mm from the starting line.

1. Calculate the retardation factor R_f, the efficiency N and the HETP H for each compound.
2. Calculate the resolution factor between the two compounds A and B.
3. Establish the relation between the selectivity factor and the R_f of the two compounds. Calculate its numerical value.

5.2 The following figure represents the results of scanning a TLC plate in normal phase (mobile phase: hexane/acetone 80/20). The three compounds have the structures A, B and C.

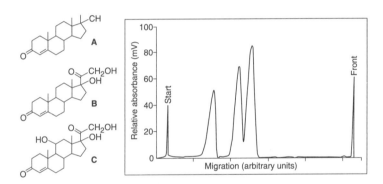

1. Indicate compounds A, B and C, from the identification of the three principal peaks of the recording.
2. What would have been the order of elution of these compounds if examined by an HPLC column containing the same types of stationary and mobile phases?
3. What would have been the order of elution of these compounds if examined by an HPLC column containing a phase of type RP-18 with a binary mixture of acetonitrile/methanol (80/20) as eluant?
4. Calculate the R_f and the HETP for the compound which migrates fastest upon the plate (use the transposed formulae of column chromatography, in particular that giving efficiency, with x as distance of migration).

The electronic balance

Before any quantitative analysis can be carried out, it is almost always necessary to weigh either the sample or to weigh compounds to prepare a standard solution. The weighing scale, one of the oldest means of quantitative analyses, is an indispensable tool that has not escaped technological progress.

Today, electronic scales have replaced scales using weights. The scale is composed of a coil and a magnet whose axes are concentric (see following figure). When an electrical current passes through the coil, it creates a magnetic field within the coil oriented along the longitudinal axis. The value of the magnetic field depends on the electrical intensity.

At rest, the coil is suspended over the magnet (1). It then reaches an equilibrium position (2) that is sensed by an opto-coupler or detection device.

When a weight is deposited on the scale, the coil reaches a new equilibrium position under the mechanical effect and that of the electromagnetic force (3). The current in the coil will then be increased automatically until the coil comes back to its initial equilibrium position as detected by the opto-coupler (4). The intensity of the electrical current, related to the mass deposited on the scale, is then digitised and sent to the microprocessor that activates the numerical display on the scale.

For analytical purposes, electronic scales are used which have a precision in the order of 1/10 or 1/100th of a mg.

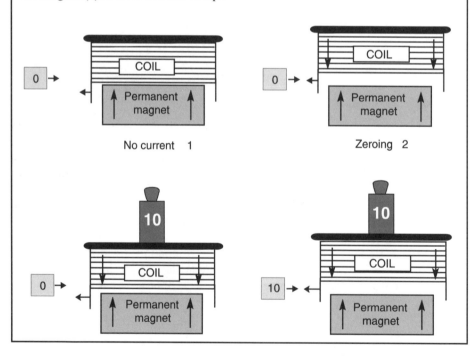

No current   1   Zeroing   2

CHAPTER 6

Supercritical fluid chromatography

Supercritical Fluid Chromatography (SFC), began to be used in industry in the 1980s, is different in many ways to GC or HPLC. Its originality stems from the nature of the mobile phase, which is a supercritical fluid. This can be used for the separation of thermolabile compounds or compounds of high molecular weight. The instrument is a hybrid between GC and HPLC. Capillary columns like those used in GC can be utilised or, alternatively, standard columns like those used in HPLC. The present trend is to use the latter. The late arrival of this technique on the instrumental market can be considered as a drawback, partly due to the fact that normalised methods whose performance is satisfactory have already been developed using standard chromatography techniques. Also, because instrumentation for SFC is more complex and expensive, this technique is still not widely employed.

6.1 Supercritical fluids

The use of supercritical fluids as mobile phases in chromatography can offer several advantages because their properties are between those of liquids and those of gases. In particular, the viscosity of a supercritical fluid is almost that of a gas (50 times lower than that of a solvent) while its solvation properties (governed by the distribution coefficients K) are similar to those of a nonpolar solvent such as benzene.

■ Transformation of a pure compound from a liquid to a gaseous state and vice versa corresponds to a phase change that can be induced over a *limited domain* by pressure or temperature. For example, a pure substance in a gas phase cannot be liquefied above a given temperature, called the critical temperature T_c, irrespective of the pressure applied to it. The minimum pressure required to liquefy a gas at its critical temperature is called the critical pressure, P_c. The boundary between gaseous and liquid states stops at the critical point C (see Fig. 6.1). Under these conditions, gaseous and liquid states have the same density.

Depending on temperature and pressure, the behaviour of a supercritical fluid is sometimes similar to that of a gas and sometimes similar to that of a liquid.

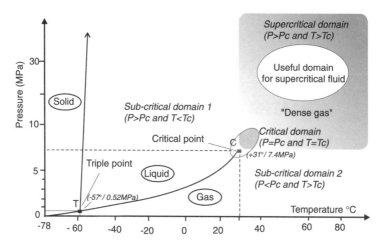

Figure 6.1—*Partial representation of the phase diagram (pressure-temperature) of carbon dioxide.* The critical point is located at 31 °C and 7.4 MPa (1 MPa $= 10^6$ Pa, or 10 bar).

■ Under laboratory conditions, carbon dioxide in its supercritical state allows the separation of unstable compounds from the matrix. It is also used in industrial processes to extract certain food products (for example, decaffeination, recovery of aromas and spices, elimination of fats). Carbon dioxide has the advantage that it can be eliminated at rather low temperatures without leaving any toxic residue. However, the use of relatively high pressures creates potential hazards in industrial installations.

6.2   Supercritical phases as mobile phases

Carbon dioxide is often used in SFC because its critical pressure and temperature are relatively easy to obtain. The critical point for this compound has the following values: $T_c = 31$ °C and $P_c = 7\,400$ kPa (Fig. 6.1). Above these conditions, the supercritical domain is obtained.

Nitrous oxide and ammonia are also used for supercritical mobile phases, albeit more rarely (N_2O, $T_c = 36$ °C, $P_c = 7\,100$ kPa; NH_3, $T_c = 132$ °C, $P_c = 11\,500$ kPa).

■ At 16 000 kPa and 60 °C, carbon dioxide has a density of 0.7 g/ml (see Fig. 6.2). Under these conditions, its polarity is similar to that of toluene and this is why the expression 'dense gas' is used in order to indicate that it is not a classical gas. Because of its low polarity, it is often customary to add an organic modifier such as methanol, formic acid or acetonitrile to the supercritical fluid.

The interesting point about these supercritical fluids is that it is possible to change their density, thus changing their solvation power, simply by modifying the pressure applied to them. A pressure gradient in SFC is equivalent to an elution gradient in HPLC or a temperature gradient in GC. Using a double gradient,

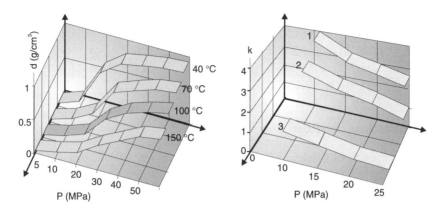

Figure 6.2—*Density for carbon dioxide as a function of pressure at four different temperatures.* At the critical point, the density of CO_2 is $0.46 \, g/cm^3$. The figure on the right represents the variation of the retention factor k for three alkaloids analysed under identical conditions and pressure, fixed downstream from the column by a restrictor ($T = 4 \, °C$, modifier: 5% water and 15% methanol; 1: codeine, 2: thebaine, 3: papaverine). As the pressure increases, k decreases.

pressure-temperature, and an organic modifier, it is possible to finely tune the retention of analytes, thus modifying the selectivity α (Fig. 6.2).

6.3   Comparison of SFC with HPLC and GC

SFC is complementary to other classical techniques of liquid or gas chromatography. The migration of the analyte is explained by a dissolution–precipitation mechanism that depends on the solvation power of the mobile phase. Thus, it is governed by the pressure that determines the density of the supercritical phase. Resistance to mass transfer between the stationary and mobile phases is less than that found in HPLC because diffusion is faster. The C factor in Van Deemter's equation is smaller so the velocity of the mobile phase can be increased (see Fig. 6.3). Moreover, because the viscosity of the mobile phase is similar to that of a gas, it is possible to use capillary columns like those used in capillary GC. However, the

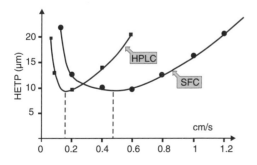

Figure 6.3—*Comparison between HPLC and SFC.* Both experimental curves have been obtained using the same column and the same compound but, in one, a classical liquid phase was used and in the other, carbon dioxide in its supercritical state was used. The HETP are comparable; however, the separation can be achieved 3 to 4 times more rapidly using SFC – a substantial gain in analysis time.

distribution coefficients are modified because of the pressure drop across the column. This causes peak broadening. Therefore, the efficiency found in capillary GC is never obtained with the SFC technique.

6.4 Instrumentation

SFC chromatographs represent hybrids between GC and HPLC instruments (Fig. 6.4). In order to deliver the supercritical fluid, syringe pumps or reciprocal pumps are used and maintained above the critical temperature using a cryostat regulated at around 0 °C. In instances where an organic modifier is used, a tandem pump is employed which has two chambers, one for the critical fluid and one for the modifier. The liquid then passes through a coil maintained above the critical temperature so that it is converted into a supercritical fluid. Stainless steel packed columns like those used in HPLC (1 to 4 mm in diameter) or fused silica capillary columns like those used in capillary GC (2 to 20 m in length, internal diameters as low as 50 μm and stationary phase film thickness of at least 1 μm) are used in SFC.

A flame ionisation detector is often used (FID) unless an organic modifier has been used. Typical HPLC detectors, such as UV/VIS, fluorescence or a specific detector based on light diffusion can also be used (cf. Fig. 7.5).

The pressure restrictor that controls the pressure in the column is installed before or after the detector, depending on its type (before for an FID, after for a UV detector).

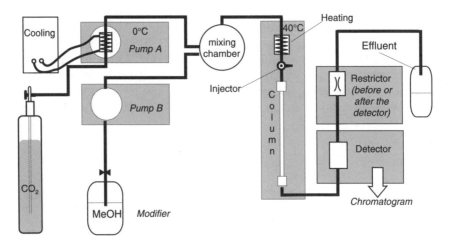

Figure 6.4—*Schematic of a supercritical fluid chromatograph.* Carbon dioxide reaches a supercritical state between the pump and the injector. A pressure regulator (restrictor) is located after the column and either before or after the detector, depending on its type. It allows the mobile phase to be kept under supercritical conditions until its exit from the column.

6.5   SFC in chromatographic techniques

The ability of SFC to control selectivity coefficients between compounds is of special interest. Its applications are mostly directed towards the separation of thermolabile compounds or compounds of high molecular weight using HPLC columns. The low viscosity of the mobile phase permits the connection of several columns in a series. This explains why several studies using SFC are directed at the analyses of oligomers (Fig. 6.5). Because it is more rapid than HPLC, SFC can be used for process analyses. The use of carbon dioxide facilitates coupling to either a mass spectrometer or an infrared spectrophotometer: the absorption bands for carbon dioxide are observed between $3\,800$ to $3\,500$ and $2\,500$ to $2\,200\,cm^{-1}$; areas in which organic compounds do not give specific absorptions.

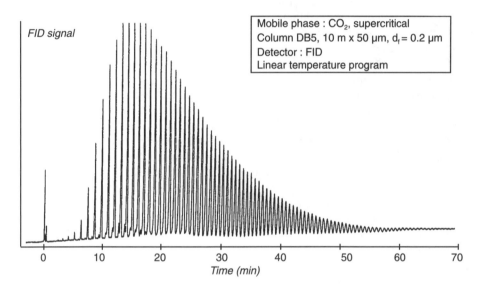

Figure 6.5—*Chromatogram of a mixture of polysiloxane oligomers analysed by SFC.* Each compound yields a unique peak in the chromatogram, thus allowing the determination of the distribution of molecular forms in a polymerisation reaction. (Reproduced by permission of FISONS Instruments.)

Trace analysis

Increasingly sensitive instruments are now used in modern chemical analysis. It is thus possible to measure analytes at very low concentrations. If preconcentration is used, *trace* or *ultratrace* amounts can be measured.

The notion of trace is related to the concentration present in the sample. A compound is considered to be in trace amount when its concentration in the environment is less than 1 000 ppm. Below 1 ppm, it is considered to be at the ultratrace level.

Obviously, these definitions do not represent strict borders. It depends on the analyte present and the matrix. Even these terms have a subjective character. Thus, if a sample of acetone used as a solvent contains 0.1% methanol, the latter will be considered as trace (1 000 ppm). However, if the sample constitutes drinking water, that quantity of methanol would be considered enormous.

In order to conduct these analyses, the detection limit of the instrument must be known. The detection limit is defined (in ppm or ppb) as the *concentration* of analyte that allows a *detectable signal* to be measured with *certainty* – for example, three times the standard deviation of the background signal or the blank. If the volume of solution needed to obtain these results is known, the preceding values can be transformed to the absolute quantities or *mole fractions* (picomole, femtomole, etc.) that are needed to obtain the signal. In general, these values are excessively small because current instruments use excessively small volumes.

For example, if 1 µl of a 1 ppb solution (1 pg) is injected into a chromatograph, this will represent 10 femtomoles for a compound with a molecular weight of 100 g/mol! The analyst must become accustomed to using prefixes that have seldom been used before: femto- (10^{-15}), atto- (10^{-18}), zepto- (10^{-21}). Thus the zeptomole contains only 602 molecules. These limits tend towards single atoms or molecules – the ultimate trace.

The detection of individual species is possible with certain methods. One can, with radioactive isotopes, measure a single atom. This, however, does not imply that methods using radioisotopes are better than those using stable isotopes. This paradox is explained by considering that in order to measure a radioactive atom, it is necessary that it decomposes *during the time of the measurement*. Thus if a radio-element has a long lifetime, the chance of observing this decomposition will be small.

Fluorescence or mass spectrometry are other means that permit detection of the signal arising from a single molecule.

If the methods used for trace analyses are the same as those normally used to measure compounds present in higher abundance, much more care is needed. Many difficulties are encountered in the measurement of ultratrace amounts. At this level, flasks and reagents appear to be more contaminated and one often has to work in a clean room.

CHAPTER 7

Size exclusion chromatography

Size exclusion chromatography (SEC) is based on a particular type of separation. The migration velocity of a given component depends on its size thus, indirectly, on its molecular mass. The instrumentation used is similar to that used in HPLC. Although its column efficiency is far from ideal, SEC has become an irreplaceable tool in the area of the separation of natural or synthetic macromolecules. However, its application in analysis is more limited: the separation of compounds according to their size is not the most efficient process for small or intermediate molecules. It is, however, a very useful process in industry for the separation of compounds with very different masses.

7.1 The principle of SEC

Size exclusion chromatography (SEC), also called *gel filtration* when the mobile phase is aqueous, and *gel permeation* (GPC) when the mobile phase is organic, is based on the differing degrees of penetration of the sample molecules into the pores of the stationary phase (Fig. 7.1). These pores must have a diameter similar to the size of the species to be separated *when these species are in solution* in the mobile phase (Fig. 7.5).

The total volume of the mobile phase in the column V_M can be split into two parts: the interstitial volume V_I (external to the pores) and the volumes of the pores V_P. V_I represents the volume of the mobile phase necessary to transport a big molecule, excluded from the pores, and $V_M = V_I + V_P$ is the volume corresponding to the elution of a small molecule that can enter the pores. The elution volumes V_E are thus comprised of V_I and V_M. For a molecule of intermediate size:

$$V_E = V_I + KV_P \qquad (7.1)$$

thus

$$K = \frac{V_E - V_I}{V_P} \qquad (7.2)$$

K, the diffusion coefficient, represents the degree of penetration of a species present in the volume V_P ($0 < K < 1$). For modern packings, V_I and V_P are essentially equal.

In practice, each stationary phase has a separation range expressed in terms of two masses; a higher mass and a lower mass above and below which it is impossible to obtain separation. Molecules with diameters greater than that of the pores are excluded from the gel (the origin of the term *size exclusion*). They flow through the column without being retained or separated ($K = 0$) and appear as a single peak in the chromatogram at position V_I (shown in Fig. 7.1). The elution volume of the smaller molecules is represented by V_M ($K = 1$). In order to expand the range of separation (elution necessarily occurs between these two volumes), it is usual to put two or three columns in a series.

7.2 Stationary phases

SEC stationary phases are usually composed of reticulated organic polymers (styrene-divinylbenzene copolymers) or minerals (hydroxylated silica) that are used as beads with diameters of 5–10 µm. Pore diameters, which can be varied during fabrication, are within the 4–500 nm range. These materials, often called *gels*, must withstand the pressure drop across the column and temperatures in the order of 100 °C in order to allow their utilisation under various conditions. Standard columns have a length of approximately 30 cm (ID = 7.2 mm).

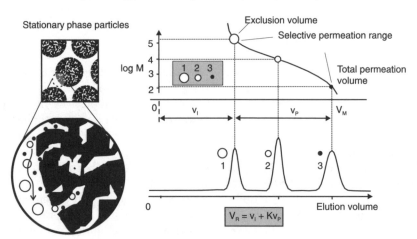

Figure 7.1—*Differential migration through a gel stationary phase.* Chromatogram of the separation of three species with different sizes in solution. The biggest molecules (1) elute faster that the molecules of medium size (2) and the small molecules (3). The elution volumes are between V_I and $V_M = V_P + V_I$.

■ If molecules are compared to vehicles, and pores to a traffic network, the principle of exclusion chromatography can easily be pictured. Trucks using the bypass around the city can move faster than cars using the smallest streets in the city!

The stationary phases include *hydrophilic* phases that are used with aqueous mobile phases (water or salts) and *lipophilic* phases that are used with organic solvents such as tetrahydrofuran or chloroform.

— Stationary phases composed of pure *polyvinylalcohols* or copolymerised with *polyglyceromethacrylates* or *vinyl polyacetates* are used to separate biological polymers that are usually contained in aqueous phases (polysaccharides, proteins). Porous silica gels containing glyceropropyl groups $[\equiv\!Si\!-\!(CH_2)_3\!-\!O\!-\!CH_2CH(OH)CH_2OH)]$ are also used for hydrophilic separations. Adsorption phenomena are weak, even for small moleules.

— Styrene-divinylbenzene gels are usually used for the separation of synthetic molecules dissolved in an organic solvent (Fig. 7.2). Tetrahydrofuran, a good solvent for most polymers, is frequently used as the mobile phase. However, chloroform, toluene or hot trichlorobenzene are also used to dissolve polymers that are not soluble in common solvents.

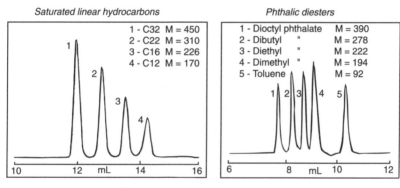

Conditions : Column, PLgel 5 µm, Pores, 5 nm, Dimensions, 300 x 7,5 mm.
Eluant, tetrahydrofurane, Flow-rate, 1 ml/min.

Figure 7.2—*Examples of separations obtained by gel permeation.* Organic compounds of small molecular weights can be separated when the stationary phase has small pores, as in this example. In the chromatogram in the right (polymer plastifiers), toluene is used to measure V_M. (Reproduced by permission of Polymer Lab.)

Each stationary phase is characterised by a calibration curve made with iso-molecular standards of known masses, M: polystyrene in THF or polyoxyethylenes, pullulanes, polyethyleneglycols in water (Fig. 7.3 and Table 7.1). The curves representing $\log(M)$ as a function of the elution volume have a sigmoidal shape. However, by mixing stationary phases, the manufacturers can provide columns for which the calibration curves are almost linear for a wide range of masses.

These curves are indicative of the size and the masses of the compounds that vary from one polymer to another.

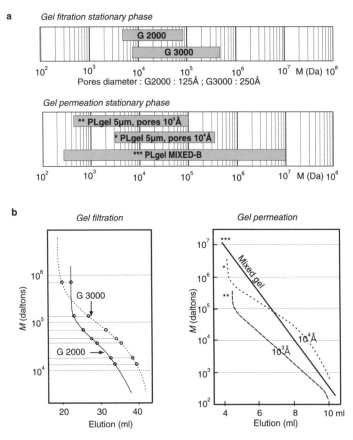

Figure 7.3—*Characteristics of stationary phases used for gel filtration or gel permeation.* a) graph indicating the mass range for each of the phases; b) calibration curves $(M = f(V))$ of these phases with different proteins and polymers of known masses.

7.3   Instrumentation

Instrumentation and methodology for SEC are similar to those used in HPLC. The columns used in SEC often have bigger volumes and, in order to increase the resolution, it is customary to use two or three columns, with different characteristics, in a series. If they are handled with care, the columns can survive for many years. Efficiency is optimal with solvents of low viscosity and when the analysis is conducted at high temperatures. The factors k are independent of the temperature.

The most commonly used detector is the differential refractometer. For polymers, the variation in the refractive index is usually independent of molecular mass. Other detectors, like photometric detectors in the UV or IR range, can also be used besides the refractometer to measure specific properties of macromolecular solutions (Fig. 7.4).

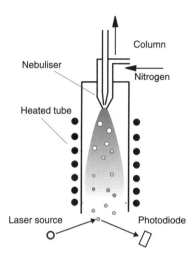

Figure 7.4—*Light diffusion detector*. Using nitrogen gas, the mobile phase is nebulised at the end of the column with a device of varying geometry. When a compound elutes from the column, the droplets under evaporation are transformed into fine particles that can diffuse light from a laser. This is called the Tyndall effect (it is similar to what is observed for a car when its lights are diffused by fog). The signal, detected by a photodiode, is proportional to the concentration of the compound. This detector can only be used for compounds that cannot be vaporised into the gas phase in the heated zone.

7.4   Domains of application

The strong points of this technique are its absence of interaction with the stationary phase, rapid dilution and the potential to recover all of the analytes. Because the technique permits the separation of nominal masses ranging from 200 to 10^7 Da, its main applications are in the analyses of synthetic and natural polymers. The choice of stationary phase for a given separation is made by examination of the calibration curve of various columns. The column of choice is that which provides a linear range over the masses of the compounds found in the sample. The calibration has to be conducted with the same type of polymers because macromolecules can have various forms ranging from pellet-like to thread-like. The data presented in Table 7.1 show the domains of application of the three gels shown in Fig. 7.3, depending on the standards that are used.

Table 7.1—Permeation ranges (Da) of three gels for various standard compounds.

Standard	G2000: 12.5 nm pores	G3000: 25 nm pores	G4000: 45 nm pores
Globular protein	5 000–100 000	10 000–500 000	20 000–7 000 000
Dextran	1 000–30 000	2 000–70 000	4 000–500 000
Polyethylene glycol	500–15 000	1 000–35 000	2 000–250 000

7.4.1   Distribution of molecular weights for a polymer

A polymer, even in its pure state, corresponds to a distribution of macromolecules with different masses. Exclusion chromatography can determine the distribution in molecular weights and the most probable mass and the mean mass. For this type of application, a calibration curve is made using macromolecules of known masses by plotting the retention times (or volumes) on the abscissa and the logarithms of the molecular masses on the ordinate. As can be seen in Fig. 7.5, a linear relationship is obtained. Using this graph, an approximate mass of the unknown can be determined by use of the retention time (volume). This assumes that the mass and the molecular volumes are directly related.

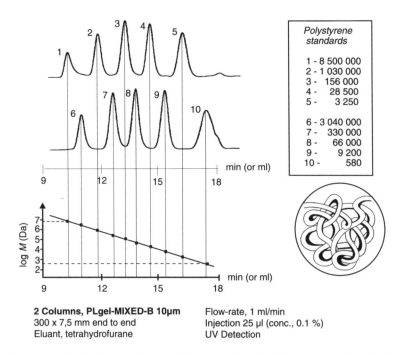

Figure 7.5—*Determination of molecular mass.* The use of a calibration curve made with standards of known molecular masses. It should be noted that the calibration curve is linear over a wide range of masses due to the use of a mixture of stationary phases. (Reproduced by permission of Polymer Lab.) The bottom right figure shows the geometry assumed by a linear polymer in solution (figure from PSS).

7.4.2   Various analyses

SEC is a useful technique for control analysis of samples with unknown compositions. These samples usually contain polymers and small molecules, as is the case in many commercial products.

For typical organic compounds that can be analysed by HPLC or GC, this technique is seldom used unless they are sugars or polysaccharides (such as starch, paper pulp, food, or beverages). Figure 7.6 shows typical examples of the application of this technique.

■ Gel filtration must not be confused with filtration itself, which produces the inverse phenomena. In filtration, the biggest molecules are retained and the smallest flow through.

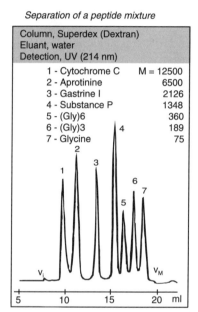

Figure 7.6—*Gel filtration chromatograms*. The separation of glucose oligomers is shown to the left. Peak number 1 corresponds to glucose ($M = 180$) and peaks are seen every approximately 20 units ($M = ca.\,2\,000$). (Reproduced by permission of Polymer Lab.) The right portion of the figure shows the separation of a mixture of proteins and glycine oligomers. (Reproduced by permission of Pharmacia-Biotech.)

Chemical analysis on the Internet

Information on different types of chemical analysis can be found on the Internet. There are numerous resources available on the Web, which include theoretical and applied concepts of instrumentation.

One of the advantages of the Internet is that information is constantly updated. Information such as spectral data banks or analytical instrumentation manufacturers can be freely obtained.

Data such as physical chemical constants, technical information, application notes and virtual expositions are also available. It is also possible to download software or to participate in discussions with interested groups or individuals.

Electronic mail is another way to obtain information from a specialist. This user-friendly system allows immediate communication and transmission of documents in many forms.

Communication can be initiated by connecting to a server such as *The Analytical Chemistry Springboard*, or the Internet site of the University of Sheffield (UK). Information on the principal sites can be found at these locations. Using Alta Vista (http://alta vista.digital.com/), most information can be found. The list below can also be used in order to explore the Web.

Sites for analytical chemistry on the Internet (December 1999):

```
http://www.anachem.umu.se/jumpstation.htm
http://www.chem.vt.edu/chem-ed/analytical/
http://analserv.chem.uva.nl/ac_inet.htm
http://www.scimedia.com/index.htm
http://www.shef.ac.uk/chemistry/web-elements-home.html
http://www.shu.ac.uk/schools/sci/chem/tutorials
http://www.chemicalanalysis.com
```

Problems

7.1 Occasionally in SEC values of $K > 1$ are observed. How might this phenomenon be accounted for?

7.2 Gel permeation chromatography is to be used to separate a mixture of four polystyrene standards of molecular mass: 9 200, 76 000, 1.1×10^6 and 3×10^6 daltons. Three columns are available for this exercise. They are prepacked with gel with the following fractionation ranges for molecular weights:

A: 70 000 to 4×10^5 daltons
B: 10^5 to 1.2×10^6 daltons
C: 10^6 to 4×10^6 daltons

How might these four polymers be separated in a single operation if it is permitted to use two of the above columns end to end?

7.3 A solution in THF of a set of polystyrene standards of known molecular mass was injected onto a column whose stationary phase is effective for the range 400–3 000 daltons. The flow rate of the mobile phase (THF) is 1 ml/min. The chromatogram below was obtained.

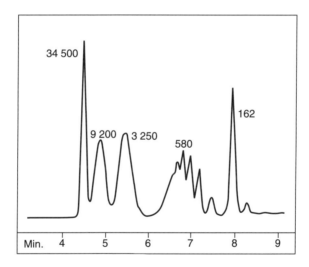

1. Plot the log of relative molecular mass vs. retention volume.
2. What is the total exclusion volume for the column (i.e. the interstitial volume), and what is the volume of the pores (the intraparticle volume)?
3. Calculate the diffusion coefficient K for the polystyrene standard whose relative molecular mass is 3 250 daltons.

CHAPTER 8

Capillary electrophoresis

Capillary electrophoresis (CE) is a very sensitive separation technique that has been developed based on the knowledge acquired from high performance liquid chromatography (HPLC). Although it is related to zone electrophoresis in the classical gel format, it is different in many ways. CE allows the separation of biomolecules with high performance where HPLC fails, and CE allows the quantitation of small molecules that cannot be separated by gel electrophoresis. Certain CE methods are a hybrid between electrophoresis and chromatography, such as electrochromatography, for which the theoretical bases are similar and will develop more fully in the next few years. High performance capillary electrophoresis (HPCE), which evolved from biochemistry laboratories, has been employed for a large number of applications developed on classical gels.

8.1 Zone electrophoresis

Zone electrophoresis, which inspired the development of capillary electrophoresis (CE) is a semi-manual technique introduced over 50 years ago. It is widely used in bioanalytical chemistry. Zone electrophoresis is based on the migration of charged species in solution under the effect of an electric field and when supported by an appropriate medium. To this effect, a strip of plastic material covered with a porous substance (a gel) is impregnated with an electrolyte buffer. The extremities of the gel-covered system are placed into independent reservoirs containing the same electrolyte and connected by electrodes to a continuous voltage supply (Fig. 8.1). The sample is deposited in the form of a transverse band – cooled and trapped between two isolating plates. Under the effect of several parameters that act jointly – voltage (500 V or more in the case of small molecules), charge, size, temperature, shape, and viscosity – the species migrate from one end to the other, generally towards the electrode or pole of opposite sign on a time scale that can vary from a few minutes to over an hour. Under these conditions, neutral species cannot be separated unless they are associated with ions of the electrolyte or with a modifier that has been introduced specifically to create differential migration velocities.

Each compound in the mixture is characterised by its relative mobility R_f, but the absence of a solvent front, compared to TLC, requires that an internal marker is used in order to measure the relative distance of migration.

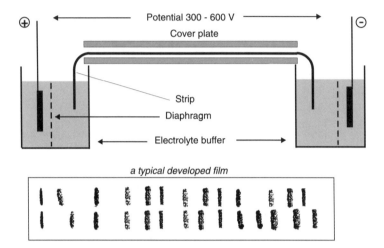

Figure 8.1—*Principle of zone electrophoresis.* Each compartment is separated by a membrane to avoid contamination of the electrolyte by secondary products formed at the electrodes. The size and the sign of the charge carried by each species depends on the chemical medium in which they are found. The experiment can be carried out at constant current, constant voltage or constant power.

Various migration media can be used. Supports can be made of polyurethane or cellulose that are covered with starch gels or, better, polyacrylamide. An electrolyte-containing sodium dodecyl sulphate (SDS) can also be used. In the latter case, the technique is known as SDS-PAGE. A filtration phenomenon that depends on the size of the pores in the gel also contributes to the migration under the electric field. When proteins, peptides, DNA or RNA fragments are analysed, specific reagents must be used for the detection system. Species present in the gel, through a contact process, are transferred onto a membrane where they are derivatised with specific reagents (Fig. 8.2); Coomassie Blue® or silver stain for visualisation. An

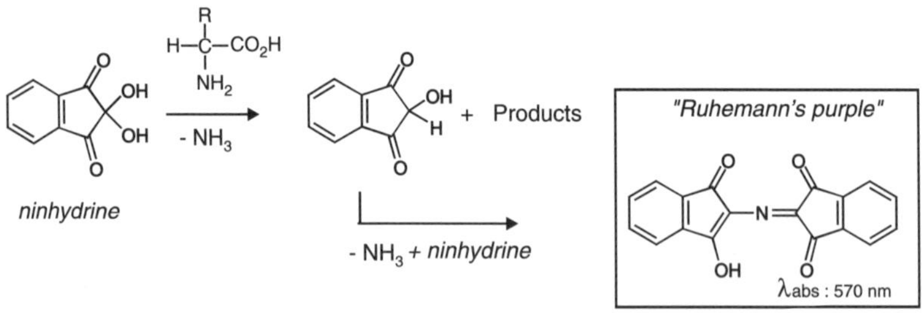

Figure 8.2—*Visualisation of proteins and amino acids by reaction with ninhydrin.* This reagent, after a series of reactions, yields a 'Ruhemann's colouration. This is one of the reagents that can be used to determine the position of compounds that have migrated on the electrophoretic gel.

electropherogram, which resembles a chromatogram, can be obtained with the aid of a densitometer that measures band intensities after derivatisation.

Zone electrophoresis is mostly used for biological applications. Peptide separation and the measurement of protein fractions from blood serum (proteinogram of albumin and α-, β- and γ-globulins) are among the better known applications. This 'TLC for biochemists' is useful for the separation of polysaccharides, nucleic acids (for DNA sequencing), proteins and other colloidal species.

8.2   Free solution capillary electrophoresis

In this more recent technique of electrophoresis, an open-ended fused silica capillary with a small inner diameter (10–150 µm) is used to replace the supported gel (Fig. 8.3). The capillary, of length L that varies between 20 and 80 cm, is filled with the same aqueous buffer solution of electrolyte as the two reservoirs. A voltage of up to 30 kV is applied to the electrodes. The intensity of the current should not exceed 100 µA (corresponding to a maximum dissipated power of 3 W) to avoid overheating the capillary, which should be contained in a thermostated enclosure.

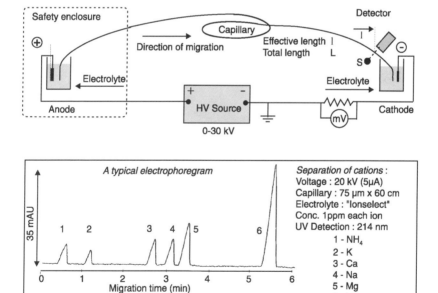

Figure 8.3—*Schematic of a standard capillary electrophoresis instrument.* The electrolyte is an aqueous ionic solution that has been filtered and degassed. It can contain several additives. There are several processes that can be used to introduce the sample into the capillary (cf. 8.4). The use of voltages above 30 kV is rare because they require special insulation. The length of the capillary (L) and the effective distance of migration (l) must not be confused since the latter is shorter by 10 or 20 cm.

The chemical species, whether they are charged positively or negatively, usually migrate towards the cathode (cf. 8.3.2). A detection system is placed near the end of the capillary. In the UV mode, for example, the capillary is inserted in the optical path between the source and the photodetector. This allows measurement of the absorbance of the solution while avoiding dead volumes. Electrochemical detection is conducted in a similar way; microelectrodes are placed within the capillary.

8.3 Electrophoretic mobility and electro-osmotic flow

Particles suspended in a liquid as solvated molecules can carry electrical charges. The size and the sign of the charge will depend on the medium as well as on the pH. This net charge is the result of the surface adsorption of ions contained in the buffer electrolyte. The migration velocities are greater for ions of comparable size if they carry more than one charge (Fig. 8.4); the separation depends on the charge-to-mass ratio of the ions. For each ion with a radius r, the *limiting migration velocity* results from the equilibrium between the electric force F, which is accociated with an electric field E acting on the particle of charge q, and the forces resulting from the viscous resistance η of the medium (Fig. 8.4).

The migration time of mineral ions depends on the limit of their equivalent conductivity.

In capillary electrophoresis, components of a mixture are separated according to two main factors: *electrophoretic mobility* and *electro-osmotic flow*. These terms apply to ions, molecules or micelles.

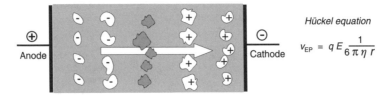

Hückel equation

$$v_{EP} = q\,E\,\frac{1}{6\,\pi\,\eta\,r}$$

Anode   Cathode

Figure 8.4—*Influence of net charge, field, viscosity and species size on the migration velocity vector of an electrolyte assumed to be immobile.* The general term cataphoresis corresponds to the displacement of ions towards the cathode whereas anaphoresis corresponds to the displacement of ions towards the anode.

8.3.1 Electrophoretic mobility – electromigration

In all electrophoresis experiments, a compound bearing an electrical charge will migrate under the influence of the electric field E. The migration velocity, also called the electrophoretic migration velocity v_{EP}, depends on the electrophoretic mobility of the ion μ_{EP}.

μ_{EP} is defined in terms of the migration velocity v_{EP} observed in an electrolyte *assumed to be immobile* and the electric field E, using equation (8.1):

$$v_{EP} = \mu_{EP} \cdot E \qquad \text{such that} \qquad \mu_{EP} = \frac{v_{EP}}{E} = v_{EP} \frac{L}{V} \qquad (8.1)$$

In the above equation, L designates the total length of the capillary and V the voltage applied across the extremities of the capillary. It will be shown later that μ_{EP} ($cm^2\ V^{-1}\ s^{-1}$) can be indirectly obtained from the electropherogram. A positive (+) or negative (−) sign is assigned to the electrophoretic mobility, depending on the cationic or anionic nature of the species: μ_{EP} is zero for neutral species. This parameter depends not only on the charge carried by the species, but also on its diameter and on the viscosity of the electrolyte.

8.3.2 Electro-osmotic mobility – electro-osmosis

Another factor that controls the migration of the solute is the *electro-osmotic mobility* μ_{EOS}, which results in movement of the electrolyte or *electro-osmotic flow*. This flow is present in gel electrophoresis to a small extent and to a greater extent in capillary electrophoresis because of the internal wall of the capillary.

The wall of the silica column is lined with silanol groups that become de-protonated when the pH is above 2. Under these conditions, a fixed polyanionic layer is formed (Fig. 8.5). However, a polycationic layer due to the electrolyte acts as its counterpart. The H_3O^+ ions and electrolyte cations are put in motion when the electrical field is applied. The net effect is to make all the species present migrate towards the cathode. This linear displacement of the electrolyte, which originates from the charge carried by the electrolyte ions and tangential applied electrical field, can be controlled or reversed by modifying the capillary inner surface, the pH, or by adding cationic surfactants.

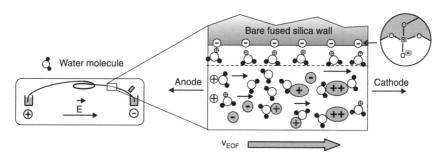

Figure 8.5—*Effect of the nature of the capillary inner wall on migration velocities.* If the inner wall has not been treated (glass or silica naturally have a negative polyanionic layer) the liquid is pumped from the anodic towards the cathodic reservoir. This is called the electro-osmotic flow. Thus, an anion can move towards the cathode. Between pH 7 and 8, v_{EOS} can increase by 35%. However, if the wall is coated with a nonpolar film (e.g. octadecyl) this flow does not exist.

μ_{EOS} is defined by a relation similar to equation (8.1). v_{EOS} represents the electro-osmotic flow (EOF) velocity of the bulk solution, i.e. the neutral molecules present in the electrolyte.

$$v_{EOS} = \mu_{EOS} \cdot E \qquad \text{such that} \qquad \mu_{EOS} = \frac{v_{EOS}}{E} = v_{EOS} \frac{L}{V} \qquad (8.2)$$

■ The electro-osmotic flow v_{EOS} must first be determined in order to calculate μ_{EOS}. v_{EOS} can be calculated by measuring the migration time t_{nm} for a _neutral marker_ to migrate over the distance l of the capillary. An organic molecule that is nonpolar at the pH of the electrolyte used and easily detected in the UV can be used as the neutral marker (e.g. mesityl oxide or benzyl alcohol). $v_{EOS} = l/t_{nm}$.

In capillary electrophoresis instruments, the electro-osmotic flow is used to impose, on all charged species in the sample, a direction of migration that is oriented from the anode towards the cathode. An increase in the electro-osmotic flow v_{EOS} decreases, at the detector, the gap in migration times of ions travelling in the same direction. The use of fused silica capillaries partially deactivated by coating the inner wall allows modulation of the electro-osmotic flow. A voltage gradient can also be used to this end.

■ When the wall is made hydrophobic by treatment with alkylsilane, it is possible to separate proteins that tend to adsorb at the surface of bare fused silica. Ultimately, it is possible to recuperate one type of ion depending on the direction of the electric field. Finally, if a cationic surfactant is added to reverse the polarity of the inner wall, it is possible to reverse the direction of the electro-osmotic flow (Fig. 8.6).

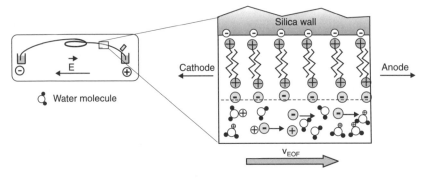

Figure 8.6—*Effect of a cationic surfactant reversing the polarity of the capillary inner wall.* Because the migration of analytes must always be in the direction of the detector, the voltage polarity of the instrument must be reversed in order for anionic species to move towards the anode, thus towards the detector.

8.3.3 Apparent mobility

Each ion has an apparent migration velocity v_{app} easily obtained from the electropherogram. If l designates the effective length of the capillary between the injector and the detector, and t_m is the migration time, then v_{app} can be obtained using the equation: $v_{app} = l/t_m$.

v_{app} depends on both the electrophoretic velocity and the electro-osmotic flow. This can be translated into the following equation:

$$v_{app} = v_{EP} + v_{EOS} \qquad (8.3)$$

The apparent electrophoretic mobility μ_{app} is defined by an equation analogous to (8.1) or (8.2) such that:

$$\mu_{app} = \frac{v_{app}}{E} = v_{app}\frac{L}{V} \quad \text{and, by consequence,} \quad \mu_{app} = \frac{l \cdot L}{t_m \cdot V} \qquad (8.4)$$

By combining the apparent mobility and the electro-osmotic flow, which is responsible for the migration of the bulk electrolyte, it is possible to calculate the migration velocity or the electrophoretic mobility of charged species. Using equation (8.3), equation (8.5) can be written as:

$$\mu_{EP} = \mu_{app} - \mu_{EOS} \qquad (8.5)$$

Alternatively, the equation below can be used:

$$\mu_{EP} = \frac{Ll}{V}\left(\frac{1}{t_m} - \frac{1}{t_{nm}}\right) \qquad (8.6)$$

8.4 Instrumentation and techniques

One of two processes can be used to introduce a few nanolitres of sample into the capillary:

— **Hydrostatic injection.** This type of injection is achieved by dipping the end of the capillary into a solution containing the sample while inducing a slight vacuum at the other end. Alternatively, a positive pressure can be applied to the sample solution.

— **Injection by electromigration.** This approach, used in gel electrophoresis, is achieved by putting the sample at a potential of appropriate polarity compared to the other extremity and dipping the capillary into it for a few seconds. In contrast to the first method, this mode of injection provokes a discrimination on the compounds present, which leads to non-representative sampling of the analytes in the sample.

8.4.1 Free solution or capillary zone electrophoresis (CZE)

This mode of electrophoresis, in which the electrolyte migrates through the capillary, is the most widely used. The electrolyte can be an acidic buffer (phosphate, citrate, etc.) or basic buffer (borate) or an amphoteric substance (a molecule that possesses both an acidic and an alkaline function). The electro-osmotic flow increases with the pH of the liquid phase, or can be rendered non-existent.

8.4.2   Micellar electrokinetic capillary chromatography (MEKC)

This technique is a variant of CZE. A cationic or anionic surfactant compound, such as sodium dodecylsulphate, is added to the mobile phase to form charged micelles. These small spherical species, whose core is essentially immiscible with the solution, trap neutral compounds efficiently by hydrophylic/hydrophobic affinity interactions (Fig. 8.7). Using this type of electrophoresis, optical purity analysis can be conducted by adding cyclodextrins instead of micelles to the electrolyte. This is useful for separating molecules that are not otherwise separable. Under such conditions, the enantiomers form inclusion complexes of different stability with cyclodextrin (cf. 3.6).

8.4.3   Capillary gel electrophoresis (CGE)

This technique represents the transposition of classical polyacrylamide or agarose gel electrophoresis into a capillary. Under these conditions, the electro-osmotic flow is relatively weak. In this approach, the capillary is filled with an electrolyte impregnated into a gel that minimises diffusion and convection phenomena. In contrast to its use for proteins that are fragile and thermally unstable, CGE is ideal for separating the more rugged oligonucleotides.

8.4.4   Capillary isoelectric focusing (CIEF)

This technique, known classically as zone electrophoresis on a support, consists of creating a stable and linear pH gradient in a surface-treated capillary that contains ampholytes. At the anode, the capillary is put into a reservoir containing H_3PO_4 solution while the cathode end is in NaOH solution. Each component migrates and is focused at the pH that has the same value as its isoelectric point (at its pI, the component net charge is zero). Then, by maintaining the electric field and using a hydrostatic pressure, these separated species are displaced towards the detector. This process, which has a very high resolution, allows the separation of peptides whose pI differs by as little as 0.02 pH units.

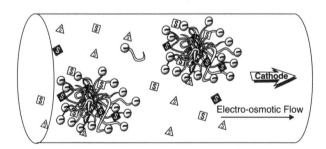

Figure 8.7—*Separation of neutral species using surfactants.* The lipophilic part of a surfactant, such as alkylsulphonate, can bind fairly easily to certain substrate species, S. The micelles that are formed in the process, although negatively charged, are carried towards the cathode by the strong electro-osmotic flow. (Adapted from Hewlett-Packard.)

8.4.5 Capillary electrochromatography (CEC)

In this electroseparation type of chromatography (Fig. 8.8), electromigration of the ions is combined with adsorption effects which are characteristic of chromatography by using a capillary column (diameter 75 µm) filled with a stationary phase of very fine particles (1–3 µm). This can be done because of the absence of a pressure drop across the column.

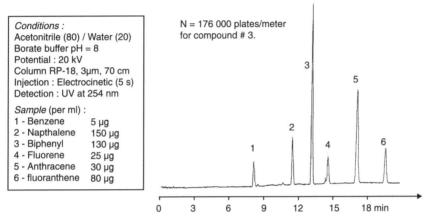

Conditions :
Acetonitrile (80) / Water (20)
Borate buffer pH = 8
Potential : 20 kV
Column RP-18, 3µm, 70 cm
Injection : Electrocinetic (5 s)
Detection : UV at 254 nm

Sample (per ml) :
1 - Benzene   5 µg
2 - Napthalene   150 µg
3 - Biphenyl   130 µg
4 - Fluorene   25 µg
5 - Anthracene   30 µg
6 - fluoranthene   80 µg

N = 176 000 plates/meter
for compound # 3.

Figure 8.8—*Electrochromatographic separation of aromatic hydrocarbons.* The movement of the mobile phase is strictly due to the electro-osmotic flow. In contrast to HPLC, no pressure is exerted at the head of the column. Separations can be carried out with a very high efficiency.

8.5 Indirect detection

One variant of absorbance detection that is widely used in HPLC can also be used in high performance capillary electrophoresis. For compounds that exhibit a very weak UV absorption, buffers such as chromate or phthalate, which have high absorption properties, can be used. Under these experimental conditions, the UV absorbance diminishes as analytes flow past the detector (due to the dilution effect of the electrolyte). This leads to negative peaks on the recorder (see Fig. 8.9).

8.6 Performance of capillary electrophoresis

The performance of capillary electrophoresis, for the separation of biopolymers, is comparable to or better than that of HPLC. The basis for separation relies on the choice of an appropriate buffer to be adapted to the analysis. Although repro-ducibility is more difficult to control, mass sensitivity is relatively high: a few thousand molecules can be detected. Sample quantity is very small and solvent and reagent consumption during an analysis is negligible (Fig. 8.10).

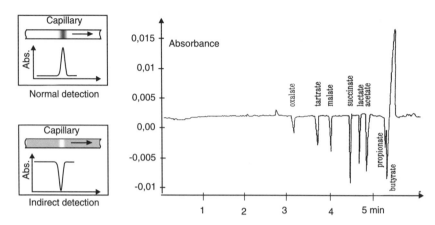

Figure 8.9—*Separation of the main organic acids in white wine*. Malic and lactic acids are indicators of the classical malolactic (fermentation process application note from TSP).

Separation parameters can be obtained from the electropherogram, analogous to chromatography (retention factor, selectivity, resolution). The theoretical efficiency of a separation N as high as 10^6 plates per metre in a column of length L can be calculated from its effective length l and the diffusion coefficient D $(cm^2\,s^{-1})$ based on the Einstein equation $(\sigma^2 = 2Dt_m)$ and the migration time t_m, or by using equations (8.7) and (8.8):

$$N = \frac{l^2}{2Dt_m} \tag{8.7}$$

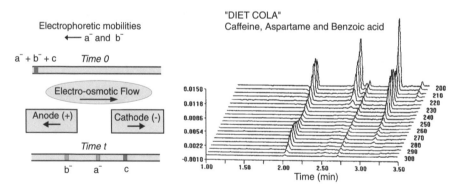

Figure 8.10—*Electrophoretic separation of three species a^-, b^- and c*. In this example, the electro-osmotic flow moving towards the cathode carries all charged and neutral species. Negatively charged species, although attracted towards the positive pole (or anode) cannot overcome the electro-osmotic flow and thus move towards the cathode. Separation of caffeine (c) from aspartame (a^-) and benzoate (b^-) in a Diet Cola sample. The presentation of data is in the form of a 3-D electrophoregram. (Reproduced by permission of TSP.)

or simply

$$N = \frac{\mu_{app}}{2D} V \frac{l}{L} \qquad (8.8)$$

Equation (8.8) shows that the efficiency of separation increases with the applied voltage. Macromolecules, whose diffusion coefficients are lower than those of small molecules, tend to give better separations (Fig. 8.11).

■ The separation of isotopes can be used to show the efficiency of separation by capillary electrophoresis. Moreover, the interfacing of a mass spectrometer to the capillary can be used for the study of biological substances (cf. Fig. 16.7).

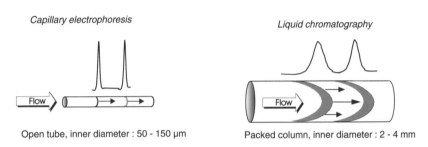

Capillary electrophoresis

Liquid chromatography

Flow

Flow

Open tube, inner diameter : 50 - 150 µm

Packed column, inner diameter : 2 - 4 mm

Figure 8.11—*Effect of diffusion on the efficiency obtained in HPLC and CE.* Diffusion increases with the square of tube diameter. This is, thus, more important in HPLC. In CE, the electrolyte is repelled by the wall leading to an almost perfect plane-like flow contrary to the usual parabolic profile obtained under hydrodynamic flow. However, other factors that depend on the difference in conductivity between the electrolyte and solutes can lead to peak deformation.

Problems

8.1 An electrophoresis analysis in free solution (CZE), calls for the use of a capillary of 32 cm and with effective length of separation 24.5 cm. The applied voltage is 30 kV. Under the conditions of the experiment the peak of a neutral marker appeared upon the electropherogram at 3 min.

1. Calculate the electrophoretic mobility, μ_{EP}, of a compound whose migration time is 2.5 min. Give the answer in precise units.
2. Calculate the diffusion coefficient under these conditions for this compound, remembering that the calculated plate number, $N = 80\,000$.

8.2 The apparatus for an experiment using a fused silica capillary is set up. The total capillary length $L = 1$ m while the effective capillary length $l = 90$ cm. The applied voltage is 30 kV. The detector is located towards the cathode and the electrolyte is a buffer of pH 5.

In a standard solution a compound has a migration time of $t_m = 10\,min$.

1. Sketch a diagram of the apparatus described.
2. From the information given, can it be deduced whether the net charge carried by the compound is positive or negative?
3. Calculate the apparent electrophoretic mobility μ_{app} of the compound.
4. If a small molecule not carrying a charge has a migration time $t_m = 5\,min$, deduce the value of the electro-osmotic flux μ_{EOS}.
5. Calculate the electrophoretic mobility of the compound μ_{EP}.
6. What is the sign of the net charge carried by the compound?
7. What would happen if the fused silica capillary was coated by trimethyl-chlorosilane?
8. Supposing that the pI of the compound was 4, what would be the sign of its net charge if the pH of the electrolyte was lowered to 3?
9. Calculate the number of theoretical plates if the diffusion coefficient of the solute is $D = 2 \times 10^{-5}\,cm^2\,s^{-1}$.
10. From the relationship between the efficiency N and the diffusion D, explain why small molecules have poorer separations than larger molecules, and why for the smaller variant these separations are much better when the capillary is narrower.

8.3 During a capillary electrophoresis experiment it is observed that if the capillary contains an acrylamide gel as well as an electrolyte then the speed of migration is slowed by the mechanical effects of filtration through the gel. This is particularly significant for larger molecules. The following relationship can be considered as proven:

$$\log M = a \cdot v + b$$

where a and b are two constants and M represents the molecular mass (Da) of a molecule migrating with a velocity v.

In an experiment designed to employ this relationship and in order to calculate the molecular mass of an unknown protein, two known standards were used; ovalbumin ($M = 45\,000\,Da$) and myoglobin ($M = 17\,200\,Da$), whose migration velocities are $1.5\,cm\,min^{-1}$ and $5.5\,cm\,min^{-1}$ respectively. In the same experiment, the unknown protein migrates at a velocity of $3.25\,cm\,min^{-1}$. Calculate its molecular mass.

8.4 Imagine that, following a capillary isoelectric focusing experiment, two polypeptides A and B are immobilised in one section of the capillary as indicated in the figure. The isoelectric point of B is superior to that of A.

Complete the figure indicating how the pH would increase at the positive end of the capillary while showing the orientation of the corresponding electric field.

8.5 The table below lists a series of proteins along with their respective isoelectric points (pI). By completing the table with positive (+) or negative (−) signs or zeros (0) indicate for each protein whether the net charge will be positive, zero or negative at the three values of pH specified.

Protein	pI	pH $=3$	pH $=7.4$	pH $=10$
insulin	5.4			
pepsin	1			
cytochrome C	10			
haemoglobin	7.1			
albumin (serum)	4.8			

Environmental analyses

Because of environmental problems, particularly those due to pollution related to human activity, there is growing public awareness of chemical analysis. In this area, the solution to problems can only arise from knowledge of the nature of the chemical pollutants, whether these are molecules, elements or a combination of atoms.

A decade ago, only a few pollutants were monitored. However, current regulations impose the monitoring of an increased number of substances at increasingly lower detection levels. Special surveillance is conducted in establishments known to release pollutants into the environment. Similarly, hazardous chemicals such as pesticides and herbicides are constantly identified and quantified. Several of these are related to air, water and soil pollution.

Most of the methods described in this book can be used in environmental applications.

Most analyses involve very low concentrations that can be considered as being at the trace or ultratrace level. Because the compounds monitored are usually found in complex and varied matrices, detection limits vary depending on the type of sample. Analytical methods must be very sensitive; methods that are easily automated and that require small samples with a minimum of pretreatment are favoured.

Examples of methods used and fields of applications are as follows:

Air pollution
— volatile hydrocarbons: gas chromatography on PLOT columns
— minerals (dust and fumes): X-ray fluorescence
— volatile organic compounds (VOC): gas chromatography on WCOT columns, coupled to mass spectrometry or infrared.

Water pollution
— heavy metals: atomic absorption or emission
— pesticides, phenols, polychlorinated biphenyls (PCB), polycyclic aromatic hydrocarbons (PAH): liquid chromatography with appropriate detectors, spectroscopic methods (fluorescence, UV/VIS).

Soil pollution
— pesticides and herbicides: gas chromatography/mass spectrometry, etc.
— heavy metals: atomic absorption or emission.

SPECTROSCOPIC METHODS

CHAPTER 9

Nuclear magnetic resonance spectroscopy

The first experiments on nuclear magnetic resonance (NMR) were conducted by the physicists Bloch and Purcell in 1945. This multifunctional spectroscopic technique has grown to be irreplaceable in many areas of chemistry, in quantitative and in structural analysis. However, it is in the latter area that NMR has shown its greatest importance. Because the technique is very efficient in gathering qualitative and structural information on molecular compounds, it is of great practical importance in both organic and biological chemistry. NMR is complementary to other spectroscopic techniques like optical or mass spectrometry, and it allows the determination of the structure and stereochemistry of the compounds studied. It can even determine the preferred conformation of molecules. The combination of computers, high performance instruments and the mastery of a chemist can yield exceptional results. For all of these reasons, NMR has become a major tool in the study of biological molecules.

NMR imaging, a more recent technique, uses the same basis to obtain a map of structured media. The map is constructed using signals obtained from a high number of qualitative and timely quantitative analyses.

9.1 General description

Nuclear magnetic resonance (NMR) is at the origin of a method widely exploited in chemistry to determine the structure of molecular compounds and materials (whether organic or not). Aside from the instruments used for routine application (shown in Fig. 9.28), NMR spectrometers are often found in research laboratories where they are considered indispensable (Fig. 9.29).

This spectroscopic method can be described using examples taken solely from organic chemistry. In fact, the determination of organic molecular structure has always been the driving force behind this technique and the numerous technical improvements that it has seen.

The data provided by this technique comprise the *NMR spectrum*. It corresponds to resonance signals emitted by the numerous atomic nuclei present in a sample (see Fig. 9.1). These signals are obtained when the sample is submitted to the joint action of two magnetic fields. The first field, produced by a magnet, has a high and constant intensity while the second is variable and approximately 10 000 times weaker. The latter field is produced by a source of electromagnetic radiation of radiofrequencies and its direction of propagation is perpendicular to the magnetic field. The NMR spectrum is related to the absorption by certain sample atoms of frequencies radiated by the electromagnetic source. Interpretation of the signals (position, appearance, intensity) leads to a variety of information from which the detailed structure of the sample molecules can be obtained. It is always easier to interpret the spectrum obtained from a pure sample.

The origin of the spectra, which differ from optical spectra, can best be understood by consideration of nuclear spins.

■ Figure 9.1 represents a one-dimensional NMR spectrum (the intensities of the peaks are not considered to be a second dimension). More sophisticated NMR studies, in two, three or four dimensions can be used to determine the position of all the atoms present in a molecule. This chapter only deals with one-dimensional (1-D) NMR.

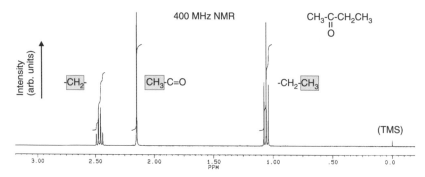

Figure 9.1—*Conventional representation of a ^1H NMR spectrum of an organic compound.* Spectrum of butanone [$CH_3(C=O)CH_2CH_3$]. Superimposed is the signal integration that allows the relative areas of each type of proton present in the spectrum to be determined. The meaning of the abscissa will be explained further on.

9.2 Spin/magnetic field interaction for a nucleus

Each atomic nucleus – as well as each particle – is characterised by a number of intrinsic parameters, including the spin $\vec{I}$. This vector quantity, which has a preferential direction, is introduced in quantum mechanics and has no classical equivalent. It explains, among other things, the behaviour of atoms in media where there is a preferential direction. The spin of the nucleus can be related to the kinetic moment $\vec{L}$ using classical mechanics and it has the same dimensions ($J \cdot s$). Thus an atom, placed in a magnetic field, will sense the orientation of the magnetic field. The spin is defined by the spin quantum number, I, which is dependent on the individual nucleus considered, and is always a positive multiple of $1/2$, including zero.

An isolated nucleus with a spin number other than zero behaves like a small magnet of magnetic moment $\vec{\mu}$ ($J \cdot T^{-1}$) where:

$$\vec{\mu} = \gamma \cdot \vec{I} \tag{9.1}$$

This *nuclear magnetic moment* $\vec{\mu}$ is represented by a vector that is colinear to $\vec{I}$ and has the same or opposite direction depending on the sign of γ, the *gyromagnetic ratio* (also called the gyromagnetic constant).

If a nucleus with a non-zero spin number, which can be compared to a small magnet, is exposed to a magnetic field $\vec{B_0}$, with an angle θ with the spin vector, $\vec{\mu}$ and $\vec{B_0}$ will become coupled. This coupling modifies the potential energy E of the nucleus. However, $\vec{\mu}$ will not necessarily align itself in the direction of the external field, in contrast to the action of a compass.

If $\vec{\mu}_z$ represents the projection of $\vec{\mu}$ on the Oz axis, which is in the direction of $\vec{B_0}$, then:

$$E = -\vec{\mu} \cdot \vec{B_0} \quad \text{where} \quad E = -\mu \cos(\theta) B_0 \quad \text{or, finally} \quad E = -\mu_z \cdot B_0 \tag{9.2}$$

According to quantum mechanics, μ_z for a nucleus can only take $2I + 1$ different values. This means that in a magnetic field B_0, the potential energy E can also only take $2I + 1$ values. The value of μ_z, the projection of the spin vector on the Oz axis, is related to the values determined by m. These values (given in $h/2\pi$ units) are given by the following series:

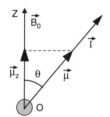

$$m = -I, -I + 1, \ldots, I - 1, I$$

Using equations (9.1) and (9.2), the $2I + 1$ allowed energy values are given by the following expression:

$$E = -\gamma \cdot m \cdot B_0 \tag{9.3}$$

For $I = 1/2$, E can have two possible values (in joules), corresponding to $m = 1/2$ and $m = -1/2$, designated by α and β:

$$E_1(\text{or } E_\alpha) = -\gamma \frac{1}{2} \frac{h}{2\pi} B_0 \quad \text{and} \quad E_2(\text{or } E_\beta) = +\gamma \frac{1}{2} \frac{h}{2\pi} B_0$$

■ This splitting in the energy level is similar to the Zeeman effect that causes separation of electronic states in a magnetic field. It is sometimes referred to in NMR as the Zeeman nuclear effect.

The projection of the spin vector onto the Oz axis essentially traces the surface of a cone of revolution with an angle between the axis and atom that can be calculated knowing $\vec{\mu}$ and $\vec{\mu}_z$. This *precession* around an axis parallel to that of the magnetic field (Fig. 9.2 and 9.4a) is characterised by a frequency that increases with the field intensity.

■ The ethereal movement of a nuclear spin around the axis is like the rotation of a spinning top around its axis normal to the field of gravity. The spinning top describes a gyroscopic movement about this direction, which is the result of the movement of a spinning top around its own axis and the coupling constant. As the movement is slowed down, the angle between the rotation axis and the vertical increases continuously. However, for a nucleus, the angle is maintained with time. This is true irrespective of the field applied because the values of the magnetic moment or its projection are quantised. Thus, for a nucleus with a spin number $I = 1/2$, the angles relative to Oz are approximately 55° and 125°. This takes into account the value of the spin ($\sqrt{I(I+1)}$), and that of the projection I. The angle 54.75° used in NMR experiments is called the *magic angle*.

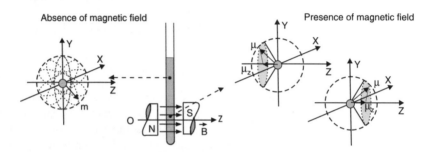

Figure 9.2—*Effect of the magnetic field on a nucleus with spin number of 1/2 for an atom of a molecule present in solution.* In the upper part of the sample tube, not influenced by the magnetic field, μ has no preferred orientation with time. However, in the portion of the tube exposed to the external field, μ traces the surface of a cone of revolution whose axis is aligned with B. Both possibilities are represented; the projection of μ is opposite or in the same direction as B.

9.3   Nuclei that can be studied by NMR

A nucleus, represented by $^A_Z X$, will have a non-zero spin number I giving an NMR signal as long as the number of protons Z and neutrons A are not both even numbers. For example, ^1_1H, $^{13}_6C$, $^{19}_9F$ and $^{31}_{15}P$ all have a spin number $I = 1/2$ while ^2_1H (deuterium, D) and $^{14}_7N$ have $I = 1$. However, nuclei such as $^{12}_6C$, ^4_2He, $^{16}_8O$, $^{28}_{14}Si$ and $^{32}_{16}S$ cannot be studied by NMR. In fact, more than half of the stable nuclei known (at least one isotope per element) yield NMR signals. However, sensitivity varies enormously depending on the nucleus. Hence the proton, also known as ^1H, or the nucleus ^{19}F, are easier to detect than ^{13}C, which is thousands of times less sensitive than the proton because of its weak natural isotopic abundance.

9.4   Bloch's theory for a nucleus of $I = 1/2$

At a microscopic level, even the smallest quantity of a compound is composed of a great number of individual molecules. Hence, the number of nuclei is so high that it is possible to reason statistically.

Let us consider a group of identical nuclei with a spin number $I = 1/2$. In the absence of an external field, the vector orientation of individual spins will be random and will vary constantly. These nuclei form a population that is considered a *degenerate* state (Fig. 9.2). When these nuclei are placed in a strong, induced magnetic field $\vec{B_0}$ (Oz orientation), an interaction between each vector and the magnetic field will be generated (cf. 9.2).

Under these conditions, two groups of nuclei with energies E_1 and E_2, as defined previously according to the projection on the Oz axis (Fig. 9.3), will evolve. The difference in energy ΔE between two states is:

$$\Delta E = E_2 - E_1 = \gamma \frac{h}{2\pi} B_0 \tag{9.4}$$

ΔE is proportional to the field B_0 (Fig. 9.3). Hence, if $B = 1.4\,\text{T}$, the energy difference for a proton will be weak: $3.95 \times 10^{-26}\,\text{J}$, or $2.47 \times 10^{-7}\,\text{eV}$. As for the ratio $\Delta E/B_0$, it strictly depends on γ and on the nucleus being studied (Table 9.1).

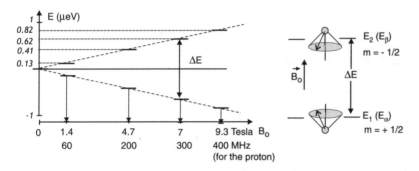

Figure 9.3—*Representation of the energy split for a nucleus with a spin number $I = 1/2$ inserted into a magnetic field.* The four values chosen for B_0 correspond, for a proton, to commercial instruments of 60, 200, 300 and 400 MHz. (B_0 represents the density of the magnetic flux in tesla: 1 T is equivalent to 10 000 gauss).

The population of nuclei in energy level E_2 is slightly less than that in energy level E_1, which is a little more stable. Population ratio (Boltzmann distribution) calculations that can be conducted using equation (9.5) for $T = 300\,\text{K}$ and $B_0 = 5.3\,\text{T}$ lead to $R = 0.999\,964$ (where $k = 8.314/6.022 \times 10^{-23}\,\text{J} \cdot \text{K}^{-1}$).

$$R = \frac{N_2}{N_1} = e^{-\frac{\Delta E}{kT}} \tag{9.5}$$

As a result, only the slight excess in the E_1 population will lead to NMR signals. By increasing the intensity of the magnetic field produced by the instrument, the difference between the two populations can be increased and, thus, sensitivity will increase.

Graphically, weak magnetisation of the sample is accounted for by using the vector $\vec{M_0}$ (macroscopic magnetisation), which is constructed using all the individual

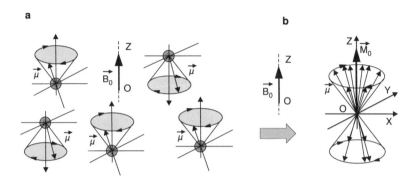

Figure 9.4—*Precession and magnetisation.* a) Snapshot showing the precession of five independent nuclei in the magnetic field; b) macroscopic magnetisation vector resulting from the individual orientations of a great number of nuclei.

$\vec{\mu}$ (Fig. 9.4b). This vector is often used to simplify the explanation of the origin of a signal observed when nuclei enter resonance.

9.5   Larmor's frequency

From an analytical point of view, a nucleus can be identified by knowledge of its gyromagnetic constant, γ, if the difference in energy separating the two populations can be determined (for $I = 1/2$). As in optical spectroscopy, this difference in energy can be determined under conditions where species can pass between one state and another. This will create a signal due to the resonance. A basic experiment consists of irradiating the nuclei present in a magnetic field with a source of electromagnetic radiation that has a variable frequency and a direction perpendicular to the external field. Absorption will occur if:

$$h\nu = \Delta E = E_2 - E_1 \tag{9.6}$$

The use of equation (9.4) leads to the fundamental rule of resonance, equation (9.7):

$$\nu = \frac{\gamma}{2\pi} B_0 \tag{9.7}$$

This general and important relationship, irrespective of the value of I, is called *Larmor's equation*. It relates the intensity of the magnetic field in which the nuclei are located to the electromagnetic radiation frequency that induces resonance; hence, a signal in the spectrum (see Table 9.1 and Fig. 9.1).

■ The radiofrequency, or frequency with which the spin vector rotates around the central axis Oz that induces a change in states, is called the *Larmor precession frequency*. The Irish physicist Larmor, whose work preceded NMR, has shown independently that ω, the angular rotation frequency around a central axis Oz, has a value of:

$$\omega = \gamma B_0$$

Table 9.1—Values of γ for the most commonly studied nuclei in NMR.

Nuclei N	γ (rad $\cdot$ s^{-1} $\cdot$ T^{-1})	Sensitivity N/^{1}H
^{1}H	$2.675\,221 \times 10^8$	1
^{19}F	2.5181×10^8	0.83
^{31}P	1.084×10^8	6.6×10^{-2}
^{13}C	$0.672\,83 \times 10^8$	1.8×10^{-4}

Since $\omega = 2\pi\nu$, the same relation is obtained as that found in NMR. The approaches are different: one leads to the frequency of the quanta that separates both energy states, and the other to the mechanical precession frequency. However, both frequencies are the same.

For a given nucleus, Larmor's frequency increases with $\vec{B}_0$. It is in the microwave region of the electromagnetic spectrum and, for a field of 1 tesla, varies from 42.5774 MHz for the $^1_1$H nucleus (proton) to 0.7292 MHz for gold, $^{197}_{79}$Au (see Table 9.2). Although instruments are built for the study of other nuclei, they are usually specifically designed for the resonant frequency of the *proton*.

Table 9.2—Resonance frequencies of various nuclei for $B_0 = 1$ T.

Nuclei N	^{1}H	^{19}F	^{31}P	^{13}C	^{15}N
frequency ν	42.58	40.06	17.24	9.71	4.32

■ Based on this principle, multi-nuclei instruments were initially built. They kept the frequency fixed while scanning the magnetic field over a wide range. This allowed the qualitative and global detection of elements by their characteristic resonance frequency (Table 9.2 and Fig. 9.5).

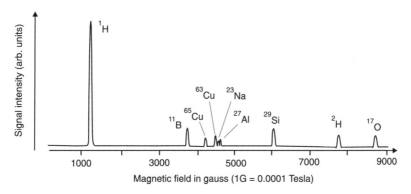

Figure 9.5—*A 5 MHz NMR spectrum of a water sample in a borosilicate glass container (magnetic field in gauss: Varian document)*. There are no commercial instruments of this type because NMR is not sensitive enough to solve many of the problems encountered in elemental analysis.

The interaction of a nucleus with the oscillating magnetic field $\vec{B}_0$, created by the electromagnetic wave, can be understood if it is assumed that it results from the composition of two half vectors, rotating in opposite directions in the xOy plane with identical angular velocities (Fig. 9.6). The vector rotating in the same direction as the precession is the only one that can interact with the nucleus.

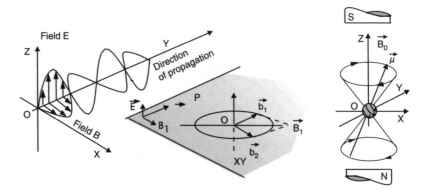

Figure 9.6—*Representation of an electromagnetic wave.* At any point on the xOy plane, the component of the magnetic wave can be dissociated into two half vectors $\vec{b}_1$ and $\vec{b}_2$ rotating with opposite velocities. Only $\vec{b}_2$ can interact with the nuclei of population E_2.

9.6 Chemical shift

The nucleus of each atom in a molecule has a particular environment in which the external applied field is perceived with slight variations. The valence electrons create a small magnetic field, opposite to the main field, thus creating magnetic shielding. These local variations induce slight resonant frequency variations as compared to what would be seen in a vacuum (Fig. 9.1).

This observation is the basis of NMR. The technique focuses on the study of a single type of nucleus instead of scanning for all nuclei, which are spread over a wide range of frequencies. Thus, to investigate the fine structure of each compound studied, the technique zooms over a small range of frequencies (for example 1 000 Hz).

The effective magnetic field $\vec{B}_{\text{eff}}$ felt by the nucleus is related to the field $\vec{B}_0$ by equation (9.8) in which σ is the *shielding constant*:

$$B_{\text{eff}} = B_0(1 - \sigma) \tag{9.8}$$

Variations in σ affect the resonant frequency of a nucleus. This phenomenon is called a *chemical shift*. As many chemical shifts appear in the spectrum as there are σ values.

■ Consequently, very large molecules in one-dimensional NMR lead to spectra that are extremely difficult to interpret because of their numerous σ values.

Using the effective magnetic field B_{eff} for a nucleus with $I = 1/2$, Larmor's relation becomes:

$$\nu_i = \frac{\gamma}{2\pi} B_0 (1 - \sigma_1) \qquad (9.9)$$

■ By theoretical calculations, it can be shown that a very high stability in time is required for the magnetic field B_0 in an instrument. For the resonance frequency of a proton, a variation of 2.3×10^{-8} T will induce a shift of 1 Hz (i.e. if the field varies by 10^{-4} T (1 gauss), the difference in frequency will be 4 259 Hz). This is why the temperature of the magnet has to be controlled to within 1/100th of a degree. The sample is also rotated in a tube with thin walls in order to reduce field inhomogeneities. Considering these constraints, the construction of a mobile NMR instrument is extremely difficult (see Fig. 9.27).

9.7 The principle of obtaining an NMR spectrum

In the *probe* area, where the sample is placed, resonance is obtained by super-imposing a weak oscillating field $\vec{B}_1$ on the field $\vec{B}_0$ generated by a coil that is fed by an AC radiofrequency. In order for $\vec{B}_1$ to have the proper space orientation, the propagation axis of the radiofrequency has to be perpendicular to the Oz axis (Fig. 9.8). The transfer of energy from the source to the nuclei occurs when the radio-frequency of the source and the frequency of precession are the same. When this occurs, nuclei in energy state E_1 can pass to energy state E_2; the relative populations become modified.

The individual vectors $\vec{\mu}$, before irradiation, are out of phase with one another and this can be represented by the vector $\vec{M}_0$ aligned in the Oz direction (Fig. 9.4). As the resonance condition is reached, all the vectors pack together and rotate in phase with $\vec{B}_1$. Hence, $\vec{M}_0$ changes direction and finally reaches an angle α with the Oz axis, which is controlled by the time and power of irradiation (Fig. 9.7). Thus $\vec{M}_0$ acquires an $\vec{M}_{xy}$ component in the horizontal plane that is maximum when $\alpha = \pi/2$, while maintaining a component $\vec{M}_z$ in the direction of the Oz axis (except if $\alpha = \pi/2$). The frequency of rotation of the magnetisation vector is equal to that of the precession movement. Under these conditions, some nuclei will proceed to the second orienta-tion allowed (in the case where $I = 1/2$). The system will slowly return to its original state after the irradiation is stopped. A coil is used to detect the component in the Oy direction (Fig. 9.8).

■ To understand what is happening at the level of the probe, it is usually assumed that an external observer is linked to a triangle rotating about the Oz axis at the same frequency of precession of the nuclei (Fig. 9.7). For example, let's assume the case of a single precession frequency. When the radiofrequency has the same value as the precession frequency, it appears to the observer that nuclei only respond to the rotating component of the field (perpendicular to Oz). The magnetisation vector starts to rotate by an angle of α. After irradiation, the individual magnetic moments lose their coherence by interaction with the spins of neighbouring nuclei faster than they can reorient to come back to the initial Boltzmann distribution. Thus, $\vec{M}_0$ loses

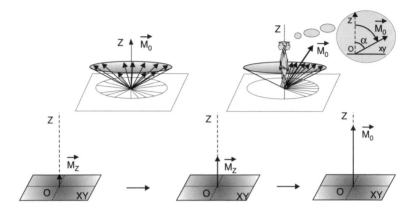

Figure 9.7—*Change of the magnetisation vector $\vec{M_0}$ with irradiation and relaxation of the system after resonance.* Schematic representation showing only the individual vectors resulting from the non-equilibrium of populations (numerical). An independent observer rotating at the same frequency as the precession would see the magnetisation vector tilted by an angle of α. Relaxation to the original position is also shown.

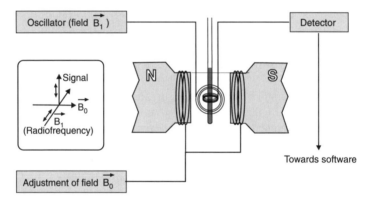

Figure 9.8—*NMR spectrometer.* Representation of the various coils around the probe, placed between the poles of the magnet, and the orientation of the different fields.

its xOy component before it regains its initial value in the Oz direction, which can take a longer period of time.

Technically speaking, the sample is irradiated for a few microseconds with an intense pulse containing all frequencies included in the domain to be sampled. This can be compared to a polychromatic radiation source (like comparing a polychromatic to a monochromatic light source). For example, when working at 300 MHz, the frequency range has to be at least 6 000 Hz in order to irradiate all the protons irrespective of their environment. Under these conditions, a small fraction of each type of proton (but not all the protons) will absorb the resonance frequency.

Obviously, it is difficult to find a schematic representation for a compound absorbing 10 different frequencies. In such a case, $\vec{M}_0$ can be dissociated into many vectors, each of which precesses around the field with its own frequency (Fig. 9.7 shows a simplified situation). As the system returns to equilibrium, which can take several seconds, the instrument records a complex signal due to the combination of the different frequencies present, and the intensity of the signal decays exponentially with time (Fig. 9.9). This damped interferogram, called *free induction decay* (FID), contains at each instant information on the frequencies of the nuclei that have attained resonance. Using Fourier transform, this signal can be transformed from the *time domain* into the *frequency domain* to give the classical spectrum.

This *pulsed wave* process provides simultaneous information on all the frequencies present. The generalisation of this process, which can be computerised and allows the study of less sensitive nuclei such as ^{13}C, has led to major developments in NMR.

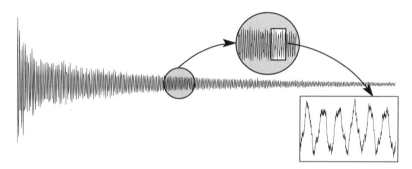

Figure 9.9—*The ^{13}C interferogram of fluoroacetone obtained with a pulsed wave instrument.* The signal $I = f(t)$ is converted by Fourier transform into the classical spectrum $I = f(\nu)$, which corresponds to the spectrum shown in Fig. 9.1. The interferogram, obtained in only a few seconds, can be accumulated tens of times before it undergoes a Fourier transformation. This results in an increase in the signal-to-noise ratio.

■ The process of obtaining a spectrum by calculations (*emission NMR*) did not originally exist. The former approach consisted of obtaining resonance by slightly modifying B_0 using a small coil wrapped around the pole piece of the magnet (Fig. 9.7). Because these instruments used to sweep the magnetic field, they operated at a single frequency (the spectrum was still reported in Hz because it is possible to establish a correlation between fields and frequencies using Larmor's equation). This technique of *absorption NMR*, now abandoned, used instruments with *continuous waves* (CW) rather than pulsed. The process used to find the signal can be related to the sequential recording of optical spectra or, in modern terms, to the location of an FM radio station. In this mode of spectrum acquisition, the quality of the spectrum only depends on a small fraction of the recording time. Modern Fourier transform spectrometers make much better use of the recording time.

9.8 Relaxation processes

At the end of the radiofrequency pulse, $\vec{M}_0$ will regain its equilibrium value and its relaxation period will depend on the medium (Fig. 9.10). This relaxation period depends both on the loss of phase coherence (relaxation time T_2), known as *transversal relaxation* (spin/spin interactions), and on the re-population of the initial state (relaxation time T_1), known as *longitudinal relaxation* (spin/network interactions). T_1 does not exceed a few seconds in solution (for ^{1}H) while it can be as long as several hours in solids. It is thus essential to dissociate both components of $\vec{M}_0$. Knowledge of the relaxation times of T_1 and T_2 can provide extremely useful information on the structure of a sample. T_1 decreases with increasing viscosity, which causes peak broadening. T_2 also affects the width of the bands. A neat liquid will yield broader signals than those obtained from dilute solutions.

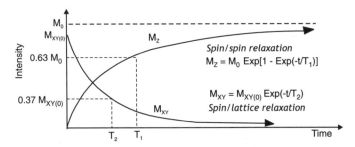

Figure 9.10—*The two processes for nuclei relaxation.* Evolution over time. T_2 decreases with increasing rigidity of the environment.

9.9   The measurement of chemical shifts

According to Larmor's relation, very weak variations in the intensity of the field will affect the frequency of resonance. It would thus be dangerous to try to compare spectra or identify compounds using the absolute frequencies obtained with different instruments. Therefore, it is usual to measure chemical shifts on the relative $\Delta\nu/\nu$ scale, which is independent of the instrument. This is done by measuring the frequency difference between the signal of a compound and that of an internal standard which serves as a reference, as well as the frequency ν_{app} of the instrument, which is imposed by design. The values obtained are expressed in parts per million (ppm). Hence, calculation of the chemical shift δ_i (ppm) that corresponds to a signal of frequency ν_i with respect to that of a reference compound (ν_{ref}) is obtained by calculating $\Delta\nu$ (it is not necessary to know ν_{ref}).

$$\delta_i = \frac{\nu_i - \nu_{ref}}{\nu_{app}} \times 10^6 = \frac{\Delta\nu}{\nu_{app}} \times 10^6 \qquad (9.10)$$

If a given compound's spectra are obtained with two different instruments with a ratio of magnetic fields B_0 equal to k and are then compared, the frequencies of the homologous signals for the two spectra are observed to have the same k value leading to unchanged values for δ. This fact has aided the generation of correlation tables based on the empirical chemical shifts as a function of chemical structure.

These tables can be used irrespective of the NMR spectrometer used (see Tables 9.5 and 9.6). Although useful, these tables are not always sufficient to allow the proper attribution of signals for a given compound. Some software programs have been developed to help in this regard.

> ■ On a spectrometer with a frequency of 60 MHz, the chemical shift of a signal with a frequency difference of 150 Hz with respect to the internal standard has a value of $\delta = 150/60 = 2.5$ ppm. If the chemical shift is recorded on an instrument at 300 MHz, it will still have a value of 2.5 ppm because the difference in frequencies will then be $150 \times 300/60 = 750$ Hz.

The reference compound used for ^{1}H and ^{13}C NMR is tetramethylsilane (TMS). This volatile and inert compound (boiling point $= 27\,^\circ$C) gives a unique signal for both types of nuclei (12 equivalent protons and four equivalent carbons). TMS is used as the reference origin on the chemical shift scale (see Fig. 9.11). In almost all cases, its resonance frequency is lower than that of all other organic compounds (δ positive). The signal corresponding to TMS is usually located on the right-hand side of the spectrum. This is in accordance with most spectroscopic methods in which the energy parameter located on the abscissa decreases as one moves from left to right. Values of δ do not exceed 15 ppm in proton NMR, while they can be in the order of 250 ppm in ^{13}C NMR.

9.10   Shielding and deshielding of the nuclei

In terms of NMR, molecules present in dilute solutions can be considered as independent entities having no noticeable interaction with neighbouring molecules. However, within a given molecule, the electronic and steric environment of each nucleus produces a shielding effect *vis-à-vis the* external field B_0.

As the screening effect increases, the nuclei are said to be *shielded*: on a continuous wave instrument operating at fixed frequency, the intensity of the field B_0 has to be increased in order to obtain resonance. Signals to the right of the spectrum are said to be resonant at *high field*. Signals observed to the left of the spectrum correspond to deshielded nuclei and are said to be resonant at *low field* (Fig. 9.11).

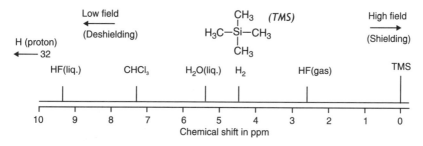

Figure 9.11—*Relationship between chemical shifts, magnetic field, and shielding effects (proton NMR).*

9.11   Factors influencing chemical shifts

Observation of a great number of NMR spectra reveals general factors which are responsible for predictable and cumulative effects on chemical shifts.

9.11.1   Effects of substitution and hybridisation

The simple replacement of a hydrogen atom by a carbon-containing group R causes a deshielding of the remaining protons. In going from RCH_3 to R_3CH, the effect reaches 0.6 ppm. The corresponding effect can be as high as 40 ppm for ^{13}C NMR. Hybridisation of carbon atoms has an even bigger influence on the position of the signals.

These hybridisation variations are caused by *anisotropy* within the chemical bonds. This is due to the non-homogeneous electronic distribution around bonded atoms to which can be added the effects of small magnetic fields induced by the movement of electrons (Fig. 9.12). Thus, protons on ethylene are deshielded because they are located in an electron-poor plane. Inversely, protons on acetylene that are located in the C–C bond axis are shielded because they are in an electron-rich environment. Signals related to aromatic protons are strongly shifted towards lower fields because, as well as the anisotropic effect, a local field produced by the movement of the aromatic electrons or the 'ring current' is superimposed on the principal field (Fig. 9.12).

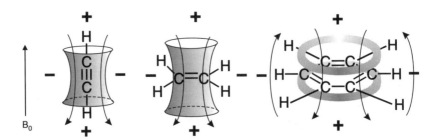

Figure 9.12—*Effects of anisotropy and local induced fields.* The presence of π bonds is shown as zones in which there is a shielding (+) or a deshielding (−) effect. Ethylenic or aromatic protons are located outside a double cone of protection.

9.11.2   Resonance and inductive effects

When the chemical shifts of the carbonyl from a ketone and an ester are compared in ^{13}C NMR, the carbon atom of the latter is observed to be less shielded than that of the ketone, which has a more electropositive nature (Fig. 9.13).

Electronic effects that modify the polarity of a bond have an influence on chemical shifts. If the position of a methyl group in an alkyl halide CH_3–X is compared to that of methane, the shift in signal increases with the electronegativity of the halogen atom (Table 9.3).

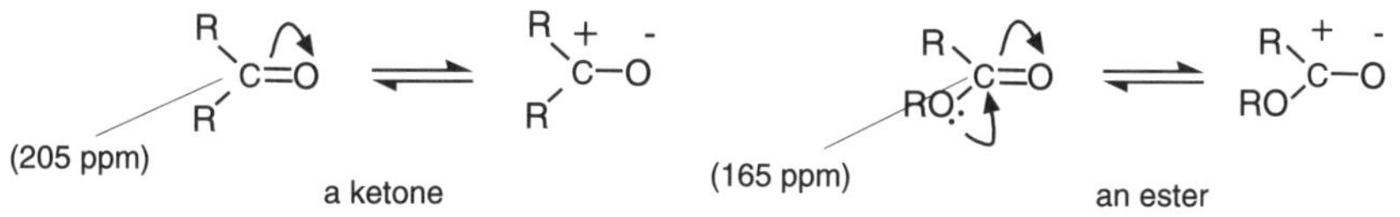

(205 ppm)   a ketone   (165 ppm)   an ester

Figure 9.13—*Effect of resonance on carbonyl-containing compounds.* Representation of the delocalisation of valence electrons in mesomeric forms of organic compounds. In ^{13}C NMR, the signal corresponding to a carbonyl in an ester is at 165 ppm, whereas it is at 205 ppm for a ketone.

Table 9.3—Influence of the electronegativity χ of a halogen on δ (TMS reference).

	$CH_3F(\chi = 4)$*	CH_3Cl $(\chi = 3.2)$	CH_3Br $(\chi = 3)$	CH_3I $(\chi = 2.6)$
δ_H (ppm)	4.5	3	2.7	2.3
δ_C (ppm)	75	30	10	−30

* electronegativity χ in eV according to the Pauling scale.

9.11.3 Other effects (solvent, hydrogen bonding, matrix)

The ^{1}H NMR spectra of organic compounds are usually obtained in an aprotic solvent at concentration levels of a few percent. The most widely used solvent is deuterated chloroform ($CDCl_3$), sufficiently polar to dissolve most organic compounds. Acetone-*d*6 (C_3D_6O), methanol-*d*4 (CD_3OD), pyridine-*d*5 (C_5D_5N) and heavy water (D_2O) are also used.

Should the compound under investigation have labile hydrogens, $H \leftrightarrow D$ exchanges can occur in certain solvents. These exchanges will cause modifications in the intensity and position of the corresponding signals. Solvent-solute associations often occur and their stability will vary with polarity. These associations can alter the resonant frequencies, thus the position, of the signals in the spectrum. Consequently, the sample concentration and solvent used must be provided with the correlation tables.

■ In chloroform/toluene mixtures, the position of the signal due to chloroform in ^{1}H NMR goes from 7.23 ppm (90% chloroform/10% toluene v/v) to 5.86 ppm (10% chloroform/90% toluene v/v). This shift towards a higher field for toluene-rich mixtures is rationalised by the presence of complexes causing the proton of chloroform to be located in the shielding zone of toluene's aromatic nuclei.

Hydrogen bonding can also have a strong influence on the electronic environment of some protons. Because of this, it is sometimes difficult to predict chemical shifts.

Finally, interactions between neighbouring molecules and viscosity will alter the resolution in the spectrum via the spin/network relaxation time.

9.12   Hyperfine structure: spin/spin coupling

As a general rule, NMR spectra have a greater number of signals than the number of nuclei with different chemical shifts. This phenomenon results from the fact that the orientation taken in the magnetic field by one nucleus is transmitted through valence electrons to the neighbouring nuclei. This coupling between nuclei rapidly reduces with distance. *Homonuclear* coupling (nuclei of the same type) or *heteronuclear* coupling (nuclei of different types) gives rise to small shifts in the signal. This hyperfine structure can provide additional information on the compound under study. Homonuclear coupling between hydrogen atoms is quite frequent. However, ^{13}C, ^{31}P and ^{19}F can also lead to heteronuclear coupling with the protons.

■ This phenomenon should not be confused with the interaction through distance between two nuclei that exchange magnetisation because the structure of the molecule is such that they are close in space, although a great number of bonds separates them. This is the *Nuclear Overhauser Effect* that causes modifications in signal intensities.

9.13   Heteronuclear coupling

9.13.1   Typical heteronuclear coupling: hydrogen fluoride

Hydrogen fluoride (HF), a molecule that contains only two atoms linked by a covalent bond, exhibits heteronuclear spin/spin coupling. For example, in the absence of coupling between the nuclei, a single signal would be observed in the ^1H NMR spectrum of HF because there is only one hydrogen atom. However, when the experiment is conducted, two signals of equal intensity are observed.

Hydrogen fluoride, when inserted into the magnetic field of the instrument, leads to a distribution of molecules because of thermal equilibrium. Because they have spin $I = 1/2$, H and F atoms give rise to four populations E_1 to E_4, that have the four following spin combinations:

$$\uparrow B_0 \qquad \uparrow H{-}F\downarrow \qquad \uparrow H{-}F\uparrow \qquad \downarrow H{-}F\uparrow \qquad \downarrow H{-}F\downarrow$$

In the absence of interaction between the H and F nuclei, populations E_1 to E_4 would have the same energy. Therefore, in ^1H NMR a single signal should appear (Fig. 9.14). However, because there is an interaction between H and F, the reality is different. Theoretical calculations indicate that in the presence of coupling, E_1 and E_2 correspond to different energy states. This leads to a splitting in the signal. The energy required to pass from states E_2 to E_3 is slightly different to that needed to pass from E_1 to E_3. Thus the spin orientation taken by the fluorine atom has an effect on the transition energy observed in ^1H NMR. Two signals are observed in the spectrum of this compound corresponding to the preceding transitions, demonstrated in Fig. 9.14.

As a general rule, couplings are defined by the distance, in Hz, between the components of the signal and by the number of bonds between the atoms that are coupled. The value of the coupling in Hz is designated by J, called the *coupling constant*. The example of HF described above leads to:

$$^1J_{FH} = 615 \text{ Hz}$$

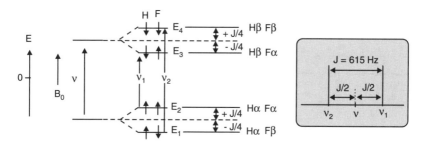

Figure 9.14—*Coupling diagram for the HF molecule in proton NMR.* a) Hypothetical situation in which there is no coupling with the fluorine atom; b) actual situation. The vertical arrows indicate transitions that differ in length by J Hz (transitions in the state of the proton do not affect the orientation of the fluorine atom; this is the reason why $E_1 \rightarrow E_3$ and $E_2 \rightarrow E_4$ are not observed).

■ The spin orientation of the fluorine atom, which has a resonance frequency very different to that of the proton, is not affected by the resonance occurring in protons: in a 2.35 T field, the resonant frequency of ^1H is 100 MHz while that of ^{19}F is 94 MHz. Signals corresponding to ^1H are separated by 615 Hz regardless of the intensity of the magnetic field B_0. The ^{19}F NMR spectrum of this molecule would have similar features. Two signals, separated by 615 Hz, would be observed due to the coupling of the proton with either orientation.

Spin/spin coupling rapidly decreases as the number of bonds between the concerned atoms increases. However, this decrease can be affected by the presence of multiple bonds (double or triple) that can propagate the spin effect through π electrons.

9.13.2   Heteronuclear coupling in organic chemistry

The presence of atoms with a spin of 1/2 such as ^{13}C, ^1H or ^{19}F within a molecule leads to numerous heteronuclear couplings that can be used for structural studies (see Fig. 9.15).

■ In ^{13}C NMR spectra where $CDCl_3$ solvent has been used, a triplet is observed because of the heteronuclear coupling between ^{13}C and the deuterium atom ^2H (D), which has a spin of 1 (Fig. 9.16).

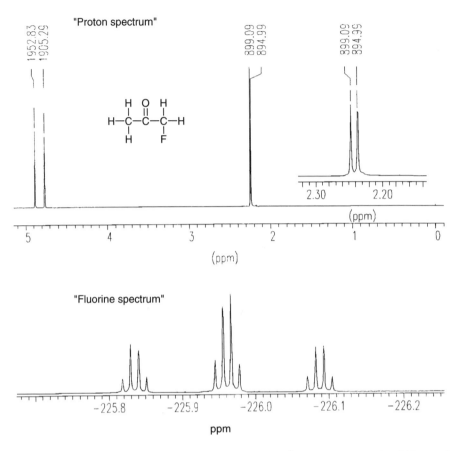

Figure 9.15—*Heteronuclear couplings. Monofluoroacetone.* Top: ^{1}H NMR spectrum in which coupling with the fluorine atom can be observed ($^2J = 47.5$ Hz, $^3J = 4.1$ Hz). Bottom: ^{19}F NMR spectrum. The single fluorine atom present in this molecule leads to a triplet from coupling with the CH_2 group and to a quartet from coupling with the three protons of the methyl group (coupling constants can be obtained using $FSiCl_3$ as a reference).

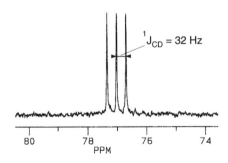

Figure 9.16—^{13}C *NMR spectrum of* $CDCl_3$ *showing the spin coupling* $^1J_{CD}$. The spectrum shows three peaks of equal intensity (the coupling $^1J_{CD} = 32$ Hz).

9.14   Homonuclear coupling

Signal splitting, as observed in the case of hydrogen fluoride, is a phenomenon often encountered in organic molecules with adjacent hydrogens.

9.14.1   Weakly coupled systems

Nuclei are said to be weakly coupled when the coupling constants are much smaller than the differences in chemical shifts between nuclei (after conversion in Hz).

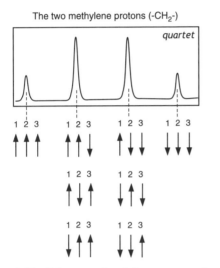
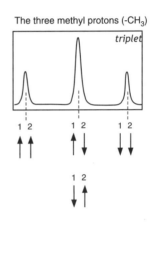

Figure 9.17—*Spin states of methylene protons and methyl protons within an ethyl group.* Represented on each row are the spin populations that produce similar effects on adjacent nuclei. For a large number of molecules, the population will be weighted according to the number of states per row in this scheme.

Butanone (Fig. 9.1) is a simple example of a molecule containing five protons (2 plus 3) on the ethyl group, which are weakly coupled to each other. These protons give rise to three peaks (a *triplet*) around 1.1 ppm and to four peaks (a *quartet*) around 2.5 ppm. The origin of these peaks in the spectrum can be assigned by their chemical shifts (Table 9.5). The triplet is due to the ethyl CH_3 group, split by its two neighbouring protons, and the quartet is due to the adjacent ethyl CH_2 group, split by its three neighbouring protons. The relative intensities of the components within the multiplet can be deduced using statistical laws (see Fig. 9.17 and Table 9.4).

As a general rule for weak coupling, if n nuclei (of spin I) are placed in the same magnetic environment and influence adjacent nuclei in the same manner, the signal observed for the latter will be composed of $2nI + 1$ peaks that are regularly spaced, i.e. $(n + 1)$ peaks if $I = 1/2$. The peak intensities in a multiplet will follow the same pattern found in each line of Pascal's triangle (see Table 9.4). However, if the group of protons is affected by neighbouring nuclei with different chemical shifts and different coupling constants, the preceding approximation is not applicable. Signal multiplicity and relative intensities of the peaks cannot be deduced as easily.

Table 9.4—Pascal's triangle and its application to NMR for $I = 1/2$.

Neighbouring hydrogens	Multiplicity	Intensity						
0	singlet	1						
1	doublet	1	1					
2	triplet	1	2	1				
3	quartet	1	3	3	1			
4	quintuplet	1	4	6	4	1		
5	sextuplet	1	5	10	10	5	1	
6	septuplet	1	6	15	20	15	6	1

■ **Nomenclature for coupled systems in ¹H NMR.** The interpretation of a spectrum of a molecule with many hydrogen atoms is simplified when signals that fall into classical cases can be observed. These particular cases are usually classified with a nomenclature that uses the letters of the alphabet, chosen in relation to the chemical shift (Fig. 9.18). Protons with identical or similar chemical shifts are designated by identical letters or neighbouring letters in the alphabet (AB, ABC, A_2B_2, etc.), while protons with very different chemical shifts are designated by letters such as A, M and X.

On this basis, an ethanal group constitutes an AX_3 system, ethoxyl an A_2X_3 system and a vinyl group is either ABC, AMX or ABX depending on the system studied.

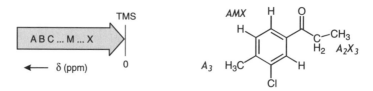

Figure 9.18—*NMR spectra nomenclature.* The ¹H spectrum of 3-chloro-4-methylpropiophenone can be considered as a group of independent signals resulting from subgroups that are easily identified.

■ Because the coupling constants $^1J_{C-H}$ in ¹³C NMR are in the order of 125–200 Hz, signal overlapping is frequent. The presence of spin coupling makes it more difficult to interpret the spectrum. It is possible in ¹³C NMR to eliminate spin coupling with protons by using a particular mode of operation (see Fig. 9.19).

9.14.2　Strongly coupled systems

When the ratio $\Delta\nu/J$ is small, the intensity of the peaks is modified. In the latter instance, new peaks appear in the spectrum and these combination peaks complicate the interpretation of spectra that contain strongly coupled protons. It is mostly

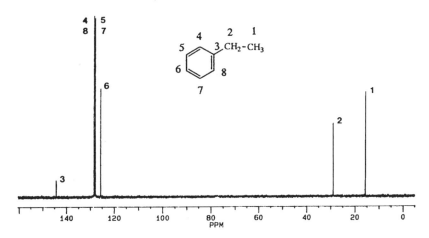

Figure 9.19—*Proton decoupled ^{13}C NMR spectrum of ethylbenzene*. Each of the carbon atoms gives a signal consisting of a singlet. These 'large band' decoupled spectra are simpler but contain less information.

because of this that new instruments operating at higher frequencies (300 to 750 MHz) have been developed. The use of superconducting magnets has allowed the development of such instruments. Because $\Delta\nu$ is proportional to the operating frequency of the instrument and J is constant, their ratio increases (Fig. 9.20). However, when frequencies above 600 MHz are used, other phenomena complicate the spectrum.

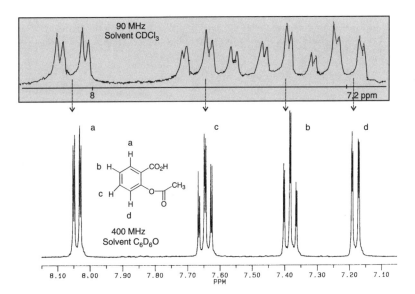

Figure 9.20—*Spectra of the aromatic protons of aspirin*. The figure shows, on a similar scale, the spectrum of aspirin obtained on two different instruments: one at 90 MHz (upper) and the other at 400 MHz (lower). It can be observed that the nature of the solvent slightly modifies the chemical shifts of the four protons. The sensitivity of NMR increases as $B_0^{3/2}$.

■ **AB system.** Two protons constitute the simplest example of a strongly coupled system when the difference in their frequencies of resonance is comparable to their coupling constant J. This system gives rise to a total of four peaks, two for each proton, separated by J Hz. However, the intensities are not equal. The chemical shifts, which can no longer be measured from the spectrum, have to be calculated using the relations shown in Fig. 9.21.

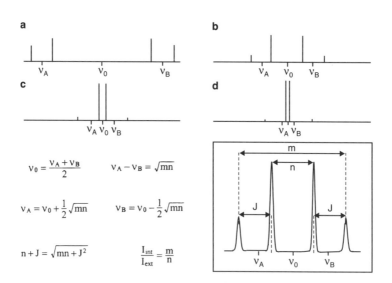

Figure 9.21—*Characteristics defining the AB system.* Figures (a) to (d) show the gradual modification of a two-proton coupled system as the ratio $\Delta\nu/J$ decreases. The general pattern and formulae used to analyse an AB system are shown.

9.15 Spin decoupling and modes of operation

NMR instruments are now equipped to facilitate spectral interpretation. Thus, it is now possible to nullify the spin-coupling effect between neighbouring nuclei. This *spin decoupling* technique stems from the belief that a nucleus in resonance does not always conserve the same spin state with time. It alternates between different states and this effects the neighbouring nuclei.

In practice, spin decoupling experiments are conducted in the following way. First, the spectrum is recorded under normal conditions. Then the spectrum is recorded while a second radiofrequency emitter irradiates at the resonance frequency of the nuclei that are to be decoupled (Fig. 9.22). This double resonance technique is used to identify nuclei which are coupled and which cause interpretation difficulties in the spectrum.

High field Fourier transform NMR ($B_0 = 17.6\,\text{T}$ for the study of ^1H at 750 MHz) has considerably expanded the scope of this method, due to the speed of acquisition and computation of spectra. It has led to the development of multidimensional NMR (2-D and 3-D) and to other experiments based on operation in a pulsed mode that allow measurement of the relaxation time of the nuclei and deduction of very useful structural information. Fourier transform NMR permits the resolution of signals that are superimposed. When analysing biopolymers, 2- and 3-dimensional NMR provide results that are comparable to those obtained by crystallography and have the advantage that representation of molecules in their natural state can be obtained.

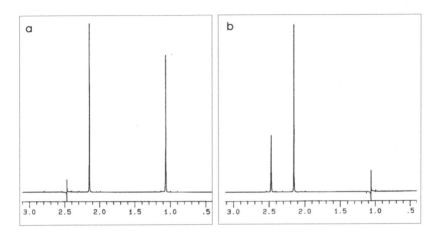

Figure 9.22—*Spin decoupling experiment on butanone.* Spectral modification: a) by irradiating the CH_2 group at 2.47 ppm; b) by irradiating the CH_3 group (of ethyl) at 1.07 ppm, compared to the spectrum shown in Fig. 9.1. For such a simple compound, these experiments are only used to illustrate the principle. On the other hand, a double resonance experiment would be useful to precisely determine the coupling in aspirin (shown in Fig. 9.20).

■ **Magnetic resonance imaging (MRI).** Proton NMR can be used to obtain images of objects containing hydrogen atoms. This important development of 3-dimensional NMR, pioneered by R. Ernst who won the Nobel Prize in chemistry in 1991, allows observation of the inside of living organisms without any penetration. In biomedical science, MRI corresponds to a non-invasive examination specially adapted for soft tissues. The total volume under observation has to be placed in a magnetic field, which requires the fabrication of very large superconducting magnets for total body instruments. The signal due to the protons of water is the most easily observed in biological tissues, which contain 90% water (a 70 kg individual contains 50 kg of water compared to 13 kg of carbon). In other instances, one can monitor the signal due to the CH_2 groups present in lipids. The final image is a map of the distribution of intensities of the same proton signal. Contrast in the image is due to variations in relaxation times of the protons. The difficulties encountered in this technique are related to the focusing or spatial selectivity needed to obtain a spectrum of a quasi-punctual volume within an object.

9.16   Fluorine and phosphorus NMR

Fluorine and phosphorous represent the two heteroatoms in organic chemistry most studied by NMR, after hydrogen and carbon.

Fluorine, consisting of 100% ^{19}F ($I = 1/2$), is comparable to the proton due to its good sensitivity. Because its electronegativity is greater than the proton, chemical shifts are distributed over a greater range (Fig. 9.23). As a consequence, it becomes possible in ^{19}F NMR to distinguish between compounds that are very similar whereas the analogous signals would be melded together in proton NMR. In particular, differences due to the stereochemistry of the molecules are important and the coupling constant J_{F-H} can be measured over a greater range than J_{H-H} (Fig. 9.15). Unfortunately, relatively few molecules naturally contain fluorine atoms. Hence, fluorine NMR spectra are usually obtained from compounds in which a fluorine atom (or CF_3 group) has been introduced in a known position. This procedure allows structural information to be obtained from the perturbations that appear in the spectra of the molecule. In summary, the fluorine atom induces only minor modification in the stereochemistry of the molecule (van der Waals' radius of fluorine is 1.35 Å instead of 1.1 for the hydrogen atom) but it produces a chemical shift comparable to that of hydroxylic group OH.

Phosphorus (^{31}P, $I = 1/2$) represents another element sensitive to NMR and consists of only one natural isotope. It has been studied since the beginning of NMR because it is an important element in the composition of inorganic compounds and has a very important role in biology.

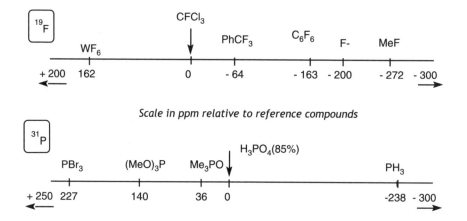

Figure 9.23—*Positions of a few NMR signals due to fluorine and phosphorus.*

APPENDIX

Quantitative NMR

Although NMR is a method essentially used for structural analysis because of the quality of the information it produces, it is also used occasionally for mixture analyses. The latter is only possible when the areas of the individual signals generated by individual components can be measured separately. Although sensitivity and precision will vary depending on the type of nuclei under study, the method has some very interesting features. The analysis can be conducted without sample preparation, without destroying the sample, without risk of polluting the instrument, and, unlike many methods like chromatography, it does not require a prior standardisation step. In organic synthesis, proton NMR can be used to calculate the yields of a reaction. In industry, many analysers based on proton NMR are used to quantitate water and other compounds containing hydrogen.

9.17   Measurement of area – application to simple analysis

Quantitative measurements in NMR are based on the area of the signals present in the spectrum. Signal areas can be produced as numerical values proportional to the area or, on less modern instruments, from the integration plots that are superimposed on the spectrum (Fig. 9.1). For the proton, the precision obtained in area measurements does not exceed 1% even if continuous wave instruments are used at slow scanning speeds. In ^{13}C NMR, it is preferable to add a relaxation reagent in order to avoid saturation related to relaxation times that alter the intensity of the signal. Using the molar ratios that are easily accessible from the spectrum, it is possible to deduce concentrations.

Consider, for example, a mixture of acetone A and benzene B (Fig. 9.24). From the ^{1}H NMR spectrum of the mixture that corresponds to a superposition of the individual spectra, a peak can be observed at $\delta = 2.1$ ppm for acetone and one at $\delta = 7.3$ ppm for benzene (relative to TMS in CDCl$_3$ solvent). The area ratio of the two peaks in this particular example – each molecule has six protons – is indicative of the molar ratio n_A/n_B of A and B (it is assumed that the response factors for A and

B are equal). If S_A represents the surface (area) of the peak due to acetone and S_B represents that of benzene, then:

$$\frac{n_A}{n_B} = \frac{S_A}{S_B} \qquad (9.11)$$

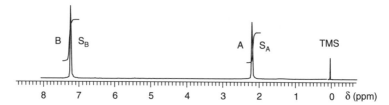

Figure 9.24—*^1H NMR spectrum of a mixture of acetone (A) and benzene (B). If $S_A = 111$ and $S_B = 153$ (arbitrary units), then $C_A = 35\%$ and $C_B = 65\%$ in mass units based on using molecular weights $M_A = 58$ and $M_B = 78$ g/mol.*

Assuming that the mixture only contains these two components, and if C_A and C_B are the concentration expressed in mass % of A and B with molar masses of M_A and M_B, then:

$$C_A + C_B = 100 \qquad (9.12)$$

and

$$\frac{C_A}{C_B} = \frac{n_A \cdot M_A}{n_B \cdot M_B} \qquad (9.13)$$

Or, by substituting into equation (9.11):

$$\frac{C_A}{C_B} = \frac{S_A \cdot M_A}{S_B \cdot M_B} \qquad (9.14)$$

Therefore,

$$C_A = 100 \times \frac{S_A M_A}{S_A M_A + S_B M_B} \qquad \text{and} \qquad C_B = 100 \times \frac{S_B M_B}{S_A M_A + S_B M_B} \qquad (9.15)$$

9.18 Samples containing compounds that can be identified

A more general case is that in which the selected peaks for each of the compounds do not correspond to the same number of protons. This can occur when the molecules do not contain the same total number of protons or, alternatively, if only a portion of the spectrum is taken to identify the compounds.

For example, if the selected peak for compound A corresponds to a protons and that chosen for compound B corresponds to b protons (A and B do not represent acetone and benzene from section 9.17), then in the recorded spectrum of the mixture of A and B each molecule of B will give a signal whose intensity is different than that of A. In such a case, equation (9.15) can be used provided corrections are inserted for these differences. The area of the selected peaks must be divided by the number of protons in the molecule in order to normalise the area to one proton of either A or B. Substituting S_A and S_B in the previous equation by the ratios S_A/a and S_B/b, equation (9.15) can be rewritten for the normalised areas as:

$$C_A = 100 \times \frac{\dfrac{S_A}{a} M_A}{\dfrac{S_A}{a} M_A + \dfrac{S_B}{b} M_B} \quad \text{and} \quad C_B = 100 \times \frac{\dfrac{S_B}{b} M_B}{\dfrac{S_A}{a} M_A + \dfrac{S_B}{b} M_B} \quad (9.16)$$

The above formula can be transposed for the situation where n constituents are visible in the spectrum. If each of the constituents has a specific area S_I with i protons for component I, then equation (9.17) is obtained, giving the mass % of each of the constituents:

$$C_i = 100 \frac{(S_I/i) \cdot M_I}{(S_A/a) \cdot M_A + (S_B/b) \cdot M_B + \cdots + (S_I/i) \cdot M_I + \cdots + (S_N/n) \cdot M_N}$$
$$(9.17)$$

■ In organic chemistry, the yield of a reaction A → B can be determined by ^{1}H NMR. This is done by recording the spectrum of the coarse material after reaction and by identifying a peak specific to the remaining reagent (A) and one specific to the product formed (B). The yield of B relative to A can be obtained by the following equation:

$$\text{Yield} = 100 \frac{(S_B/b)}{\left(\dfrac{S_A}{a} + \dfrac{S_B}{b} \right)} \quad (9.18)$$

9.19   Internal standard method

A more general situation relates to the measurement of the concentration (in g/l or %) of a compound present in a sample. To quantify a single component in solution, it is not necessary to know the nature of all the other compounds that are present or to identify *all* of them in the NMR spectrum. It suffices to identify a signal generated by the compound of interest.

In order to quantify compound X (with mass M_X) in a mixture, a quantity p mg of reference compound R (with mass M_R) is added to P mg of the sample containing compound X. The internal reference standard R is chosen so that the signal it generates does not interfere with the signal used to quantify compound X (Fig. 9.25). The concentration in mass % of R, C_R, is equal to:

$$C_R = 100 \times \frac{p}{P + p} \quad (9.19)$$

Then, in the NMR spectrum of the mixture containing the internal standard:

— a peak belonging to compound X is chosen (area S_X for x protons)
— a peak belonging to reference R is chosen (area S_R for r protons).

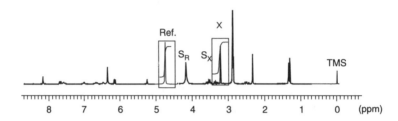

Figure 9.25—*Spectrum of a sample into which an internal standard R has been added.* Peak X belongs to the compound to be quantified and peak R to the internal reference compound.

Considering the molar ratio n_X/n_R of X and R, equation (9.20) can be obtained as before:

$$\frac{C_X}{C_R} = \frac{n_X}{n_R} \frac{M_X}{M_R} \qquad (9.20)$$

Assuming that:

$$\frac{n_X}{n_R} = \frac{S_X/x}{S_R/r} \qquad (9.21)$$

then the mass concentration (in %) C_X of compound X can be obtained by:

$$C_X = 100 \times \frac{p}{P+p} \times \frac{(S_X/x) \cdot M_X}{(S_R/r) \cdot M_R} \qquad (9.22)$$

9.20 Standard additions method

The previous method can be improved by preparing a number of standard solutions that contain the sample to be analysed plus increasing quantities of that same compound. Using the NMR spectra obtained from these solutions, a plot can be made of the peak areas used for quantification. The intercept of the plot with the concentration axis will give the required sample concentration (Fig. 9.26).

One of the difficulties encountered with this method is in maintaining the stability and the sensitivity of the NMR instrument while the standard solutions are being analysed.

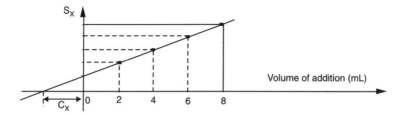

Figure 9.26—*NMR plot for the standard additions method.* The unknown concentration corresponds to the abscissa segment between the origin of the axis and the intersection with the calibration curve.

9.21   Analysers using pulsed NMR

Pulsed wave Fourier transform NMR also permits routine quantitation yet is under-used by quality control analysts. For example, water and other organic compounds like lipids can be quantified using this technique, which gives the concentration of hydrogen in the sample. The corresponding instruments do not plot the usual NMR spectrum but measure total intensity of the FID after the irradiation period and its time decay (see Figs 9.27 and 9.28). The rate of proton relaxation provides information on the environment of the hydrogen atoms. It thus becomes possible to distinguish between protons that are present in a solid and those that are part of a liquid.

In food technology, it is possible after standardisation to measure the concentration of major constituents like water and lipids in diverse samples.

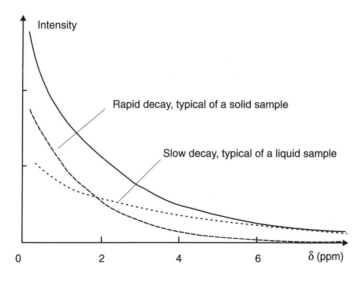

Figure 9.27—*Decay or FID for a solid/liquid sample.* The amplitude of the FID signal, which relates the quantity of protons as a percentage of the total mass, can be distinguished from relative measures based on T_2 that allow the percentage of solids in a sample to be determined by convolution of the curve envelope and its constituents.

Figure 9.28—*^1H NMR instrument used for routine analyses.* Automatic analyser based on pulsed NMR used to quantify water and lipids in food technology. (Reproduced with authorisation of Bruker.)

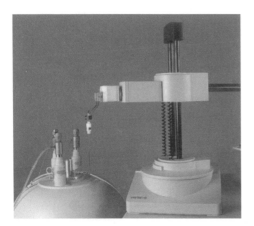

Figure 9.29—*NMR magnets and samples.* Robotic introduction (on the left) of a sample in solution placed in an NMR tube within a magnetic field generated by a superconducting coil and maintained at liquid helium temperature. (Reproduced by permission of Varian.) Electromagnet (on the right) of sizeable volume used for a special kind of sample – the human body (part of an MRI instrument from the SMIS Society).

Table 9.5— NMR chemical shifts of the principal types of protons in organic molecules (reproduced by permission of Spectrometry Spin & Techniques).

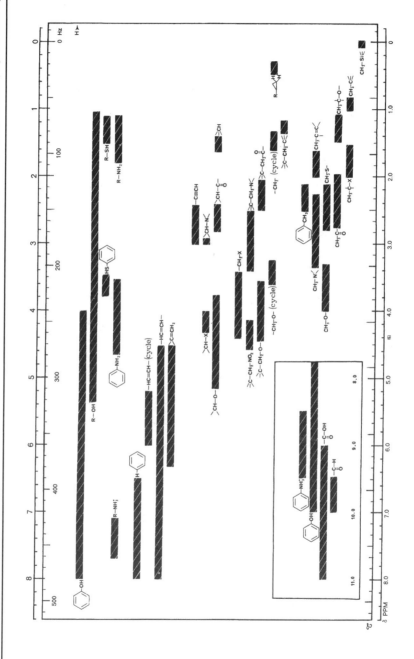

Table 9.6—Correlation table of organic functional groups in ^{13}C NMR.

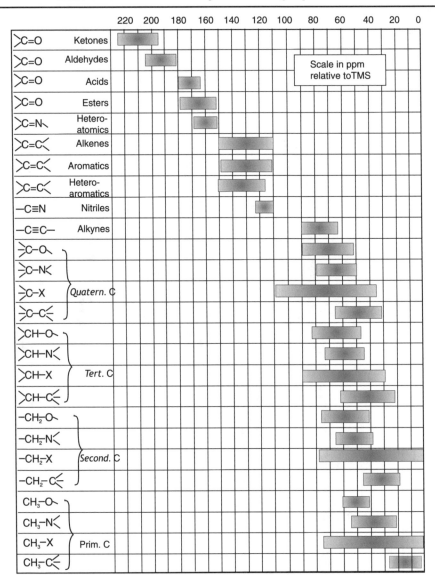

Problems

9.1 Often in tables the values describing a nucleus' magnetic moment are represented as lying upon an axis parallel to $\vec{B}$, the applied magnetic field. Therefore, if for a proton, $\mu_z = 1.41 \times 10^{-26}$ J · T^{-1}, calculate the constant γ for the proton.

9.2 If $T = 300$ K, calculate the ratio of the populations N_{E1}/N_{E2} for a proton in an NMR spectrometer where the applied magnetic field $B = 1.4$ T. Make the same calculation for the case where the field $B = 7$ T, given that $\gamma = 2.6752 \times 10^8 \, \text{rad T}^{-1} \text{s}^{-1}$.

9.3 In the ^{1}H NMR spectrum from a 200 MHz spectrometer the scale along the abscissa is represented by 1 ppm = cm.
1. What is the distance between two signals with a separation of 7 Hz?
2. If it is known that $\gamma_H/\gamma_C = 3.98$, what would happen to the distance calculated above if a ^{13}C spectrum were to be recorded on the same apparatus?

9.4 1. Calculate the chemical shift, δ, in ppm of a proton (^{1}H), whose NMR signal is displaced by 220 Hz with respect to TMS (the field of the spectrometer is 1.879 T).
2. The resonance signal for a proton is displaced by 90 Hz with respect to TMS when measured on a 60 MHz spectrometer. What would happen to this displacement if an apparatus of 200 MHz was employed?
3. What would be the corresponding chemical shift of the same proton with both of these spectrometers?

9.5 Two isomers A and B share the same molecular formula, $C_2HCl_3F_2$. What is the structure of each of these isomers when their proton NMR spectra, as recorded upon a 60 MHz apparatus, are the following:

The spectrum of isomer A comprises two doublets at 5.8 ppm and 6.6 ppm respectively ($J = 7$ Hz), while that of isomer B contains a triplet at 5.9 ppm ($J = 7$ Hz).

Consider now that a third isomer is present. Under the existing conditions, describe what its spectrum must be.

9.6 Numerous NMR experiments make use of the 'Magic Angle' technique which corresponds to the angle made by the spin vector and an axis parallel to the direction of the magnetic field in which it is held. What is the value of this angle in degrees?

9.7 25 mg of vanillin ($C_8H_8O_3$) is added to 100 mg of an unknown organic compound. The ^{1}H integration curve displayed upon the spectrum of the new mixture can be described as follows: one proton of the original compound corresponds to a value of 20 mm while a single proton of vanillin corresponds to 10 mm. What is the molecular weight of the unknown compound? Remember: H = 1, C = 12, O = 16 g/mole.

9.8 If the ratio of the magnetogyric constants γ_F/γ_H is 0.4913, calculate the distance which would separate the TMS signal from that originating from an atom of fluorine, given that the scale upon the spectrum is represented by 1 ppm = 2 cm. (The apparatus is a 200 MHz spectrometer.)

9.9 Following the reduction of acetone in isopropanol by means of a hydride the ^{1}H NMR spectrum of the crude, isolated product reveals that the reaction has not completed. The integration curve of the remaining acetone corresponds to a value of 24 (arbitrary units), while that of isopropanol is equivalent to 8. Calculate the molar yield of this transformation.

9.10 From the ^1H NMR spectrum below, which corresponds to a mixture of bromobenzene, dichloromethane and iodoethane, calculate the percentage composition of the three components. The values (without units) given in brackets on the integration curve are proportional to the areas of the corresponding signals.

Given: $H = 1$, $C = 12$, $O = 16$, $Cl = 35.5$, $Br = 80$ and $I = 127\,g/mol$.

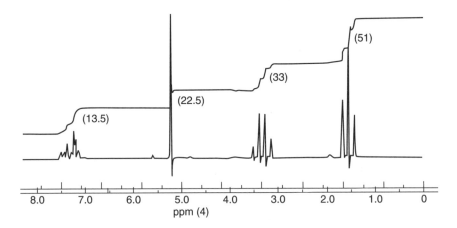

CHAPTER 10

Infrared spectroscopy

Analytical infrared (IR) spectroscopy includes several methods that are based on the absorption (or reflection) of electromagnetic radiation with wavelengths in the range of 1 to 1000 µm. This spectral range can be divided into three groups: near IR (1 to 2.5 µm), mid IR (2.5 to 25 µm) and far IR (beyond 25 µm). The frequency range that is the most accessible and the richest in providing structural information is the middle range. The absorption bands present in this frequency domain form a molecular fingerprint, thus allowing the detection of compounds and the deduction of structural details.

Infrared analyses are conducted on dispersive (scanning) and Fourier transform spectrometers. Non-dispersive industrial infrared analysers are also available. These are used to conduct specialised analyses on predetermined compounds (e.g. gases) and also for process control allowing continuous analysis on production lines. The use of Fourier transform has significantly enhanced the possibilities of conventional infrared by allowing spectral treatment and analysis of microsamples (infrared microanalysis). Although the near infrared does not contain any specific absorption that yields structural information on the compound studied, it is an important method for quantitative applications. One of the key factors in its present use is the sensitivity of the detectors. Use of the far infrared is still confined to the research laboratory.

10.1 Spectral representation in the mid infrared (IR)

All infrared spectrometers generate data that are contained in the *infrared spectrum* (see Fig. 10.1). The spectrum represents the ratio of transmitted intensities with and without sample at each wavelength. This intensity ratio is called *transmittance* (T) can be replaced by percent transmission (%T) or by *absorbance* $A = \log(1/T)$. If the experiment is conducted using reflected or diffuse light, *pseudo-absorbance* units are used (cf. 10.10.2). Finally, it is common to report wavelengths in terms of *wave number* $\bar{\nu}$ (cm^{-1} or *kaysers*) knowing that:

$$\bar{\nu}_{cm^{-1}} = \frac{1}{\lambda_{cm}}$$

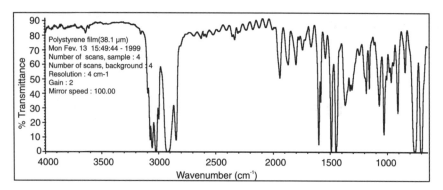

Figure 10.1—*Mid infrared spectrum of a film of polystyrene.* Typical representation of an infrared spectrum with a linear scale abscissa in cm^{-1} and % transmittance as the ordinate (in order to better visualise the right-hand portion of the spectrum, it is customary to change scales around $2\,000$ cm^{-1}; see Fig. 10.12). The abscissa in cm^{-1} (or kaysers) is linear in energy ($E = hc/\lambda$), that decreases from left to right (from high to low energies).

10.2   Origin of absorption bands in the mid infrared

In this spectral range, the absorption bands originate from the interaction of the electrical vector of the electromagnetic wave with the electrical dipole of non-symmetrical bonds. If the dipole corresponding to a chemical bond oscillates at the same frequency of that bond, then the electrical component of the wave can transfer its energy to the bond. It is essential for this energy transfer that the mechanical frequency of the bond be the same as that of the electromagnetic radiation (Fig. 10.2). This simplified approach can be used to rationalise that there is neither coupling with the electromagnetic wave nor absorption of energy in the absence of a permanent dipole. Consequently, molecules such as O_2, N_2 and Cl_2 with non-polar bonds are 'transparent' in the mid IR.

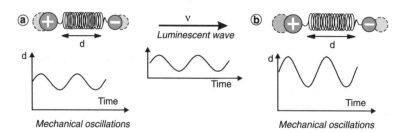

Figure 10.2—'*Mechanical' interpretation of the interaction between a light wave and a polar bond.* Upon interaction of a bond with light of the same frequency, the amplitude of oscillation is changed, not the mechanical frequency.

10.3   Rotational–vibrational bands in the mid infrared

In isolation, the total mechanical energy of a molecule is the sum of three quantised terms that are related to the rotational energy, vibrational energy, and electronic energy:

$$E_{\text{Tot}} = E_{\text{Rot}} + E_{\text{Vib}} + E_{\text{Elec}} \qquad (10.1)$$

These terms have very different values and can vary independently. This is called the Born–Oppenheimer principle.

■ An absorption band located at $1\,000\,\text{cm}^{-1}$ corresponds to a photon energy $E = hc\bar{\nu} = 0.125\,\text{eV}$. This energy is too weak (3 or 4 eV) to cause transitions between the different electronic energy levels, thus E_{Elec} is not involved in the mid IR.

In practice, for compounds in the condensed (liquid) phase or the solid phase, spectral lines such as those presupposed by the quantum aspect of equation (10.1) are not observed. In most cases, spectral bands with tens of cm^{-1} are present (Fig. 10.1). This can be rationalised by dipole–dipole interactions and by the solvation of molecules which perturbs the energy level of individual bonds.

It is noteworthy that the width of an absorption line is inversely proportional to the lifetime of the excited state (Heisenberg's uncertainty principle). Hence, for gases, the lifetime is long and the absorption lines are sharp. However, the lifetime is short for compounds in the condensed phase and band broadening occurs. Except for very simple molecules, no instrument allows the observation of individual lines.

10.4   Simplified model for vibrational interactions

The *harmonic oscillator* is used as a simple model to represent the vibrations in bonds. It includes two masses that can move on a plane without friction and that are joined by a spring (see Fig. 10.3). If the two masses are displaced by a value x_0 relative to the equilibrium distance R_e, the system will start to oscillate with a period that is a function of the *force constant* k $(\text{N} \cdot \text{m}^{-1})$ and the masses involved. The frequency, which is independent of the elongation, can be approximated by equation (10.2) where μ (kg) represents the reduced mass of the system. The term 'harmonic oscillator' comes from the fact that the elongation is proportional to the exerted force while the frequency ν_{Vib} is independent of it.

$$\nu_{\text{Vib}} = \frac{1}{2\pi}\sqrt{\frac{k}{\mu}} \qquad \text{where} \qquad \mu = \frac{m_1 m_2}{m_1 + m_2} \qquad (10.2)$$

For energy to be absorbed by this moving system, it must be in phase and in the form of an oscillation of the same frequency. This phenomenon is similar to that by which the amplitude of a swing is increased.

The mechanical vibrational energy of this model E_{Vib} can vary in a continuous fashion. After a weak elongation x_0 relative to the equilibrium distance R_e, the energy is given by:

$$E_{\text{Vib}} = \frac{1}{2} k x_0^2 \qquad (10.3)$$

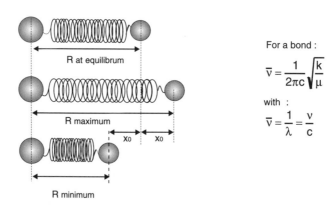

For a bond :

$$\bar{\nu} = \frac{1}{2\pi c} \sqrt{\frac{k}{\mu}}$$

with :

$$\bar{\nu} = \frac{1}{\lambda} = \frac{\nu}{c}$$

Figure 10.3—*Harmonic oscillator representation of a diatomic molecule.*

The previous model can be used on a molecular scale to represent chemical bonds. However, it is necessary to introduce the quantum aspect for species at the atomic level. Because a chemical bond of vibrational frequency ν can absorb radiation of identical frequency, the energy will increase by the quantum $\Delta E = h\nu$. According to this theory, the simplified relation giving possible energy values E_{Vib} is given by:

$$E_{\text{Vib}} = h\nu(V + 1/2) \qquad (10.4)$$

where $V = 0, 1, 2, \ldots$ is known as the vibrational quantum number. It can only vary by one unit ($\Delta V = +1$, 'single quantum' transition). The different values obtained using equation (10.4) are separated by the same interval $\Delta E_{\text{Vib}} = h\nu$ (Fig. 10.4).

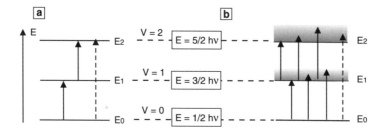

Figure 10.4—*Vibrational energy levels of a bond.* a) For isolated molecules; b) For molecules in the condensed phase. The transition from $V = 0$ to $V = 2$ corresponds to a weak harmonic band. Because of the photon energy involved in the mid IR, it can be calculated that the first excited state ($V = 1$) is 10^6 times less populated than the ground state. Harmonic transitions are exploited in the near IR.

At room temperature, molecules reside in the ground state $V = 0$, thus having an energy $E_{(Vib)0} = 1/2h\nu$ (zero point energy). Besides the normal transition where $\Delta V = +1$, a weak transition corresponding to $\Delta V = +2$, 'forbidden' by theory, is observed when the fundamental absorption is very strong (this is the case, for example, for valence vibrations in organic compounds containing ketones or aldehydes).

10.5   Real model for vibrational interactions

The harmonic oscillator model does not take into account the real nature of chemical bonds, which are not perfect springs. The force constant k decreases if the atoms are pulled apart and increases significantly if they are pushed close together. The vibrational levels, instead of being represented by a parabolic function as in equation (10.3), are contained in an envelope. This envelope can be described by the Morse equation (Fig. 10.5):

$$E_{Vib} = D(1 - e^{-\beta x})^2 \qquad (10.5)$$

where D represents the sum of $E_{(Vib)0}$ and the dissociation energy E_D, β is a constant and x corresponds to the increase in interatomic distance from its equilibrium value at rest (for $V = 0$). This function, which resembles a parabola for low energy values (Fig. 10.5), tends toward the limiting value D when x is large. Quantum theory is the only one that can precisely describe vibrational energies. It uses correction terms due to the anharmonicity of the oscillations observed experimentally.

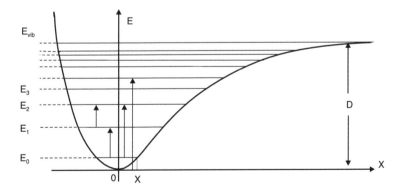

Figure 10.5—*Morse diagram*. The segments within the boundaries of the envelope show the possible energy levels of the bond. This envelope also explains the significant increase of the bond length x as its energy approaches the level of the dissociation energy.

10.6   Rotational bands of compounds in the gas phase

Low energy photons in the far IR can only modify the term E_{Rot}. This leads to pure rotational spectra that can be easily studied for small diatomic gases. However, in the mid IR, photons have sufficient energy to modify E_{Vib} and E_{Rot}. This leads to vibrational–rotational spectra (Fig. 10.6). Each vibrational transition is accompanied by tens of individual rotational transitions. The molecule becomes an 'oscillating rotor' for which energy E_{VR} approximately corresponds to the following values, where J ($J = 0$, 1, 2, 3, ...) and V ($V = 0$, 1, 2) are the rotational and vibrational quantum numbers, respectively.

$$E_{VR} = h\nu_{VR} = h\nu_{Vib}(V + 1/2) + h\nu_{Rot} J(J + 1) \tag{10.6}$$

This application of equation (10.1) can be used to determine the possible absorptions:

$$\Delta E_{VR} = (E_{VR})_2 - (E_{VR})_1 = \Delta E_{Vib} + \Delta E_{Rot} \tag{10.7}$$

Hence, the transitions observed in carbon monoxide (CO) correspond to $\Delta V = +1$ and $\Delta J = \pm 1$ (Figs 10.6 and 10.7). Consequently:

$$\Delta E_{Vib} = h\nu_{Vib} \quad \text{and} \quad \Delta E_{Rot} = h\nu_{Rot} 2(J + 1) \quad \text{or} \quad \Delta E_{Rot} = h\nu_{Rot} \times (-2J)$$

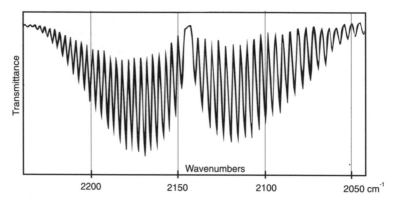

Figure 10.6—*Vibrational–rotational bands of carbon monoxide (P = 1 000 Pa)*. The various lines illustrate the principle of the selection rules (see Fig. 10.7). In this case, $\Delta V = +1$ and $\Delta J = \pm 1$. Branch R can be seen on the left-hand side of the spectrum while branch P is on the right. The distance between the rotational bands allows the moment of inertia I of the molecule to be calculated. I is not constant due to the anharmonicity factor.

Differences in energies can be converted into wavenumbers ($\bar{\nu} = E/hc$ and $\nu/c = \bar{\nu}$). At room temperature, many rotational levels, J, are populated:

R band ($\Delta J = +1$) $\bar{\nu}_{VR} = \bar{\nu}_{Vib} + \bar{\nu}_{Rot} \times 2(J + 1)$ (10.8)

P band ($\Delta J = -1$) $\bar{\nu}_{VR} = \bar{\nu}_{Vib} - \bar{\nu}_{Rot} \times 2J$ (10.9)

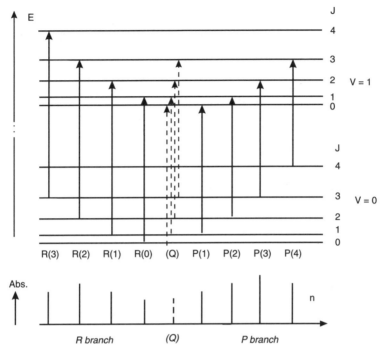

Figure 10.7—*Representation of the rotational and vibrational energy levels and conversion into the vibrational–rotational spectrum (at bottom).* The fundamental vibration corresponds to $V = +1$ and $J = +1$. The vibrational–rotational band corresponds to all the allowed quantum leaps. If the scale of the diagram is in cm^{-1} the arrows correspond to the wavenumbers of absorption. The R branch corresponds to $\Delta J = +1$ and the P branch to $\Delta J = -1$. They are located on each side of the Q band, which is absent in the spectrum ($\Delta J = 0$ corresponds, here, to a forbidden transition).

10.7 Characteristic bands for organic compounds

Infrared spectroscopy, before its present expansion as a quantitative method, was always used as a semi-empirical method for structural analysis. This application stems from the extension of the above rules to organic polyatomic molecules.

It has been empirically observed that a correlation exists between the position of band maxima and the presence of organic functional groups or structural features within the skeleton of molecules (see table 10.1 at end of chapter). If each type of functional group absorbs at a particular wavelength, it can be concluded that the value of the force constant k does not vary significantly for a given type of bond. Moreover, the reduced mass of a given group tends towards a limiting value as the molecular mass increases.

However, absorption bands observed in the lower part of the spectrum, below 1500 cm^{-1}, are numerous and vary with each compound. They are due to deformations in the chemical bonds and in the skeleton and it is difficult to assign all of these bands.

■ A molecule involving n interacting atoms can be defined by $3n$ co-ordinates called *degrees of freedom*. As 3 co-ordinates are required to define the centre of gravity (the translation of the molecule is described by 3 degrees of freedom), the remaining $3n - 3$ co-ordinates are called internal degrees of freedom. The description of the rotation of the molecule usually requires 3 degrees of freedom. Therefore, the remaining $3n - 6$ degrees of freedom correspond to internal vibrations within the molecule. A molecule containing 30 atoms will have 84 modes of vibration. Fortunately, these vibrational modes are not all active because a variation in the dipole moment is required for absorption. Such a high number of levels leads to complex spectra.

The most widely known molecular movements are stretching vibrations (symmetrical and asymmetrical) and bending vibrations (Fig. 10.8).

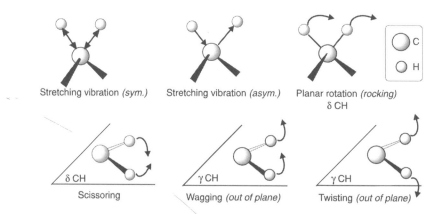

Figure 10.8—*Molecular vibrations in CH_2*. Characteristic stretching and bending vibrations in plane and out of plane. In IR spectroscopy, the position and intensity of the absorption bands are modified by molecular associations, solvent polarity, etc.

10.8 Instrumentation

Mid infrared spectroscopy began commercially in the 1940s with *single beam* spectrometers allowing the point-by-point recording of the transmission spectrum. Today, instruments fall into two categories: dispersive spectrometers that function in a *sequential mode* and Fourier transform spectrometers that are capable of *simultaneous analysis* of the full spectral range using interferometry. The first category uses a monochromator with a motorised grating that can scan the range of frequencies studied. The second is based on the use of a Michelson interferometer or similar device coupled to a specialised computer that can calculate the spectrum from the interferogram obtained from the optical bench. However, the light sources and optical materials are identical for all instruments.

10.8.1 Dispersive spectrometers

The components of a single or double beam scanning spectrometer are shown schematically in Fig. 10.9a and b.

The initial *single beam* dispersive spectrometers that did not, at the time, produce digitised spectra (this would have allowed for baseline correction) were soon replaced by *double beam* spectrometers. This more complex arrangement can directly yield the spectrum corrected for background absorption. The use of two distinct but similar optical paths, one as a reference and the other for measurement, allows the alternate measurement of the transmitted intensity ratios at each wavelength.

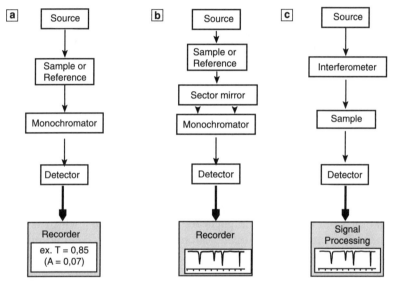

Figure 10.9—*Schematic diagram of various infrared spectrometers.* a) Single beam model: its principle is still used for measurements at a single wavelength; b) double beam model; c) single beam Fourier transform instrument. Contrary to UV/VIS spectrometers, the sample is placed immediately after the light source. Since photon energy in this range is insufficient to break chemical bonds and degrade the sample, it can be permanently exposed to the full radiation of the source.

The radiation coming from the light source is split in two by a set of mirrors (Fig. 10.10). For small intervals in wavelength defined by the monochromator, light that has alternately travelled both paths (reference and sample) arrives at the detector. A mirror (chopper) rotating at a frequency of approximately 10 Hz orients the beam towards each of the optical paths. The comparison of both signals, obtained in a very short time period, can be directly converted into transmittance. Depending on the spectral range and on various constraints (rotation of the monochromator grating, response time of the detector) the spectrum can be recorded within minutes. The detector in this type of instrument is only exposed to a weak amount of energy in a given period of time because the monochromator only transmits a narrow band of wavelengths.

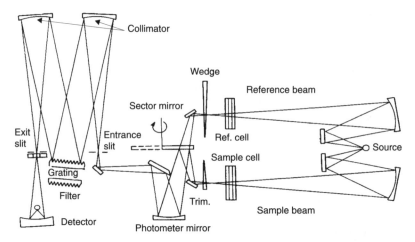

Figure 10.10—*Double beam spectrometer with a rotating mirror.* Model IR 435 (1986, reproduced by permission of Shimadzu).

10.8.2   Fourier transform infrared spectrometer (FTIR)

Fourier transform infrared spectrometers first appeared in the 1970s. These single beam instruments, which differ from scanning spectrometers, have an interferometer of the Michelson type placed between the source and the sample, replacing the monochromator (Figs 10.9c and 10.11).

Irradiation from the source impacts on the *beam splitter* which is made of a germanium film on an alkali halide support (KBr), for the mid-infrared. The semi-transparency of this device allows the generation of two beams, one of which falls on a fixed mirror and the other on a mobile mirror whose distance from the beam splitter varies. The two beams, later recombined along the optical path, travel through the sample before hitting the detector, which measures the global light intensity received. This is a multiplexing phenomenon applied to optical signals. The heart of the Michelson interferometer is the moving mirror that oscillates in time between two extreme positions. When its position is such that the path travelled by both beams is the same length, the light composition of the beam exiting the interferometer is identical to that which enters. However, when the mobile mirror is out of this particular position, the composition of the light beams varies due to the loss of phase between the two beams. The transmitted signal, which varies with time, is recorded as an *interferogram*, $I = f(\delta)$, where δ represents the path difference between both beams (Fig. 10.11). An electronic interface controls the optical bench and data acquisition. During the displacement of the mirror, an ADC converter linked to the detector samples thousands of points across the interferogram. Each of the values recorded corresponds, to the second term of a formidable linear equation of which the unknowns are the amplitudes of all the wavelengths (assumed to be a finite number) present in the signal. A specialised microprocessor calculates, using these thousands of values, a giant matrix using a rapid Fourier transform algorithm developed by Cooley. These computations lead to the amplitude of each wavelength in the spectral band. Using a resolution factor imposed by the calculation, a classical representation of the spectrum can be obtained; $I = f(\lambda)$ or $I = f(\bar{\nu})$.

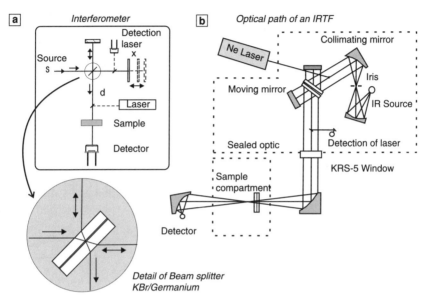

Figure 10.11—*Optical arrangement of a Fourier transform IR spectrometer.* a) A 90° Michelson interferometer including the details of the beam splitter (expanded view); b) optical diagram of a single beam spectrometer (based on a Nicolet model). A weak intensity HeNe laser (632.8 nm) is used as an internal standard to measure precisely the position of the moving mirror using an interference method (a simple sinusoidal interferogram caused by the laser is produced within the device). According to the Nyquist theorem, at least two points per period are needed to calculate the wavelength within the given spectrum.

To obtain a sample spectrum equivalent to the one obtained on a double beam spectrometer, two spectra of transmitted intensities are recorded: the first without sample (absorption background), and the second with sample. The conventional $\%T$ spectrum can be obtained from the two preceding measurements (Fig. 10.12).

> ■ The French mathematician J-B. Fourier (1768–1830) has become even more famous with the growth of microcomputers. The principle of his calculations, published in an article on the propagation of heat in 1908, is now applied in many scientific software packages for the treatment of spectra (acoustical and optical) and images. It is said that Fourier developed these calculations when Napoleon's army asked him to improve the dimensions of cannons.
> In general terms, a Fourier transform is a mathematical operation that allows the user to go from one measurement domain to another (for example, from time to wavelength).

The FTIR method, which has significantly altered traditional methods of obtaining IR spectra, has several advantages:

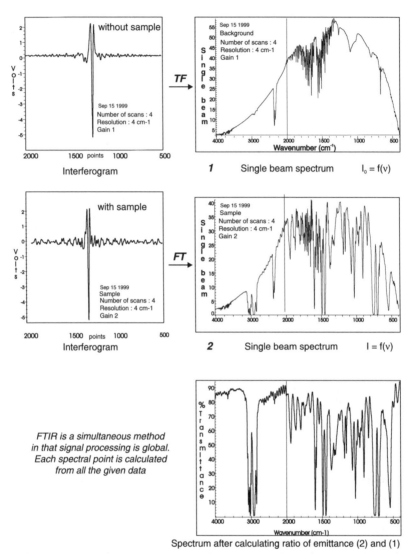

Spectrum after calculating ratio of emittance (2) and (1)

Figure 10.12—*Sequence of events necessary to obtain a pseudo-double beam spectrum with a Fourier transform IR spectrometer.* The instrument records and stores in its memory two spectra representing the variation of I_0 (blank) and I (sample) as a function of wavenumber (emission spectra 1 and 2 above). Then, it calculates the conventional spectrum, which is identical to that obtained on a double beam instrument, by calculating the ratio $T = I/I_0 = f(\lambda)$ for each wavenumber. Atmospheric absorption (CO_2 and H_2O) is thus eliminated. The figure illustrates the spectrum of a polystyrene film.

— there is no stray light, and replacement of the entrance slit by an iris yields a brighter signal so that the detector receives *more energy* (advantage of multiplexing)

— the signal-to-noise ratio is *much higher* than with the sequential method because it can be increased by accumulating several scans (called 'Fellgett's advantage')

— wavelengths are calculated with a *high precision* which facilitates the comparison of spectra

— the resolution is constant throughout the domain studied.

This method will soon replace the sequential method for IR spectrometers (Fig. 10.13).

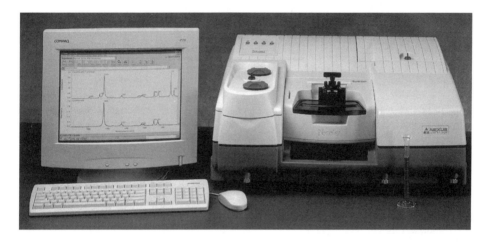

Figure 10.13—*Computer controlled IR spectrometer*. Many models of FTIR instruments permit the adaptation of an infrared microscope for the study of microsamples model NEXUS (1999), reproduced by permission of Nicolet).

10.8.3 Non-dispersive IR photometers

Besides the instrument previously described, many spectrometers have been developed for target applications. Generally simple and robust, these instruments do not record a spectrum but instead are designed to quantify one or many compounds. Almost one hundred gases or volatile compounds can be quantified using this approach. Some models of these IR instruments incorporate interchangeable filters to isolate a particular wavelength for measurement (Fig. 10.14). These instruments are mainly used for applications of routine control analysis in the laboratory or for on-site control and regulation.

Currently, the most widely used IR analysis of this type is the measurement of carbon monoxide and dioxide from car exhausts, and this is carried out in almost all commercial garages. CO is measured by recording the absorbance at $2\,170\,cm^{-1}$ while CO_2 is measured at $2\,350\,cm^{-1}$. Figure 10.14b shows one of these instruments, together with other inventive solutions.

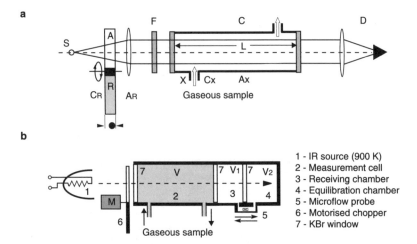

Figure 10.14—*Two models of gas analysers.* a) The light coming from the source S passes through a interference filter F chosen as a function of the wavelength to be measured. Cell C contains the sample to be measured. In order to maintain the simplicity of a single beam instrument with the advantages of a double beam spectrometer, a reference cell R containing a known concentration of the same type of gas to be measured can be placed alternatingly in the optical path with a second cell A filled with nitrogen. By comparison of the absorbance measured by detector D with and without cell R, it is possible to obtain the concentration of the gas within the sample (adapted from a British Servomex, instrument). b) The light beam from the source, after passing through the analytical cell, goes through cell V1 which contains the reference gas (i.e. to quantify CO, V1 and V2 contain this gas). Measurement of the gaseous flow between V1 and V2 is an indication of the pressure variation between both chambers and thus an indication of the IR radiation absorbed by the gas present in V1: the intensity of the beam reaching V1 will be attenuated if the same gas is present in cell V. A beam chopper is necessary to obtain a repetitive pulsed signal (adapted from a Siemens instrument).

10.9   Optics, sources and detectors used in the mid infrared

10.9.1   Optical materials

Traditional optical materials used in visible and near IR do not transmit in the mid IR. Sequential first generation instruments used large dimension prisms made of sodium chloride (useful up to $650\,cm^{-1}$) or sodium bromide (useful up to $400\,cm^{-1}$). The optical dispersion of these fragile and water-sensitive materials was variable with wavelength. They have now been abandoned and replaced with metallic surface-coated gratings to ensure reflection. However, a single grating is not sufficient to cover the whole spectral range of instruments currently used. Optimal dispersion is achieved when the number of elementary mirrors per cm corresponds approximately to the value of the wavenumber. It is therefore necessary to use two gratings with different line spacing or *pitch*. These are used in association with interference filters to avoid the reflection overlap of higher orders.

10.9.2 Light sources

Several types of sources are used in the mid IR. These are either a large filament or a hollow rod 3 to 4 cm long ($\phi = 3$ mm) made of zirconium oxide or rare earth oxides (Nernst source) heated by an internal resistor (the Joule effect). These sources radiate a power of a few hundred watts and have a lifetime of several years. A rod of silicium carbide can also be used (i.e. Glowbar^{TM}). It is more robust and preferable for wavelengths of above 15 μm. A potential of a few volts is usually applied to these devices. When heated to 1 500 °C without a protective shield, they emit radiation over a large domain ranging from visible to thermal IR (Fig. 10.15). The radiative intensity varies enormously with temperature and wavelength. A maximum is observed at $\lambda = 3\,000/T$ (λ in μm and T in Kelvin – adaptation of Stefan's law). In sequential model spectrometers, the entrance slit is programmed as a function of the wavelength to compensate for the difference in intensity between both extremities of the spectrum.

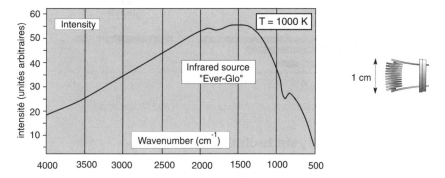

Figure 10.15—*Typical emission spectrum of a mid IR source and at right design of a simple source.* The emission spectrum varies depending on the model used. Current sources are stable, last for many years and produce sufficient radiation intensity from one end of the spectrum to the other.

10.9.3 Detectors

The most widely used detection mode relies on the thermal effect of IR radiation. Because of absorption by the sample and the range of wavelengths involved, the radiation intensity that reaches the detector is weak.

The oldest detectors were *thermocouples* consisting of a double junction of two metals (bismuth and antimony). One metal served as a reference and the other was placed in the optical path (Fig. 10.16). These devices are very sensitive but their time response is too slow for them to be used with Fourier transform instruments.

Current detectors use a deuterated triglycerine sulphate (DTGS) crystal or lithium tantalate (LiTaO$_3$) sandwiched between two electrodes from which they receive the radiation. This allows monitoring of the rapid modulation of the radiation intensity. Under the effect of a potential difference, the crystal becomes *pyroelectric*. It is polarised and acts as a dielectric whose degree of polarisation varies with the radiation. Because of its small thermal mass, linear response and size, this detector is popular in FTIR instruments. The signal is sent to an A/D convertor.

The last category of detectors is the *photovoltaic* detectors. These non-thermal detectors are made of a photoconductive film that is an alloy of mercury, cadmium and tellurium (MCT) or indium/antimony (In/Sb) deposited on an inert support. These detectors are usually cooled down to liquid nitrogen temperature (77 K) in order to increase their sensitivity.

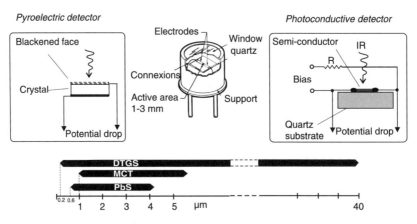

Figure 10.16—*Two types of detectors used in mid IR spectroscopy.*

10.10 Sample analysis techniques

Many types of sample can now be studied in the mid IR. Numerous devices have been inverted to meet the needs of qualitative and quantitative analyses. Samples can now be analysed by transmission or by reflection.

10.10.1 Transmission analysis

Cells with a minimum optical path of a few centimetres are used for *gas* analysis (Fig. 10.17). If the absorbance under these conditions is too weak, a cell using mirrors to increase the path length can be used. Cells with an optical path of several metres represent a complex and cluttered arrangement. Filliform gas cells with a volume of a few tens of microlitres (l = 10 cm with diameter < 1 mm) are used for the hyphenated technique of GC-IR. These cells use reflecting side walls.

Liquids are usually analysed with cells which have dismountable IR windows. For qualitative analysis, a droplet of the sample is compressed between two NaCl or K Br disks without a divider. However, for quantitative analysis, either *Infrasil*™ quartz cells (with an optical path from 1 to 5 cm) or cells that have a variable or fixed width, generally smaller than 1 mm (see Fig. 10.17), can be used. In the mid-infrared, the latter consist of two KBr or NaCl windows with a spacer. The optical path length must be calibrated and periodically controlled.

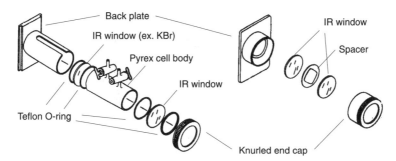

Figure 10.17—*Cells used for mid IR spectroscopy*. a) Gas cell with direct optical path (reproduced with permission of Wilmad); b) expanded view of a liquid cell with a fixed path-length (from Eurolabo document).

For *solids*, many approaches can be used:

— a few milligrams of sample may be dissolved in paraffin oil (NUJOL^{TM}), which has only three interfering bands. The spectrum of the sample is first analysed outside these interfering bands. A second spectrum can be recorded in hexachlorobutadiene, which is transparent in the absorption area of paraffin.

— the sample is crushed along with dry KBr in a mortar. The mixture is then compressed into a disc using either a hydraulic press with a pressure of 5 to 8 t/cm^2 or manually. The resulting translucent fritted disc corresponds to a dispersion of the sample in a solid matrix. Diffusion losses of the radiation can be reduced when the solid is crushed into a fine powder. Not all solids are compatible with this crushing technique.

— if the solid can be dissolved in a suitable solvent, then the same method as for liquids can be used. Unfortunately, no solvents are transparent over the entire range of wavelengths employed in the mid IR. This process allows measurements to be performed at all wavelengths except those absorbed by the solvent.

■ For solids that do not lend themselves to the above procedures (such as strongly absorbing, non-crushable or insoluble solids; certain elusive molecules), photoacoustic detection is a possibility if an FTIR is available. This is a very expensive, seldom used accessory that includes a detector placed directly into the compartment with the sample (see Fig. 10.18).

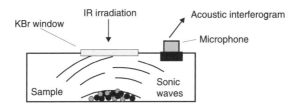

Figure 10.18—*Photoacoustic detection device*. The spectrum is obtained in the photoacoustic units PAS $= 100 I_s/I_{ref}$ (where I_s is the sample signal intensity and I_{ref} is the reference intensity). The reference consists of activated charcoal. The spectrum resembles an absorption spectrum.

The sample is submitted to radiation modulated by the interferometric device. Temperature variations within the sample are translated into pressure variations within the cell. A microphone is used to detect the sound (this is called photo-acoustic detection). The acoustic interferogram obtained from the pressure waves is converted into a classical spectrum.

10.10.2   Reflection analysis for solid samples

When the incident beam reaches the interface of a second medium with a greater refractive index, the beam, depending on the incident angle, will be totally reflected as on a mirror or partially reflected after penetrating the medium by approximately one half wavelength (2 to 10 μm for the mid IR). The sample absorbs part of this radiation.

Many techniques are based on this principle and can be used for the analysis of all types of samples. The spectrum obtained from reflected light is not identical to that obtained by transmittance. The spectral composition of the reflected beam depends on the variation of the refractive index of the compound with wavelength. This can lead to *specular reflection, diffuse reflection* or *attenuated total reflection*. Each device is designed to favour only one of the above. The recorded spectrum must be corrected using computer software.

— **Specular Reflection.** The specular reflection accessory can only be used with FTIR spectrometers. It permits the reflected light to be measured in a direction symmetrical to that of the incident beam (Fig. 10.19). This approach is mostly used for transparent samples which have a reflective surface (such as polymer films, enamel, certain coatings).

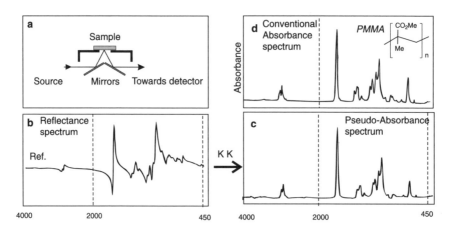

Figure 10.19—*Specular reflection device for FTIR.* a) Optical path in a specular reflection device at a fixed angle; b) raw signal obtained by specular reflection of a sample of methyl polymethacrylate (Plexiglas); c) application of Kramers–Kronig calculation (transformation of reflectance into pseudo-absorbance); d) conventionally obtained transmission spectrum, presented in absorbance units, for comparison.

a *Diffuse reflection*

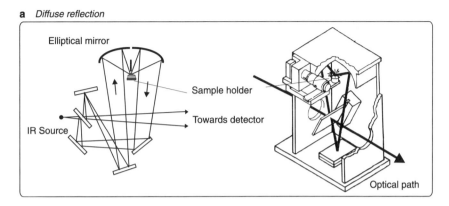

b *Attenuated Total Reflection*

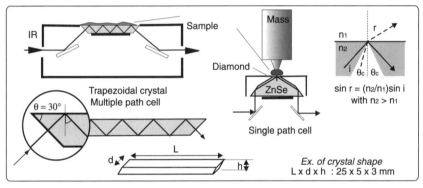

c *Benzoic acid*

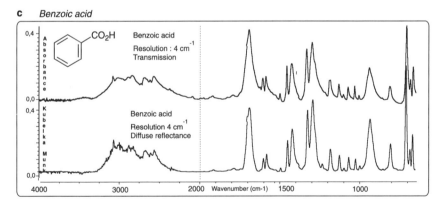

Figure 10.20—*Devices allowing the study of samples by reflection.* a) Diffuse reflection device; b) attenuated total reflection (ATR) device; c) comparison of the spectra of benzoic acid obtained by transmission (KBr disc) and by diffuse reflection using the Kubelka–Munk correction. The depth of penetration of the IR beam depends on the wavelength. The absorbance for longer wavelengths would be overestimated if no correction was applied.

The reflected light, which corresponds to only a weak fraction of the incident radiation absorbed, represents the variation of the compound's refractive index with wavelength. By comparing, after recording the spectrum, the specular reflection I of the sample and the total reflection I_0, which is obtained at each wavelength by replacing the sample with an aluminium mirror, the instrument can calculate the reflectance spectrum $R = I/I_0 = f(\lambda)$. Finally, by a mathematical transformation using Kramers–Kronig (K–K) theory, it is possible to obtain a pseudo-absorption spectrum equivalent to one obtained by transmission.

— **Diffuse Reflection.** Using a set of flat and elliptical mirrors, this device can measure a sufficient amount of light diffused by a sample dispersed in KBr powder (Fig. 10.20). By comparing the diffused reflection obtained with neat KBr, a result resembling the transmission spectrum is obtained. Kubelka–Munk's correction can be used to improve the spectrum.

— **Attenuated Total Reflection (ATR).** The principle of ATR devices is to impose, on the optical beam, multiple reflections at the interface between a parallel-epipedal crystal ($25 \times 10 \times 2$ mm) with a high refractive index (e.g. Ge, CaF_2, ZnSe or KRS-5) and the sample (Fig. 10.20). Because of the angle of incidence and the difference in refractive indices of the media, the radiation penetrates the sample by half a wavelength at each reflection. The succession of many total reflections attenuated in this way leads to accumulation of absorption. The final spectrum, after it has been corrected for the penetration depth, is identical to the spectrum obtained by transmission. This approach can be used for powders and liquids. In some devices, the crystal is submerged in the solution being analysed.

10.11 Infrared microscopy

Thanks to the increased sensitivity of detectors, good reflectance spectra can be obtained for small samples such as those examined under an optical microscope. Focusing the beam down to almost the exact size (a few μm) allows the composition of samples that have a microstructure. The instrument includes a spectrometer coupled to an optical microscope (Fig. 10.21).

■ **Optical fibre probe** – these devices include two shielded optical fibres (made from vitreous oxide) that are transparent in the IR and can collect and transport light over a distance of 1 to 2 m. A double reflection ATR-type probe with a crystal (ZnSe) is located at one end of the fibre. The advantage of this device is that the probe can be introduced directly into the sample, which allows rapid control analysis in environments that are difficult or aggressive yet in the proximity of the spectrometer. Optical yield decreases with the length of the fibre: the beam, located off-axis to the fibre, undergoes thousands of reflections which causes a significant loss in light intensity.

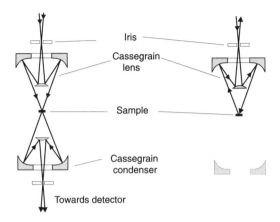

Figure 10.21—*Optical path of the IR beam in an IR microscope*). The sample can be examined in transmission or in reflection mode. This accessory can be installed on a spectrometer that has a deflected beam.

10.12 Archiving spectra

Because they can precisely measure wavelengths, FTIR spectrometers permit spectra and file storage using an instrument format and certain rules so that spectra can be treated by instruments from other manufacturers. Each data point in the spectrum can be stored as an ACSII file. However, the software normalises the raw data and lowers the spectral resolution to obtain fewer calculated values (one point every 8 or 4 cm^{-1} – giving less than 1 Koctet/spectrum). The inverse operation can be used to give a proper corrected image of the spectrum.

With recent rapid increases in the size of computer hard disks and memory, another format, JCAMP.DX (Joint Committee of Atomic Molecular and Physical Data), has become more widespread. It preserves all the numerical values in the original spectrum as well as information regarding the spectrum in an ASCII file. One of the advantages of this format is that it allows all comparison algorithms to be used for further identification of the spectrum (Fig. 10.22).

■ Besides the latter approach to storing data, it is possible in IR as in many other analytical techniques to export a spectrum with its annotations stored in memory as a pictorial image to a report via a word processor. It is using this procedure that all spectra shown in this chapter have been integrated into the text.

■ GC-FTIR requires a large number of spectra to be stored over a very short time interval. The gas phase exiting the column is sampled at very short time intervals throughout the elution. The resolution of the spectra is lowered in order to occupy less memory. The data are then used to reconstruct a pseudo-chromatogram called Gram–Schmidt after the mathematicians whose method is used to treat the data.

```
##TITLE=polystyrene (film 38.1µm)##JCAMP-DX=4.24  $$ Nicolet v. 1
##DATATYPE=INFRARED SPECTRUM ##ORIGIN= ##OWNER=IUT Le Mans
##DATE=1999/12/16 ##TIME=13:09:20
##DATA PROCESSING=Ratio against background
##XUNITS=1/CM ##YUNITS=TRANSMITTANCE
##FIRSTX=399.192596 ##LASTX=3999.639893 ##FIRSTY=0.954292 ##MAXX=3999.639893
##MINX=399.192596 ##MAXY=0.955734 ##MINY=-0.002079 ##XFACTOR=1.000000
##YFACTOR=1.000000E-009 ##NPOINTS=1868 ##DELTAX=1.928467 ##XYDATA=(X++(Y..Y))
399.193 954291712 955734016 932161088 926776192 914560704 910738240
410.763 912457024 912845696 892691072 878786240 868191424 865718208
422.334 850579072 835266432 823590912 825456000 815852416 803264768
.....(missing values)...,..)
3982.284 848554240 845863744 843439040 841975232 839750656 838881536
3993.854 838110336 834689344 830800000 830396352
##END=
```

Figure 10.22—*Example of a JCAMP.DX file read by a word processor.* This file corresponds to the spectrum in Fig. 10.1. The header contains information about the spectrum and the data points are organised in sequences of 6 values. Only a few values have been retained.

10.13 Comparison of spectra

Identification of compounds can be facilitated using generalised or specialised *spectral libraries* (which contain polymers, solvents, adhesives and organic molecules classified by functional group) offered by various societies or publishers (e.g. Aldrich, Sigma, Sadtler and Hummel). The spectra, stored in absorbance values, are normalised using the most intense band, which is given an absorbance equal to one. To obtain a reliable comparison, the archived spectra must have been obtained from compounds in the same physical state as the unknown (e.g. gas phase if comparing to GC-FTIR).

The library search is a mathematical comparison of the unknown compound's spectrum with that of all reference compounds in the database. The aim of the comparison is to find the spectrum that most resembles that of the unknown compound. At the end of the search, the computer software makes a list of all spectra that resemble the unknown spectrum. The software lists the spectra relative to the unknown, along with a reliability or correlation index, irrespective of the library used. Because the object of the library search is to help the analyst and not act as a substitute for him/her, the analyst must manually examine the results. The best approaches to identification are interactive approaches in which the analyst can define *filters* to reduce the field of investigation. Several different algorithms are used for comparison and can lead to different spectra listings.

■ The algorithm that uses *absolute difference* is rapid and simple. For each of the j spectra in the library and n points along the abscissa, the sum S_j is calculated as the absolute difference between the ordinate of the unknown and that of the corresponding spectrum j in the library (equation (10.10)). Then the summations S_j are ranked. The results are presented along with a correlation index.

$$S_j = \sum_{i=1}^{n} |y_i^{\text{ref}(j)} - y_i^{\text{inc}}| \qquad (10.10)$$

The parameters used in the equation can be replaced by their squares or by the squares of their differences. The algorithm is chosen in order to minimise or increase the value of certain factors (e.g. intensity differences, background, differences in band position).

10.14   Calibration of cell thickness

The possibility of obtaining precise absorbance measurements and the storage of numerical data have given a new dimension to quantitative analysis by IR. This is especially true in situations where correlations can be made between the spectra and the composition or physico-chemical properties of the sample. This method is widely used because it is easier in the mid IR than UV/VIS to find absorption bands specific to compounds within a mixture.

For solid samples dispersed in a KBr disc which have a thickness that cannot be precisely measured, an internal standard is used (e.g. calcium carbonate, naphthalene, sodium nitrite). This reference is added in equal quantity to all standards and to the sample.

Absorbance measurements for solutions are carried out in cells with a small optical path d to minimise absorption of the solvents, none of which are transparent in this spectral domain. Uncertainty in the value of d is related to the fragility of the material used for the cell windows and their construction. Therefore, it is necessary to periodically calibrate the optical path.

For cells with a small thickness, the interference pattern method is used. This method consists of measuring the transmittance of the empty cell between two wavenumbers $\bar{\nu}_1$ and $\bar{\nu}_2$. It can be seen in Fig. 10.23 that beam S_2 has gone through a double reflection on the internal wall of the cell. Thus, if the angle of incidence is normal to the cell wall and if $2d = k\lambda$, the addition of both intensities is observed (the two beams S_1 and S_2 are in phase). A modulation of the main beam S_1 will be observed as a function of wavelength. This modulation is in the order of a few percent.

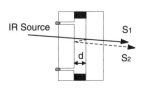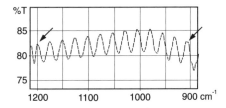

Figure 10.23—*Measurement of cell thickness using the interference pattern.* On the left is shown the reflections on the inner wall of a cell (for clarity, the incident angle of the beam on the cell is shifted from the normal by a small angle). To the right is shown part of a spectrum recorded with an empty cell. The spectrum shows 12 interference fringes between the two arrows. Calculation using this data leads to a cell path-length of $d = 204\,\mu m$.

If proper software is not available to calculate the thickness of the cell using a Fourier transform, the following equation can be used. In this equation, N represents the interference between $\bar{\nu}_1$ and $\bar{\nu}_2$ (in cm^{-1}):

$$d_{(cm)} = \frac{N}{2(\bar{\nu}_1 - \bar{\nu}_2)} \qquad (10.11)$$

■ Some analysts still consider that dispersive double beam instruments are preferable to FTIR instruments for quantitative analysis because they are the only ones that compare simultaneously the transmitted intensities in both reference and sample beams.

10.15   Raman diffusion

It has been demonstrated in sections 10.3 and 10.4 that absorptions observed in the mid IR are due to vibrational and rotational energy modifications in the molecules when they are submitted to radiation from a source.

There is another optical method which studies these energy modifications and produces a spectrum that contains almost the same information as that obtained in the mid IR: Raman. In this technique, a solution of the sample in a solvent such as water is irradiated by intense, monochromatic laser light in the visible region. The composition of the beam diffused by species present in the sample is analysed at 90° to the incident beam. In this process, bands called Stokes lines are observed beside the incident beam, at greater wavelengths. If the differences between these bands and the wavelength of the incident beam are expressed as wavenumbers, the values obtained correspond to the difference in rotational and vibrational energy levels obtained by absorption spectroscopy (Fig. 10.24).

Although this method is as old as infrared absorption spectroscopy, Raman diffusion has been plagued by its lack of sensitivity. This situation has recently changed with the advent of laser excitation sources and Fourier transform computations. However, the cost of Raman instruments is still relatively high. This method, which can have advantages over absorption spectroscopy, is not yet widely used.

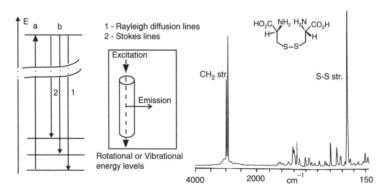

Figure 10.24—*Raman diffusion.* Origin of emission bands; cell geometry; Raman spectrum of L-cystine.

Table 10.1—Correlation Chart in the mid-Infrared

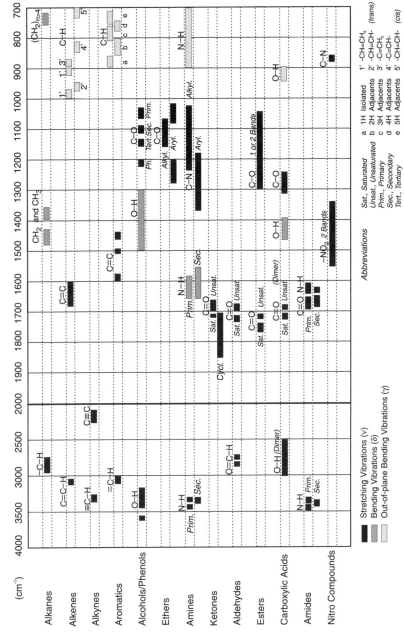

Abbreviations			
Sat., Saturated			
Unsat., Unsaturated			
Prim., Primary			
Sec., Secondary			
Tert., Tertiary			

a	1H	Isolated	1'	-CH=CH₂

a 1H Isolated
b 2H Adjacents
c 3H Adjacents
d 4H Adjacents
e 5H Adjacents

1' -CH=CH₂
2' -CH=CH- (trans)
3' =C=CH₂
4' -C=CH-
5' -CH=CH- (cis)

Stretching Vibrations (ν)
Bending Vibrations (δ)
Out-of-plane Bending Vibrations (γ)

Problems

10.1 1. What is the energy possessed by a radiation of wavenumber $1\,000\ \text{cm}^{-1}$?
2. Transform $\lambda = 15\ \mu\text{m}$ into units of cm^{-1} and then into m^{-1}. What is the wavelength corresponding to a wavenumber of $1\,700\ \text{cm}^{-1}$?
3. At the maximum of an absorption band, the transmittance is only 5%. What is the corresponding absorbance?

10.2 If chloroform (trichloromethane) exhibits an infrared peak at $3\,018\ \text{cm}^{-1}$ due to the C–H stretching vibration, calculate the wavenumber of the absorption band corresponding to the C–D stretching vibration in deuterochloroform (experimental value $2\,253\ \text{cm}^{-1}$).

10.3 A ketone possesses an absorption band with a peak centred around $1\,710\ \text{cm}^{-1}$. From this information deduce a value for the force constant of the C=O double bond.

10.4 The bond between the two atoms of a diatomic molecule is characterised by a force constant of $1\,000\ \text{N/m}$. This bond is responsible for a vibrational absorption at $2\,000\ \text{cm}^{-1}$. Accepting that the energy of radiation is transformed into vibrational energy, estimate a value for the length of the bond at the maximum separation of the two atoms.

10.5 The examination of an absorption band located around $2\,900\ \text{cm}^{-1}$, expressed by a sample of HCl in the gaseous state, reveals that the band is the result of a superimposition of two forms of vibration, one of which is clearly more intense than the other. These two series are separated by an approximate distance of $2\ \text{cm}^{-1}$. How might this phenomenon be interpreted? Use calculations to illustrate your answer.

10.6 Knowing that the fundamental frequency of carbon monoxide is $2\,135\ \text{cm}^{-1}$ when the spectrum is recorded with tetrachloromethane as the solvent, calculate the force constant of this molecule under these conditions.

10.7 The following experiment is used to determine the vinyl acetate (VA) level in an ethylene vinyl acetate (EVA) commercial packaging film.
Infrared spectra of packaging films with known vinyl acetate contents are recorded. The absorbance peak at $1030\ \text{cm}^{-1}$ used to determine the content of the vinyl acetate was measured by the baseline method ($A = \log(I_0/I)$. The following results were obtained.

(EVA)	% VA	A_{1030}	A_{720}	$d(\mu\text{m})$
1	0	0.01	1.18	56
2	2	0.16	1.55	80
3	7.5	0.61	1.49	82
4	15	0.36	0.45	27

1. Taking into account the film thickness, determine, from the data in the table, the best line $A_{1030} = f(\%\,\text{VA})$, using linear regression, for a film thickness of 1μm.

2. Explain why the polyethylene peak at $720\,\text{cm}^{-1}$ may be chosen as an internal standard, then calculate the ratio A_{1030}/A_{720} for the four films.

3. Using both above methods calculate the vinyl acetate content (%) of an unknown EVA film (giving $d = 90$μm, $A_{1030} = 0.7$ and $A_{720} = 1.54$).

10.8 1. In supposing that the signal to noise ratio obtained for an FTIR is 10 for a signal corresponding to a single scan, what would be the expected value following 16 scans?

2. The resolution (in cm^{-1}) is given by the expression $R = 1/\Delta$, where Δ represents the difference of the maximum optical path between the two beams (in cm) when the mirror glass is moving. What must be the displacement of the mirror relative to the central position to obtain a resolution of $1\,\text{cm}^{-1}$? (The central position of the mirror is such that δ, the difference between the optical paths, is zero.)

3. To locate the position of the moving mirror with great precision, we superimpose in the apparatus a laser source of monochromatic radiation ($v = 15\,800\,\text{cm}^{-1}$), which permits a computer to pick up a point of the interferogram each time the laser light is extinguished.

   What, in μm, is the value of the displacement dx of the mirror between two successive extinctions of the laser beam's light?

CHAPTER 11

Ultraviolet and Visible absorption spectroscopy

The interaction of electromagnetic radiation with matter in the domain ranging from the close ultraviolet to the close infrared, between 180 and 1,100 nm, has been extensively studied. This portion of the electromagnetic spectrum, called UV/Visible because it contains radiation that can be seen by the human eye, provides little structural information except the presence of unsaturation sites in molecules. However, it has great importance in quantitative analysis. Absorbance calculations for compounds absorbing radiation in the UV/Visible using Beer–Lambert's Law is the basis of the method known as colorimetry. This method is the workhorse in any analytical laboratory. It applies not only to compounds that possess absorption spectra in that spectral region, but to all compounds that lead to absorption measurements.

This area of analytical chemistry includes a great number of instruments that range from colour comparators and other visual comparison devices to automated spectrophotometers that can carry out multicomponent analysis. Liquid chromatography and capillary electrophoresis have accelerated the development of improved UV/Visible detectors, which are at the origin of the current mode of acquiring chromatograms, accompanied by the possibility of identification and quantification of compounds.

11.1 General concepts

The spectral range of interest can be subdivided into three ranges: the near UV, the Visible and portions of the near Infrared (185–400, 400–700 and 700–1100 nm, respectively). Most commercial spectrometers have a spectral range of 185 to 900 nm. Absorption by materials in the atmosphere beginning at 185 nm is the limiting factor for the lower wavelength in this working range. About 10 to 20 nm can be gained at the shorter wavelength end by recording the spectrum under vacuum – the domain of the far UV.

These instruments allow a spectrum to be obtained which is a plot of absorbance (cf. 11.11) as a function of wavelength (Fig. 11.1). Wavelengths are expressed in nanometres (nm), the recommended unit in this spectral domain.

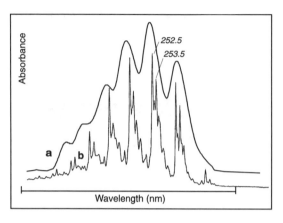

a Benzene in solution
b Benzene vapour
c Iodine vapour

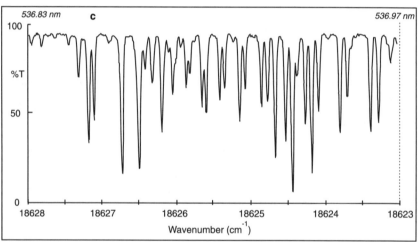

Figure 11.1—*Three different aspects of spectra obtained in the UV/Visible region.* a) Band spectrum of a compound in solution (most frequent case); b) spectrum showing fine structure; c) line spectrum obtained with a high resolution instrument. Only 0.14 nm separates the two sides of this spectrum. Absorbance can be measured with up to 6 units of certainty with some instruments, however, high values are not as reliable.

As will be discussed later, the position of absorption maxima for a given compound can vary by a few nm depending on pH or solvent. The determination of wavelengths in cm^{-1} is reserved for higher resolution spectra of compounds with absorption bands (e.g. 110 cm^{-1} separates 300 and 301 nm).

Few, yet large, absorption bands are obtained when compounds are studied in the condensed phase, whether pure or in solution. However, when the same compound is studied under reduced pressure in the gas phase, the spectrum is seen to have fine structure. Only very simple molecules give a line spectrum in the gas phase.

11.2 Molecular absorptions

The impact of a UV/Visible photon on an isolated molecule modifies the term E_{elec} in equation (11.1), determining the quantification and variation of its total mechanical energy. This electronic perturbation is also accompanied by modification in the terms E_{rot} and E_{vib} that correspond to the transition. Many such transitions are possible within the same molecule (Fig. 11.2). These transitions are related to the excitation of valence electrons.

$$\Delta E_{tot} = \Delta E_{rot} + \Delta E_{vib} + \Delta E_{elec} \qquad (11.1)$$

■ In the condensed phase, due to interactions of molecules with neighbouring species, no single instrument, regardless of its resolution, can record a spectrum in which individual transitions are seen. In the best case for simple molecules, fine structure corresponding to envelopes of transitions can be observed (Fig. 11.1).

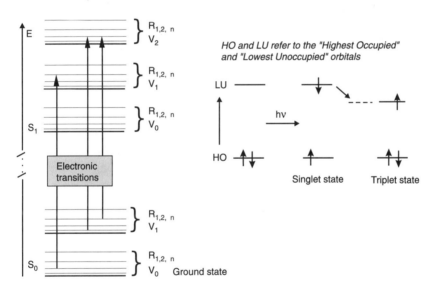

Figure 11.2—*Energy diagram showing the rotational, vibrational and electronic states of a molecule.* On the scale of this diagram, the rotational and vibrational levels should be comparatively closer than the electronic levels because their relative values are in the order of 1 000:50:1. The framed section shows, in the form of an energy diagram, the absorption of a photon. This figure represents the promotion of an electron from an occupied orbital (HOMO) to an unoccupied orbital (LUMO) giving rise to a singlet, which rapidly evolves into a more stable triplet state.

After absorption, the energy can relax by emission of photons (Fig. 11.3) or by non-radiative transitions. The most frequent radiative transitions include *phosphorescence* and *fluorescence*. These are the basis of other methods of analysis that will be discussed in Chapter 12.

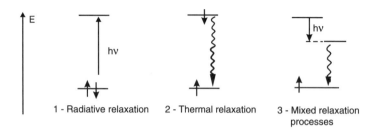

1 - Radiative relaxation   2 - Thermal relaxation   3 - Mixed relaxation processes

Figure 11.3—*Processes showing the relaxation of an excited state to its ground state.* Because transitions are almost instantaneous, the interatomic distance does not change (this is the *Franck–Condon principle*).

11.3 Origin of absorption in relation to molecular orbitals

In common organic compounds, considering the bonds between light elements (H, C, N, O) for example, the observed transitions involve the σ, π and non-bonding n electrons (Fig. 11.4). Because modifications in the polarity of chemical bonds occur during these transitions, the resultant spectra are often called *charge transfer spectra*.

Each transition is characterised by its wavelength and its molar absorption coefficient ε $(1\,mol^{-1}\,cm^{-1})$ at this wavelength (see 11.11).

11.3.1 $\sigma \rightarrow \sigma^*$ transition

The great stability of σ bonds within organic compounds is reflected by the significant energy difference between bonding and non-bonding orbitals (Fig. 11.2). The transition of an electron between the bonding σ and non-bonding σ^* molecular orbital requires a lot of energy, thus the transition appears in the far UV. This is the reason for saturated hydrocarbons like hexane or cyclohexane that only contain σ bonds being transparent in the conventional UV range. For example, *hexane*: $\lambda_{max} = 135\,nm$ $(\varepsilon = 10\,000)$.

■ Cyclohexane and heptane are used as solvents in the near UV because they are transparent above 200 nm (at this wavelength, the absorbance is equal to 1 for a thickness of 1 cm). Unfortunately, their solvation power is insufficient to dissolve many polar compounds.

Similarly, the transparency of water in the near UV ($A = 0.01$ for $l = 1\,cm$ at $\lambda = 190\,nm$) is due to the fact that there can only be $\sigma \rightarrow \sigma^*$ and $n \rightarrow n^*$ transitions in this compound.

11.3.2 $n \rightarrow \sigma^*$ transition

The promotion of an n electron to the σ^* orbital in O, N, S, and X-containing compounds is observed in the near UV with moderate intensity. This transition is seen around 180 nm for alcohols, 190 nm for ethers or halogen derivatives and

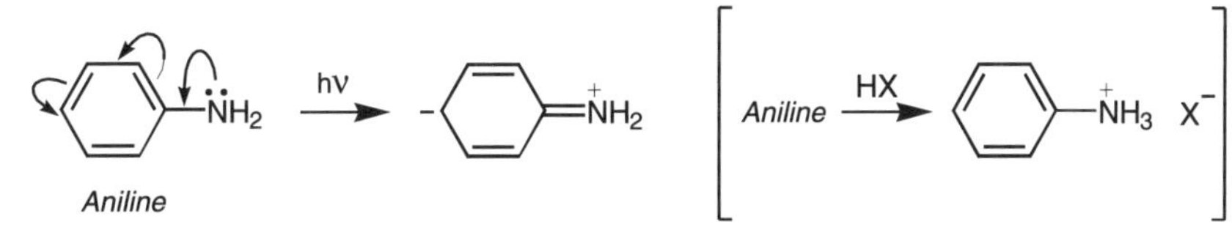

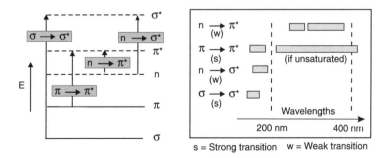

Aniline

Figure 11.4—*The n → σ* transition in primary amines.* This transition corresponds to an increase in the 'weight' of the mesomeric form. Absorption corresponding to this transition disappears when a proton-containing hydrogen halide is added. This effect is due to the formation of a quaternary ammonium salt that mobilises the electron pair of the nitrogen atom, which is necessary for this transition (see formula in brackets).

around 220 nm for amines (see Fig. 11.4). For example, methanol: $\lambda_{max} = 183$ nm ($\varepsilon = 50$); ether: $\lambda_{max} = 190$ nm ($\varepsilon = 2\,000$); ethylamine: $\lambda_{max} = 210$ nm ($\varepsilon = 800$).

11.3.3 $n \rightarrow \pi^*$ transition

This weak transition is due to the promotion of an electron from the non-bonding molecular orbital n to an anti-bonding π^* orbital. This transition is usually observed in molecules that contain a heteroatom as part of an unsaturated system. The most common of these bands corresponds to the carbonyl band at around 270 to 295 nm, which can be easily observed. The molar absorption coefficient for this band is weak. The nature of the solvent influences the position of absorption bands because the polarity of the bond is modified during absorption. For example, ethanal: $\lambda_{max} = 293$ nm ($\varepsilon = 12$ in ethanol as solvent).

11.2.4 $\pi \rightarrow \pi^*$ transition

Compounds with an isolated ethylenic double bond lead to a strong absorption band around 170 nm, where the position of this band depends on the presence of hetero-atom substituents. For example, ethylene; $\lambda_{max} = 165$ nm ($\varepsilon = 16\,000$).

Figure 11.5—*Comparative diagram showing the most common transitions in simple oxygen or nitrogen-containing compounds.* These four types of transitions are often shown on a single energy diagram to indicate their relative position, or to indicate the spectral regions involved.

■ **d → d* transition.** Numerous inorganic salts with electrons in **d** orbitals lead to weak transitions in the visible region that are responsible for their colour. Hence, solutions containing salts of metallic titanium $[Ti(H_2O)_6]^{+++}$ or copper $[Cu(H_2O)_6]^{++}$ are blue. Potassium permanganate will yield solutions that are violet, and so on.

11.4 Donor–acceptor association

A compound that is transparent within a spectral domain when in its isolated state can sometimes absorb when in the presence of a species with which it can interact through a *donor–acceptor* relationship (D–A). This phenomenon is related to the passage of an electron from a bonding orbital of the donor (which becomes a radical cation) to an unoccupied orbital of the acceptor (which becomes a radical anion), which has a close energy level (Fig. 11.6). The position of the absorption band in the spectrum is a function of the *ionisation potential* of the donor and the *electron affinity* of the acceptor. The value of ε for these transitions is usually large.

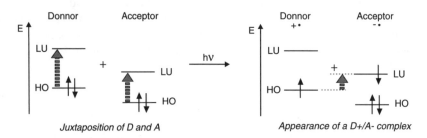

Figure 11.6—*Energy diagram for a donor–acceptor interaction.* The excited state is assumed to be essentially in the ionic form.

11.5 Isolated chromophores

Chromophores (or *oxochromes*) are small groups of atoms responsible for characteristic absorptions. By extension, the chromophore in a molecule corresponds to the site responsible for the electronic transition. A *chromogene* is a species formed by a skeleton on which many chromophores can be found. For a series of molecules containing the same chromophore, the position and intensity of the absorption bands are constant (Table 11.1). When a molecule contains several isolated chromophores separated by at least two single bonds, the overlapping of individual effects is observed. If the chromophores are adjacent to one another, a different situation results.

Table 11.1—Characteristic chromophores of several nitrogen-containing groups.

Name	Chromophore	λ_{max}(nm)	ε_{max}(l mol^{-1} cm^{-1})
amine	$-NH_2$	195	3 000
oxime	$-NOH$	190	5 000
nitro	$-NO_2$	210	3 000
nitrite	$-ONO$	230	1 500
nitrate	$-ONO_2$	270	12
nitroso	$-N=O$	300	100

11.6 Solvent effects: solvatochromism

The position, intensity and shape of absorption bands of a compound in solution will vary with the solvent (see Fig. 11.7). These changes reflect the physical solute–solvent interactions that modify the energy difference between the ground and excited states. Study of the displacements and of the variation of absorption band intensities with solvent can be used to determine the type of transition.

11.6.1 Hypsochromic effect (*blue shift*)

In some transitions, the polarity of the chromophore is weaker after absorption of radiation. One case of this is the n → π* absorption due to the carbonyl present in ketones in solution. Before absorption, the C^+–O^- polarisation stabilises in the presence of a polar solvent whose molecules will be clustered around the solute because of electrostatic effects. Thus, the n → π* electronic transition will require more energy and its maximum will be displaced towards a shorter wavelength, contrary to what would be observed in a nonpolar solvent. This is the *hypsochromic effect*. Because the excited state is readily formed, the solvent shell around the

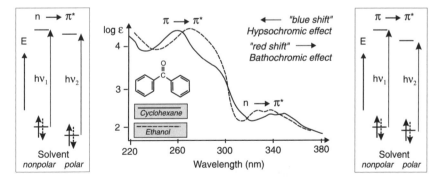

Figure 11.7—*Absorption spectrum of benzophenone in cyclohexane and ethanol.* Note that the bathochromic and hypsochromic effects are reversed for the two types of transitions, depending on the nature of the solvent.

carbonyl does not have time to reorient itself and bring about stabilisation after absorption of the photon. The same effect is observed for n → $\sigma*$ transitions. This effect is usually accompanied by a variation in ε.

11.6.2   Bathochromic effect (*red shift*)

For less polar compounds, the solvent effect is weak. However, if the dipole moment of the chromophore increases during the transition, the final state will be more solvated. This is the case for $\pi \to \pi*$ transitions in ethylenic hydrocarbons with a slightly polar double bond. A polar solvent has the effect of stabilising the excited state, which favours the transition. A shift towards greater wavelengths is observed unlike the spectrum obtained in a nonpolar solvent. This is the *bathochromic effect*.

11.7   Chromophores in conjugated systems

If several chromophores within the same molecule are separated by a single bond, this forms a conjugated system. It results in a strong modification in the spectrum compared to the overlap of effects produced by isolated chromophores. Spatial delocalisation of electrons increases with the number of carbon atoms involved in the conjugated system and this causes a decrease in the energy difference between the ground and excited states (see Table 11.2). The absorption spectrum is displaced towards longer wavelengths (*bathochromic effect*) with an increase in intensity (*hyperchromic effect*).

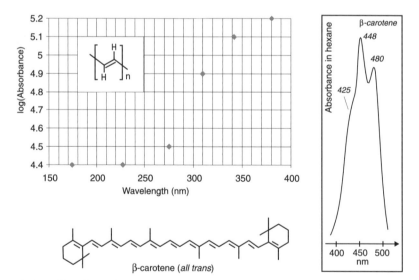

Figure 11.8—*Effect of several conjugated double bonds.* This effect is at the origin of colour in many natural compounds with conjugated chromophores. The orange colour of 'all trans' β-carotene shown above is due to the presence of 11 conjugated double bonds.

■ The absorption of aromatic compounds produces more complex spectra than that of ethylenic compounds. The $\pi \rightarrow \pi^*$ transitions result in the presence of fine structure in the spectrum. The spectrum of benzene vapour, obtained by depositing a droplet in a quartz cell of 1 cm pathlength, is an excellent test to evaluate the resolution of instruments in the near UV (Fig. 11.1). Substitution on the benzene ring produces modifications in the shape of the absorption bands.

Table 11.2—*Displacement of absorption maximum for a family of conjugated polyenes.* Values of λ_{max} for a family of disubstituted conjugated polyenes with differing numbers of conjugated double bonds. The general structure is given on the right.

n	λ_{max} (nm)	$\varepsilon_{max} 10^{-4}$ ($1\,mol^{-1}\,cm^{-1}$)	Structure
1	174	1.6	
2	227	2.4	
3	275	3	
4	310	7.7	
5	342	12.2	
6	380	14.7	

11.8 Woodward–Fieser rules

Structural analysis from electronic spectra yields little information because of their relative simplicity. In the 1940s, however, before the advent of more powerful identification techniques, UV/VIS visible spectroscopy was used for structural identification. The study of a great number of spectra of various molecules has revealed correlations between structures and the positions of absorption maxima. The most widely known empirical rules, due to Woodward, Fieser and Scott, involve unsaturated carbonyls, dienes and steroids. Using incremental tables based on various factors and structural features, it is possible to predict the position of the $\pi \rightarrow \pi^*$ absorption bands in these conjugated systems (Table 11.3). Agreement between the calculated values and the experimentally determined position of absorption bands is usually good, as can been seen by the following four examples:

λ_{max} Obs.	231	234	217	386	nm
Calc.	229	234	215	385	nm

Table 11.3—*Correlation table in UV/Visible spectroscopy.* Woodward–Feiser–Scott rules for the calculation of the absorption maxima for enones and dienones (precision of 3 nm).

Type of structure concerned: (chemical formula)

$$-C{=}C{-}C{=}C{-}C{=}O$$

with substituent X

solvent: methanol or ethanol

Basic structure (chemical formula)

X = H	207 nm
X = alkyl	215 nm*
X = OH, O-alkyl	193 nm

open chain or ring of 6 C
(ring of 5 C: 202 nm)

Increments:
- each additional conjugated double bond   30 nm
- exocyclic character of a C=C double bond   5 nm
- dienic homoannular character   39 nm
- for each substituent, add (in nm):

positions	α	β	γ	δ
alkyl	10	12	18	18
−Cl	15	12		
−Br	25	30		
−OH	35	30		50
−O-alkyl	35	30	17	31
−O-acyl	6	6	6	6
−N(R)$_2$		95		

Solvent increments:
- water   +8 nm
- chloroform   −1 nm
- ether   −7 nm
- cyclohexane   −11 nm

11.9   UV/Visible spectrophotometers

A UV/VIS spectrophotometer consists of three components: the *source*, the *dispersive system* (combined in a *monochromator*) and a *detector*. These components, which can be used independently to design a system appropriate for a desired application, are typically integrated into the same instrument to make spectrophotometers for chemical analysis. The sample can be placed in the optical path before or after the dispersive system (see Figs 11.2 and 11.3) and recorded spectra can be treated using a number of different computer algorithms.

Because the spectra obtained from compounds in solution do not contain fine structure, it is not necessary to use spectrophotometers with high resolution. However, it is important to be able to precisely measure the absorbances over a range of several units. The simplest instruments, called spectrocolorimeters, are used for routine quantitative applications. Higher performance instruments designed for the best possible resolution are used in other areas besides analysis.

11.9.1   Light sources

Two light sources are commonly used in this spectral domain. An *incandescent lamp* made from a tungsten filament housed in a glass envelope is used for the Visible portion of the spectrum, above 350 nm. For that portion of the spectrum below this region, a medium pressure *deuterium arc lamp* is used (medium pressure must be applied to obtain an emission continuum).

■ A deuterium arc lamp has two electrodes, bathed in an atmosphere of deuterium, between which a metallic screen pierced with a hole of 1 mm in diameter is placed. The discharge current creates an intense arc at the level of this hole, which is close to the anode. Under electron bombardment, deuterium molecules dissociate, emitting a continuum of photons over the range of 160 to 400 nm (Fig. 11.9).

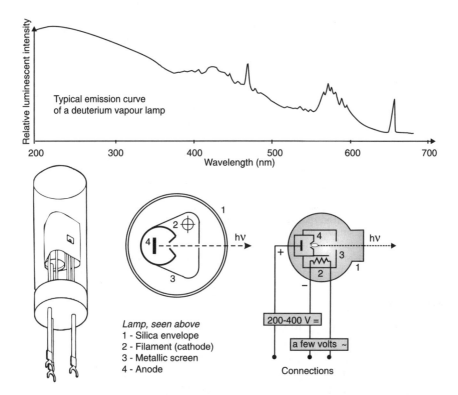

Figure 11.9—*Emission spectra of a deuterium lamp*. Side, top view (reproduced by permission of ISTC, USA) and the different electrical circuit of this lamp. A voltage of 3 to 400 V is used to trigger this lamp. The anode is in the form of a plate, while the cathode is a filament to which a few volts are applied.

11.9.2   Dispersive systems and monochromators

Light emitted by the source is dispersed by a planar or concave grating with approximately 1 200 lines per mm. For scanning spectrophotometers, the grating is integrated into an assembly called a monochromator, which extracts a narrow spectral band. The wavelength in these instruments is varied by pivoting the grating (Fig. 11.10). Optical paths with long focal lengths (0.2 to 0.5 m) yield the best resolution.

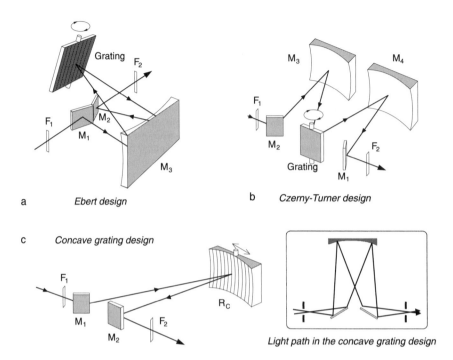

a   *Ebert design*

b   *Czerny-Turner design*

c   *Concave grating design*

Light path in the concave grating design

Figure 11.10—*Grating monochromators.* a) Ebert design incorporating a single concave spherical mirror, M_3, which compensates for aberrations and yields an excellent quality of image; b) Czerny–Turner design, similar in conception, incorporating two spherical mirrors, M_3 and M_4; c) design with a concave grating G_C allowing simultaneous dispersion and focusing of the radiation. The spectral bandwidth of these monochromators depends on the width of the entrance and exit slits, F_1 and F_2.

11.9.3 Detectors

The detector measures the light signal at a given wavelength. It is by nature a single channel device that converts the light intensity selected either by the monochromator exit or by the position, in the case of a spectrograph, into an electrical signal. In the latter case, the use of a large number of detectors in the format of a *diode array* allows simultaneous multichannel detection (Fig. 11.1). Two types of detectors exist: photomultiplier tubes and semiconductors (e.g. silicon photodiodes and charge transfer devices (CCD/CID)).

The photomultiplier tube – a very sensitive device that has a linear response over seven decades – has for a long time been the most widely used detector in spectrophotometers. Its efficiency depends on the yield of the photocathode, which varies with wavelength (e.g. 0.1 e$^-$/photon at 750 nm), and with the signal gain provided by the dynode cascade (e.g. gain of 6×10^5). With such values, the impact of 10 000 photons per second produces a current of 0.1 nA.

■ As for the human eye, it is difficult for a photomultiplier tube to precisely compare two light intensities, one from the reference beam and the other from the sample, when they are very different. For this reason, it is desirable to have the absorbance of solutions no higher than 1 (see section 11.15). On the other hand, for an instrument with a stray light of only 0.01% (measured in % transmittance), an increase in solution concentration will not create significant variations in the signal up to 4 absorbance units.

In routine spectrophotometers, photomultiplier tubes are replaced by photodiodes (Fig. 11.11), which have excellent sensitivity, linearity and dynamic range. The photoelectric threshold, in the order of 1 eV, allows detection up to wavelengths of 1.1 μm. In *diode array* systems, each rectangular rectangular diode (15 μm × 2.5 mm) is associated with a capacitor. The electronic circuit sequentially samples the charge of each capacitor. While a photomultiplier tube measures the instant intensity in watts, a diode measures the emitted energy in joules over a time interval.

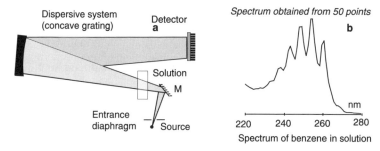

Figure 11.11—*Multichannel detection.* a) Multichannel detection with a diode array located in the focal plane. The light beam is diffracted by the concave dispersive system after travelling through the sample. Note the absence of an exit slit; b) spectrum of a 1:1 000 solution of benzene in methanol. This spectrum represents a typical spectrum without smoothing and is obtained with commercial photodiodes (*note*: in contrast to mid IR spectroscopy, interferometry followed by Fourier transform has led to few commercial achievements in this area).

11.9.4 Single beam, single channel type spectrometers

Many routine measurements are conducted at fixed wavelengths using simple photo-colorimeters equipped with wideband, interchangeable colour filters. An *analytical blank* (or control) containing the solvent and reagents for the analysis (without the sample to be measured) is first placed in the optical path, then this is replaced by the solution to be analysed. These instruments, for a slight additional cost, can be fitted with a grating and an electronic compensation mechanism for light source intensity variations (Fig. 11.12). The compensation device is known as a *split-beam*. Part of the light beam is diverted before it reaches the sample, permitting stabilisation of the source intensity (it is not a true reference beam). These spectrometers produce absorbance data by alternate measurements between the two cells (control and sample) and lead to the desired concentration.

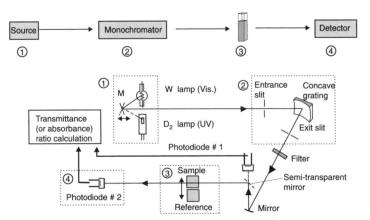

Figure 11.12—*Schematic and optical path of a single beam spectrophotometer equipped with electronic regulation (Hitachi U-1000).* Measurements in solution are often carried out at a fixed wavelength after a calibration curve has been plotted. The use of higher performance double beam UV/Visible spectrometers is not necessary for these measurements in which the spectrum is not recorded. On the other hand, quantitative measurements from mixtures represent a different type of analysis.

11.9.5   Single beam, multichannel type spectrophotometers

This type of instrument resembles a spectrograph because it allows the simultaneous measurement of all wavelengths. It uses an array of up to 2 000 miniaturised photodiodes (Fig. 11.13). A full spectrum can be recorded in milliseconds with this type of simultaneous acquisition detector, each diode measuring the light intensity over a small interval of wavelength. The resolution power of these diode-array spectrometers is limited (usually 1 to 2 nm) by the size of the diodes.

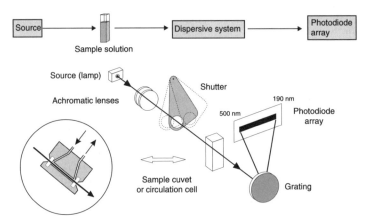

Figure 11.13—*Schematic and optical path showing the principle and simplified view of a diode array spectrophotometer.* The shutter is the only mobile piece in the assembly, allowing subtraction of the background signal (dark current) without any light intensity striking the photodiodes. This inverted optical design allows the sample to be exposed to the exterior light. These instruments are widely used as detectors in liquid chromatography (cf. 3.7).

11.9.6   Double beam optical spectrometers (scanning type)

The best instruments in this area are still the double beam spectrometers in which one of the beams passes through the sample while the other passes through the reference. Two rotating mirrors, called *choppers*, which are synchronised with the displacement of the grating, allow the comparison of transmitted light at the detector of the two beams with the same wavelength (Fig. 11.14). Amplification of the modulated signal allows the elimination of stray light. Electronics associated with the photomultiplier tube adjust the sensitivity as the inverse of the light intensity perceived by the detector. A simpler set-up can be used, consisting of a semi-transparent mirror and two paired photodiodes. The signal corresponds to the voltage needed to maintain a constant detector response (i.e. a *feedback* system). These instruments are characterised by a fast scanning speed (30 nm/s) and by the possibility of measuring absorbances up to several units.

■ Double beam spectrophotometers allow differential measurements to be made between the sample and the analytical blank. They are preferable to single beam instruments for measurements in problematic solutions. For high performance instruments, the bandwidth can be as low as 0.01 nm.

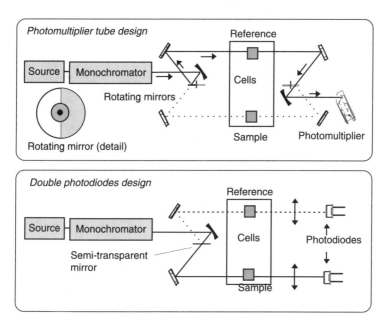

Figure 11.14—*Optical path between the monochromator exit and the detector for two double beam instruments (rotating mirror model and semi-transparent mirror model).* Instruments with rotating mirrors are similar to those used in IR spectrophotometers. However, the light beam from the source goes through the monochromator before it hits the sample. This minimises photolytic reactions that could occur if the sample is exposed to the total radiation from the source. The optics of instruments with two detectors are simpler and only one mirror, semi-transparent and fixed, is necessary to replace the delicate mechanisms of synchronised, rotating mirrors.

To allow routine remote measurements on an entire batch without the necessity to sample, or to monitor the concentration of a product on a fabrication line, immersion probes are used (Fig. 11.15). These devices incorporate two fibre optics to conduct the light to the sample and retrieve the light after absorption in the medium under investigation.

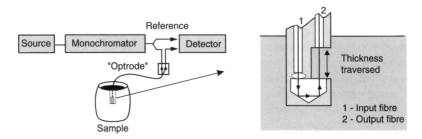

Figure 11.15—*Principle of a spectrophotometer fitted with an immersion probe.* Monochromatic light from the spectrophotometer is guided toward an immersion cell and then brought back to the detector. The reference beam is also guided by a fibre optic.

11.10   Quantitative analysis in the UV/Visible

The spectral domain of the UV/Visible is well known because it includes the visible part of the spectrum and is widely used in quantitative analysis. Measurements are based on the Beer–Lambert law, which relates the absorption of light under certain conditions to the concentration of a compound.

It is not necessary that the compound contain a chromophore as long as derivatisation is carried out before measurement to ensure absorption of the light. Through derivatisation, it becomes possible to quantify a chemical species that has no significant absorption because it is weak or, alternatively, because it lies in the same spectral domain as interferences.

To this effect, the measurement of absorbance is preceded by a chemical transformation (derivatisation) that has to be specific, total, rapid, reproducible and yield a UV/VIS absorbing derivative that is stable in solution. This is the principle of colorimetric tests.

■ The term colorimetry comes from the fact that initial measurements in this spectral domain, well before the invention of spectrophotometers, were carried out with white light without any optical instrument. Visual comparison of the sample colour with that of a reference solution of known concentration was then performed.

Two situations can be encountered with the use of this method (Fig. 11.16):

— *The constituent A to be quantified is present in a mixture with a variable quantity of compound B that absorbs in the same spectral range*: direct measurement of the absorbance due solely to compound A is impossible because of the super-position of the spectra of A and B (Fig. 11.16a). To remedy this situation, compound A is totally transformed by chemical reaction into compound C with an absorption band removed from that of B (Fig. 11.16, curves a and b).

— *Compound A does not possess a chromophore*: here again a chromophore can be incorporated into A, by transforming this substance into compound C, following the same principle (Fig. 11.16, curves c and d).

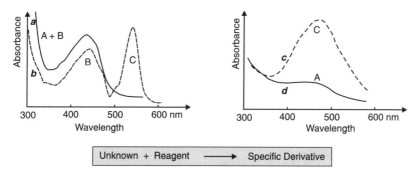

Figure 11.16—*Two situations frequently encountered in colorimetry*. A compound present in a mixture can be quantified by colorimetry if it is chemically transformed (derivatised) prior to the measurement of absorbance.

11.11   Beer–Lambert's law

Beer, a nineteenth-century German physicist, gave his name to a law that allows the calculation of the quantity of light transmitted through a defined thickness of a compound in solution in a non-absorbing matrix. His work is often associated with that of the French mathematician Lambert, who laid down the basis for photometry in the eighteenth century. The result is *Beer–Lambert's law*, shown here in its current form:

$$A = \varepsilon l C \tag{11.2}$$

where A is the absorbance, an optical parameter without units, measured with a spectrophotometer; l is the thickness of solution through which incident light is passed; C is the molar concentration; and ε is the molar absorption coefficient $(\text{lmol}^{-1}\,\text{cm}^{-1})$ at a given wavelength. The molar absorptivity of a compound depends on the wavelength used, the temperature and the nature of the solvent. Its value is usually known for the wavelength at the absorption maximum. This value, which corresponds to the absorbance of a solution with a 1 M concentration and a thickness of 1 cm, can vary over a wide range (0 to 200 000). If m is the quantity of compound per litre and M its molar mass (expressed in g), then equation (11.2) becomes:

$$A = \varepsilon l \frac{m}{M} \tag{11.3}$$

This formula is based on Lambert's hypothesis that the intensity I of monochromatic radiation is decreased by dI (i.e. negative) as it passes through a thickness dx of a material with an absorption coefficient k at the chosen wavelength (Fig. 11.17). Thus,

$$-\frac{dI}{dx} = kI_x \tag{11.4}$$

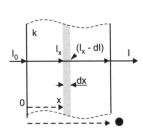
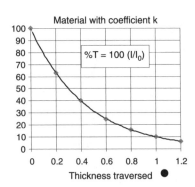

Figure 11.17—*Absorption of light by a homogeneous material and representation of % transmittance as a function of the material's thickness.*

If I_0 represents the incident intensity of the radiation before it passes through a medium of thickness l that has an *absorption coefficient* k, the transmitted intensity I will be given by the integrated form, equation (11.6), of the previous equation:

$$[\ln I_x]^I_{I_0} = -k[x]^l_0 \tag{11.5}$$

$$\ln \frac{I}{I_0} = -k \cdot l \tag{11.6}$$

$$I = I_0 \exp(-kl) \tag{11.7}$$

In 1850, Beer applied equation (11.7) to a dilute solution of a compound dissolved in a transparent medium. He proposed that k was proportional to the molar concentration C of the compound (Fig. 11.18). This is why equation (11.7) is better known in the form (11.2) or (11.3) in which the absorbance A is represented by one of the following equations:

$$A = \log \frac{I}{I_0} \quad \text{or} \quad A = \log \frac{1}{T} \tag{11.8}$$

where T is the transmittance defined as follows:

$$T = \frac{I}{I_0} \quad \text{or} \quad \%T = \left(\frac{I}{I_0}\right) \times 100$$

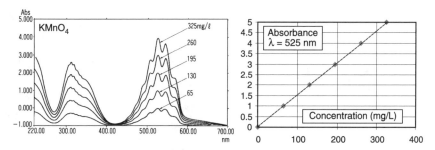

Figure 11.18—*Beer–Lambert's law.* Representation of absorbance and a plot showing the linearity of absorbance for solutions containing an increasing amount of potassium permanganate.

■ In colorimetry, it is preferable to measure the absorbance of a chromophore at the higher wavelength to reduce the risk of superimposition with absorption bands related to other compounds. When absorbance measurement is preceded by a chemical reaction, it is conceivable that the exact structure of the compound whose absorbance is being measured is not known. Nevertheless, if it is assumed that the reaction is quantitative, its molar absorption coefficient can be calculated using the molar concentration of the compound that has been derivatised.

11.12   Visual colorimetry

Visual colorimetry, which has been used for over two centuries, is a simplified form of instrumental colorimetry. Because of its low cost, it is used on a daily basis and its precision is often surprising.

■ Numerous tests using strips that change colour when dipped into a medium represent current applications of colorimetry. However, the result obtained by visual examination of the reflected light is related to reflectometry more than to transmission colorimetry (see Fig. 11.18). These selective tests, which are ready to use and do not require an instrument, are complementary to established methods. Because they yield immediate results, they are useful for all sorts of semiquantitative analyses.

Simple colorimeters, used for routine analysis, are visual comparators related to Nessler tubes. The latter are made of flat-bottomed glass tubes with a volumetric mask and filled with reference solutions of varying concentrations (and possibly derivatising agents). The solution to be analysed, placed in an identical tube, is inserted beside the series of standard solutions.

The tubes are placed on a background of white light and examined for transparency. The concentration of the solution to be analysed falls between the two standard solutions that bracket it.

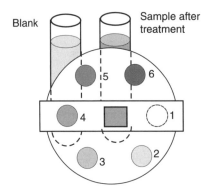
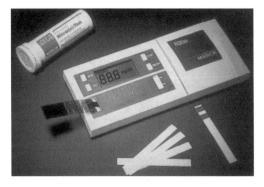

Figure 11.19—*Visual colorimetry.* To the left, a rotating filter comparison system. By using two cells, one of which contains the sample and the other the blank, it is possible to take into account the normal coloration of the two tubes. Observation is made by comparing the transparency for a specific filter against white light. To the right, a portable reflectometer that allows comparisons to be made without the human eye. (Reproduced by permission of Merck.)

■ Nephelometry, a technique that differs from colorimetry, also uses Beer–Lambert's law. This method involves the formation of a precipitate and, based on the absorbed light at a given wavelength, allows the concentration of the precursor to be determined.

For example, insoluble barium salt is added to quantify sulphate ions. The absorbance of the precipitate of barium sulphate is measured after it has been stabilised with a hydrosoluble polymer such as Tween.

11.13 Absorbance measurements

Light impinging on a sample can be transmitted, refracted, reflected, scattered or absorbed. Beer–Lambert's law, which is only related to the fraction absorbed, is applicable under the following conditions:

— the light used must be monochromatic
— the concentration must be weak
— the solution must not fluoresce and must be homogeneous
— the solute must not undergo photochemical reactions
— the solute must not form variable associations with the solvent.

Experimentally, a calibration curve $A = f(C)$ is constructed using solutions of known concentration of the compound. The solutions undergo the same treatment as the sample. This curve is often a straight line for dilute solutions. It allows determination of the concentration C_X of the unknown sample.

In many instances, a single reference solution of concentration C_R can be used. This concentration is chosen such that the absorbance A_R is close to or slightly above that of the unknown solution A_X (see Fig. 11.20).

The following formula permits calculation of C_X:

$$C_X = C_R \frac{A_X}{A_R} \qquad (11.9)$$

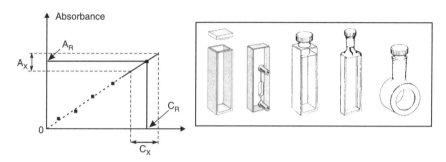

Figure 11.20—*Colorimetric calibration curve and classical quartz cells for absorption measurements.* If a single standard solution is used, it is assumed that the calibration curve is a straight line that passes through the origin. The precision of the result will improve if the unknown concentration is close to that of the reference solution (i.e. the results will be determined more or less by interpolation not by extrapolation).

Beer–Lambert's law is additive. This implies that absorbance A, measured in a cell of thickness l for a mixture of two compounds 1 and 2 in solution in the same solvent, will be identical to the absorbance measured after light has passed through two cells of the same thickness l placed in series, the first containing compound 1 and the last containing the other compound, 2 (compounds must be at the same concentrations as in the initial mixture). By giving indices 1 and 2 to compounds 1 and 2 respectively, the following additive relation is obtained:

$$A = A_1 + A_2 = l(\varepsilon_1 C_1 + \varepsilon_2 C_2) \qquad (11.10)$$

■ **Isobestic points.** Consider compound A, which can be transformed by a first order reaction into compound B. Assume that the absorption spectra obtained under the same conditions of concentration cross over at a point I when they are superimposed (Fig. 11.21). That is, the absorbances of the two solutions are the same for the wavelength at point I. Consequently, the coefficients ε_A and ε_B are identical. In this type of experiment, A is initially pure and at the end of the experiment B is pure. For all the intermediate solutions, mixtures of A and B can be prepared but the global concentration does not change ($C_A + C_B$ = constant). This leads to the following relationship:

$$A_1 = \varepsilon_A l C_A + \varepsilon_B l C_B = \varepsilon l(C_A + C_B) = \text{constant}$$

All spectra of the mixtures A + B will pass, over the course of time, through point I, called the isobestic point, where the absorbance of A will be constant.

Families of concurrent curves are observed for a coloured indicator as a function of pH or as a function of reaction kinetics.

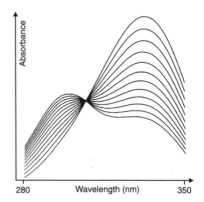

Figure 11.21—*Isobestic point.* Alkaline hydrolysis of methyl salicylate at 25 °C. Superposition of the successive spectra recorded between 280 and 350 nm at 10 min intervals.

11.14   Confirmation analysis (purity control)

Classical calculation of the concentration of an analyte using equation (11.8) will yield erroneous results if the sample contains an impurity that absorbs at the same wavelength. Thus, a *confirmation analysis* is often required.

For a given pure compound, the ratio of absorption coefficients at two different wavelengths is constant and characteristic of this compound. Therefore, if one or two additional absorbance measurements at a few nanometres separation are made, absorbance ratios of the solution can be obtained and compared to those which have been established from a reference solution. If these ratios are different, it can be concluded that there is an impurity in the sample.

■ It is possible from this principle to control peak homogeneity in liquid chromatography. For this, a UV detector allowing simultaneous measurements of absorbance at different wavelengths is necessary (such as a diode array detector). Any variation of the absorbance ratio at the two predetermined wavelengths during elution will confirm that compounds are co-eluting from the column (Fig. 11.22). This method will not work, however, if there is perfect co-elution.

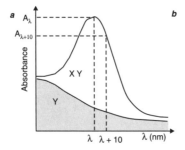
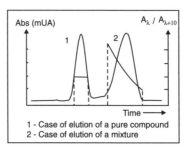

Figure 11.22—*Confirmation analysis.* a) Spectrum of a mixture (X + Y) and spectrum of Y in the same range (grey region); b) chromatogram illustrating two peaks, the first of which is a unique compound and the second, shifted in time, represents a co-elution. Calculation of the absorbance ratio during elution permits the determination of the purity of eluted compounds. These variations are usually shown by the chromatographic software as coloured zones.

11.15   Distribution of relative errors due to instruments

Some spectrophotometers allow the measurement of absorbance over a dynamic range of 4 to 6 decades. However, elevated values of absorbance are less reliable because they correspond to very weak transmitted intensities ($I/I_0 = 10^{-6}$ for $A = 6$). For most instruments, there are three independent causes of error that can affect transmittance (Fig. 11.23):

— *background noise of the radiation source*. This term, ΔT_1, is considered to be constant and independent of T: $\Delta T_1 = k_1$ (curve 1)

— *background noise from the photomultiplier tube*. This term, ΔT_2, varies as a function of T and its complex relationship is shown in the following equation (curve 2):

$$\Delta T_2 = k_2 \sqrt{T^2 + T} \tag{11.11}$$

— *reflection and scattering of light* in the optical path of the instrument. This term is proportional to T: $\Delta T_3 = k_3 T$ (curve 3).

These sources of errors are additive, which causes a variation of the total error in concentration which has a minimum generally located around $A = 0.7$. For quantitative measurements, it is desirable to adjust dilution factors so that absorbances are in this optimal domain.

Beer–Lambert's law can be used to relate the relative error in concentration C to the relative error in transmittance T.

$$\frac{\Delta C}{C} = \frac{1}{\ln T} \frac{\Delta T}{T} \tag{11.12}$$

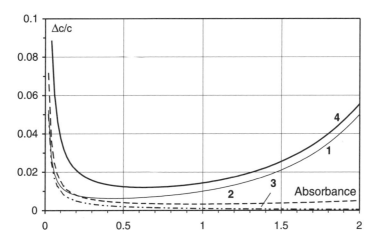

Figure 11.23—*Average curves representing each of the instrumental causes of error in absorbance measurements and the global curve (4) resulting from their sum (see text).*

11.16   Baseline correction

Some samples of natural origin contain micellar particles in suspension that cause, by light scattering (the Tyndall effect), a supplementary absorbance that varies with wavelength.

Morton and Stubbs' method (and the method using derivative curves, cf. 11.18) allows an efficient baseline correction if it is assumed that absorption varies linearly in the measurement range. Thus, in the situation represented in Fig. 11.24a, it is essential for quantification to correct absorbance A_2 at the maximum λ_2 by subtracting the values indicated as x and y:

$$A = A_2 - x - y \tag{11.13}$$

To determine x and y, a reference spectrum is obtained from the pure compound at two wavelengths, λ_1 and λ_3, such that the absorption coefficients are the same (Fig. 11.24b). In this spectrum, the same absorbance A' is obtained for λ_1 and λ_3, and an absorbance A'_2 at the maximum λ_2. The ratio between these two absorbances is $R = A'_2/A'$. Then, for the sample spectrum, the ratio of absorbances is determined at the same two wavelengths. The value of x can thus be deduced from the figure:

$$x = (A_1 - A_3)\frac{(\lambda_3 - \lambda_2)}{(\lambda_3 - \lambda_1)} \tag{11.14}$$

From R and the ratio A_2/A_3 of the sample, y can be determined because the absorbance ratio (after corrections for x and y) has the value R (Fig. 11.24):

$$R = \frac{A_2 - (x + y)}{A_3 - y}$$

$$y = \frac{RA_3 - A_2 + x}{R - 1} \tag{11.15}$$

This equation, which is independent of the concentration of the reference compound in solution, yields satisfactory results. The method only necessitates that a spectrum of the pure compound is available.

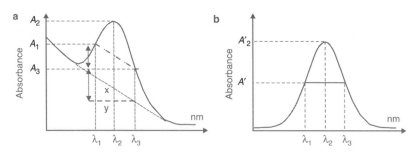

Figure 11.24—*Illustration of calculations obtained using the Morton and Stubbs' method.*

11.17 Multicomponent analysis (MCA)

When a mixture containing compounds with known spectra is analysed, it is possible to determine the mixture's composition. According to the additivity law (equation (11.10)), the spectrum of the mixture corresponds to the weighted superposition of each of the individual constituents. The following section recalls the classical calculation method. It is rarely used as such because it is typically incorporated into software.

11.17.1 Basic algebraic method

Given a mixture of three components a, b and c in solution (with concentrations C_a, C_b, C_c), it is possible to measure the absorbance of this solution at three different wavelengths λ_1, λ_2 and λ_3, giving A_1, A_2 and A_3. Since the absorption coefficients of the three components at each of these wavelengths is known (9 values in total, from ε_a^1 to ε_c^3) it is possible to write the following system of simultaneous equations using the additivity law (it is assumed that the optical path is 1 cm):

$$\text{at } \lambda_1 \qquad \varepsilon_a^1 C_a + \varepsilon_b^1 C_b + \varepsilon_c^1 C_c = A_1$$

$$\text{at } \lambda_2 \qquad \varepsilon_a^2 C_a + \varepsilon_b^2 C_b + \varepsilon_c^2 C_c = A_2$$

$$\text{at } \lambda_3 \qquad \varepsilon_a^3 C_a + \varepsilon_b^3 C_b + \varepsilon_c^3 C_c = A_3$$

The solution to this mathematical system, which corresponds to a 3×3 matrix, allows the determination of the unknown concentrations C_a, C_b, C_c.

$$\begin{bmatrix} \varepsilon_a^1 & \varepsilon_b^1 & \varepsilon_c^1 \\ \varepsilon_a^2 & \varepsilon_b^2 & \varepsilon_c^2 \\ \varepsilon_a^3 & \varepsilon_b^3 & \varepsilon_c^3 \end{bmatrix}^{-1} \times \begin{bmatrix} A_1 \\ A_2 \\ A_3 \end{bmatrix} = \begin{bmatrix} C_1 \\ C_2 \\ C_3 \end{bmatrix} \tag{11.16}$$

This approach is most efficient when the spectra of each individual component are significantly different. Otherwise, it is vulnerable to errors in measurements. In practice, with instruments which have diode array detectors, the software utilises many tens of data points. The system to be resolved becomes overdetermined and, under these conditions, yields very precise results.

11.17.2 Multiwavelength linear regression analysis (MLRA)

Mixture analysis has given rise to several methods, made possible with the range of spectrometers available. Computers dedicated to the spectrometer optical platform include software that can treat a great number of data points obtained from sample spectra and standard solutions.

For example, a method is given below that can improve the results of an analysis of a two-component sample using linear regression (Fig. 11.25).

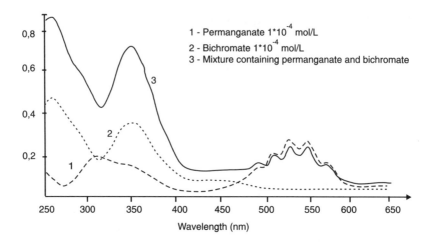

Figure 11.25—*Multicomponent analysis.* Spectra of a 10^{-4} M permanganate solution, a 10^{-4} M dichromate solution and a mixture containing 0.8×10^{-4} M permanganate with 1.8×10^{-4} M dichromate (from Bianco *et al.*, *J. Chem. Educ.* 1994, **66**(2), p. 178).

The instrument uses three recordings stored in memory: a spectrum of the sample (composed of the two compounds to be analysed) and a spectrum of each of the pure components in the same spectral domain (reference solutions of known concentrations).

For each wavelength, the law of additivity of absorbances allows the absorbance of the mixture of a and b (which must not react with one another) to be given by the following expression:

$$A = \varepsilon_a l C_a + \varepsilon_b l C_b \tag{11.17}$$

For each of the two reference spectra, assuming a cell thickness of $l = 1$ cm, the following equations can be written:

– for component a $\qquad\qquad A_{\text{ref.a}} = \varepsilon_a C_{\text{ref.a}}$ $\hfill$ (11.18)

– for component b $\qquad\qquad A_{\text{ref.b}} = \varepsilon_b C_{\text{ref.b}}$ $\hfill$ (11.19)

These expressions permit the determination of the absorption coefficient ε for each of the pure compounds for a given wavelength. Thus, equation (11.17) can be rewritten as:

$$A = \frac{A_{\text{ref.a}}}{C_{\text{ref.a}}} C_a + \frac{A_{\text{ref.b}}}{C_{\text{ref.b}}} C_b \tag{11.20}$$

Dividing the first term by $A_{\text{ref.a}}$, for each wavelength:

$$\left(\frac{A}{A_{\text{ref.a}}}\right) = \frac{C_a}{C_{\text{ref.a}}} + \frac{C_b}{C_{\text{ref.b}}}\left(\frac{A_{\text{ref.b}}}{A_{\text{ref.a}}}\right) \tag{11.21}$$

The first term of equation (11.21) is thus a function of the ratio of absorbances appearing in the second term. The calculated values lie on a straight line whose slope and intercept allow the determination of C_a and C_b. The precision of the results increases with the number of points used.

11.17.3   Deconvolution

Many types of data treatment software allow the composition of a mixture to be obtained from its spectra. Kalman's least squares filter is one of the most widely known of these methods. Using successive approximations, it automatically finds the spectra of the sample solution by addition of the spectra of each compound contained in the spectral library (i.e. by additivity of the absorbances) and use of weight coefficients. These are called deconvolution methods (Fig. 11.26).

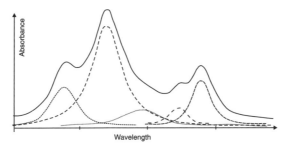

Figure 11.26—*Example of spectrum deconvolution.* The experimental spectrum (uppermost curve) is treated in order to obtain the spectra of the five individual components. This method assumes that the spectra of the individual components are known.

11.18   Derivative spectrometry

Derivative spectrometry improves the precision of analysis when compounds with very similar spectra are present in the mixture. The derivative spectral curves, in the mathematical sense, amplify the weak variations in the slopes of the initial spectrum (called the *zeroth order spectrum*). These derivative plots show greater variations than when in absorbance units (Fig. 11.27).

The principle can be extended to higher derivatives (*nth* derivatives) from the curve of the *first derivative*, $dA/d\lambda = f(\lambda)$, and is accessible on almost all spectrometers.

The spectrum of the *second derivative* shows regions of zero slope corresponding to inflection points in the zeroth order spectrum (Fig. 11.27). Use of the fourth derivative is even better at showing weak variations in absorbance from the initial spectrum.

Many types of analyses would be improved by using this principle because the use of a derivative spectrum is no more difficult than the use of an absorbance spectrum. The amplitude between the minimum and maximum on the *nth* derivative curve is

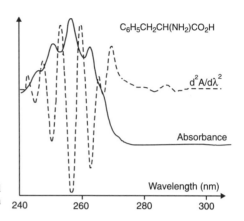

Figure 11.27—*Derivative curve.* UV spectrum and second derivative spectrum of phenylalanine (from Perkin–Elmer document).

proportional to the value of the absorbance of the solution. A calibration curve can be obtained from several standard solutions of varying concentrations to which the same mathematical principles are applied.

This process can be useful in three cases, where the absorbance can be erroneous:

— if the spectrum of the sample solution suffers from an increase in absorbance due to the presence of a uniform background, the derivative curve will not be affected because it is only sensitive to variations in the slope of the zeroth order spectrum (Fig. 11.28).

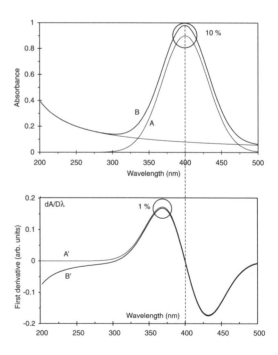

Figure 11.28—*Effect of light scattering on a spectrum recorded in absorbance and on its derivative spectrum.* This figure shows a comparison of the derivative spectra corresponding to the spectrum A of a compound in solution without scattering and to the same spectrum B in the presence of scattering. It can be seen that the effect of scattering is in the order of 10% in absorbance units but only about 1% in the derivative spectrum (modelled spectra).

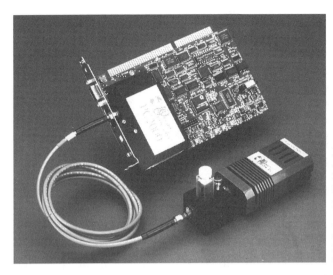

Figure 11.29—*A PC-plug-in spectrophotometer*. This instrument is a fully integrated system for colorimetric analysis. The source is linked by an optical fiber to the diode array detector fixed onto the card (Model CHEM2000 reproduced by permission of Ocean Optics Europe.)

— if light scattering is present in the sample solution, producing an increase in background absorption that increases towards lower wavelengths, the derivative spectrum will be only weakly affected.

— the absorbance of an analyte barely visible in the original spectrum because it is swamped in normal absorption mode from the matrix becomes detectable using the derivative spectrum.

Figure 11.30—*Microplate reader*. This device, indispensable for laboratories involved with immunochemical tests, is a spectrometer with an optical path that allows sequential absorbance measurements (for reaction kinetics) along the axis of each well of the microplate. (Multiscan instrument, reproduced by permission of Labsystems.)

Problems

11.1 Calculate the energy of a mole of photons corresponding to a wavelength of 300 nm.

11.2 Calculate the absorbance of an organic dye ($c = 7 \times 10^{-4}$ mol l^{-1}), knowing that the molar absorptivity $\varepsilon = 650$ mol l^{-1} cm^{-1} and that the length of the optical path of the cell used is 2×10^{-2} m. What would happen to the absorbance if the cell used was of double its present thickness?

11.3 A 1.28×10^{-4}M solution of potassium permanganate has a transmittance of 0.5 when measured in a 1 cm cell at 525 nm.
1. Calculate the molar absorptivity coefficient for the permanganate at this wavelength.
2. If the concentration is doubled what would be the absorbance and the percentage transmittance of the new solution?

11.4 A polluted water sample contains approximately 0.1 ppm of chromium, ($M = 52$ g mol^{-1}). The determination for Cr(VI) based upon absorption by its diphenylcarbazide complex ($\lambda_{max} = 540$ nm, $\varepsilon_{max} = 41\,700$ l mol^{-1} cm^{-1}) was selected for measuring the presence of the metal. What optimum path-length of cell would be required for a recorded absorbance of the order of 0.4?

11.5 Paints and varnishes for use on exteriors of buildings must be protected from the effects of solar radiation which accelerate their degradation (photolysis and photochemical reactions). Given that: $M = 500$ g mol^{-1}; $\varepsilon_{max} = 15\,000$ l mol^{-1} cm^{-1} for $\lambda_{max} = 350$ nm, what must be the concentration (expressed in g l^{-1}) of a UV additive M such that 90% of the radiation is absorbed by a coating of thickness 0.3 mm?

11.6 By employing the empirical rules of Woodward–Fieser, predict the position of the maximum absorption of the compounds whose structures are seen below.

11.7 To determine the concentrations (mol/l) of Co(NO$_3$)$_2$ (A) and Cr(NO$_3$)$_3$ (B) in an unknown sample, the following representative absorbance data were obtained.

A (mol/l)	B (mol/l)	510 nm	575 nm
1.5×10^{-1}	0	0.714	0.097
0	6×10^{-2}	0.298	0.757
Unknown	Unknown	0.671	0.330

Measurements were made in 1.0 cm glass cells.

1. Calculate the four molar absorptivities: $\varepsilon_{A(510)}$, $\varepsilon_{A(575)}$, $\varepsilon_{B(510)}$ and $\varepsilon_{B(575)}$.
2. Calculate the molarities of the two salts A and B in the unknown.

11.8 The determination of phosphate concentration in a washing powder requires first the hydrolysis of the tripolyphosphate components into the phosphate ion or its protonated forms. Then a quantitative colorimetric method based upon the absorption of the yellow complex of phosphate and ammonium vanado-molybdate can be used. A series of standard solutions was prepared. The complex was then developed in 10 ml aliquots of these solutions by adding a 5.0 ml aliquot of an ammonium vanadomolybdate solution.

   Measurements were made in a glass cell of 1.0 cm pathlength at 415 nm. The following absorbance data were obtained.

Solutions	Concentration (mmol/l)	Absorbance
S_0	0.0	0
S_1	0.1	0.15
S_2	0.2	0.28
S_3	0.3	0.4
S_4	0.4	0.55
S_5	0.5	0.7

1. Produce a calibration curve from these data.
2. By the method of least squares, derive an equation relating absorbance to phosphate concentration.
3. An unknown solution obtained by hydrolysis of 1 g of detergent treated in an identical way gave an absorbance of 0.45. Determine the phosphate concentration of this solution.

11.9 The concentration of the sulphate ion in a mineral water can be determined by the turbidity which results from the addition of excess $BaCl_2$, to a quantity of measured sample. A turbidometer used for this analysis has been standardised with a series of standard solutions of $NaSO_4$. The following results were obtained:

Standard solution	Conc. SO_4^{2-} (mg/l)	Reading of turbidometer
S_0	0.00	0.06
S_1	5.00	1.48
S_2	10.00	2.28
S_3	15.00	3.98
S_4	20.00	4.61

1. In supposing that a linear relationship exists between the readings taken from the apparatus and the sulphur ion concentration, derive an equation relating readings of the turbidometer and sulphate concentration (method of least squares).
2. Calculate the concentration of sulphate in a sample of mineral water for which the turbidometer gives a reading of 3.67.

The creators of modern colorimetry

Visual colorimetry, probably the oldest analytical method, was last used by the Greeks and Romans. This method found its scientific basis in 1729 when Pierre Bouguer theorised that 'if a given width of glass absorbs half of the light emitted by a source then double the width will reduce the light by one quarter its initial value'.

Some 30 years later, Jean-Henri Lambert (1728–1777) produced the first mathematical relationship: 'the logarithm of the decrease in light intensity (today we would say the inverse of the transmittance) is equal to the product of the capacity of the medium times its thickness'.

Finally, in 1850, Auguste Beer established the relationship between concentration and optical density (the expression is now termed absorbance). This led to the Beer–Lambert law.

Among all the devices that have been invented for visual colorimetry, one of the most original was described by Jules Duboscq in 1868. Now redun-dant, the Duboscq Comparator used the Beer–Lambert law to determine the concentration of the species by comparing the different thicknesses through which light is passed. The device superimposes, in a small visual channel using a system of total reflecting prisms, the light intensities that have travelled through two cells, one containing the sample (C_X) and the other containing a known standard (C_R). The observer adjusts the heights, h_X and h_R, of the solutions until the coloration of both cells is identical. Under these conditions, the absorbances are equal. If $A_R = A_X$, then:

$$\varepsilon h_R C_R = \varepsilon h_X C_X$$

and thus the concentration C_X can be determined.

This visual colorimeter was used until 1960. The observer adjusts the transmitted intensities in both paths by moving the containers of the two solutions to be compared along the two transparent bars.

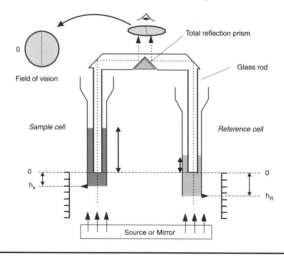

CHAPTER 12

Fluorimetry

Certain compounds, whether present in solution or in solid state (as molecular or ionic crystals) emit light when they are excited by photons in the visible or near ultraviolet domain of the spectrum. This phenomenon, called luminescence, is the basis of fluorimetry, a very selective and sensitive analysis technique. The corresponding measurements are made with fluorimeters or spectrofluorimeters and, for chromatographic applications, with fluorescence detectors.

The fluorescence of organic molecules and ions in solution is a photoluminescent process that decreases extremely rapidly when the excitation ceases, in contrast to phosphorescence, which latter has a much slower decrease and is seldom used for analysis. Finally, chemiluminescence, the emission of light during a chemical reaction, is involved in some analytical applications.

12.1 Fluorescence and phosphorescence

Certain compounds, when excited in solution by visible or near ultraviolet radiation, re-emit all or part of this energy as radiation. According to Stokes law, the maximum of the spectral emission band is located at a higher wavelength than that of either the incident radiation or the excitation band maximum (see Figs 12.1 and 12.2). After excitation, the intensity of the emitted light decreases (decays) exponentially according to equation (12.1), which relates the instantaneous intensity to time:

$$I_f = I_0 \exp(-k_f t) \tag{12.1}$$

The emission can be classified as *fluorescence*, which has a very rapid decrease in intensity, or *phosphorescence*, where emission decay is much slower. A compound can both fluoresce and phosphoresce.

The *lifetime* of fluorescence (or phosphorescence), τ_0, is defined using the rate constant k_f as $\tau_0 = 1/k_f$. By this definition, the residual intensity according to equation (12.1) is only 36.7% of the initial intensity at τ_0; 63.3% of the initial species have relaxed to a non-emissive state. Because fluorescence lifetimes are only a few nanoseconds, fluorimeters require measurements to be made at the same time as excitation, i.e. in a stationary state. However, phosphorescence, which has a lifetime of many seconds, is usually studied after excitation.

These analytical methods are very sensitive. The detection limit of a compound in solution that fluoresces is often 1 000 times lower than the detection limit in the UV/Visible. Therefore, proper use of these techniques requires a good knowledge of the phenomenon in order to avoid errors.

■ Certain gases present at trace levels in the earth's atmosphere are quantified by fluorescence. This is done by excitation using a very short pulse (1 μs) of a powerful laser (LIDAR device).

12.2   Origin of fluorescence

Fluorescence from molecular compounds can be understood in a similarl way to UV/visible absorption by considering energy exchanges between frontier orbitals designated by HOMO and LUMO (cf. 11.2 and 11.3). Molecules initially in the ground electronic state S_0 are instantaneously excited by absorption to a vibrational level V_i of the excited electronic state S_1 (Fig. 12.1). Molecules relax down to the V_0 vibrational level of this electronic state S_1 very rapidly (in 10^{-12} s) by *internal conversion* without the emission of photons. It is from this energy level that fluorescence occurs (in 10^{-9} to 10^{-7} s) causing molecules to return to a vibrational level of the ground electronic state S_0.

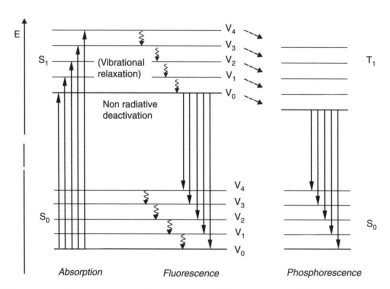

Figure 12.1—*Energy diagram comparing fluorescence and phosphorescence.* Short arrows correspond to internal conversion without the emission of photons. Fluorescence is an energy transfer between states of the same multiplicity (spin state) while phosphorescence is between states of different multiplicity. The situation is more complex than that shown by this Jablonski diagram.

Over the course of fluorescence, which accompanies energy relaxation, the molecule can keep part of the energy it received in the form of vibrational energy of the ground state. This excess vibrational energy is dissipated by collisions or other non-radiative processes called *vibrational relaxation*. The emission of lower energy photons is also possible and gives rise to fluorescence in the mid infrared.

The diagram in Fig. 12.1 shows an apparent symmetry that exists between the absorbance and fluorescence of many compounds. This symmetry is often observed when the absorbance and fluorescence spectra are overlaid (Fig. 12.2).

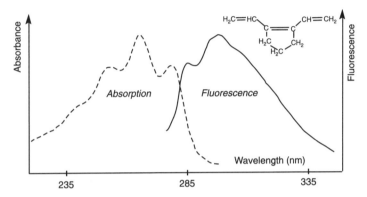

Figure 12.2—*Representation on overlaid graphs of the absorption and fluorescence spectra of a triene.* The apparent symmetry of the traces can be rationalised considering Fig. 12.1 (figure from *Tetrahedron* 1993, H. Jacobs *et al.*, p. 6045).

■ Fluorescence is not useful simply for chemical analysis. For example, a fluorescent additive that sticks to textile fibres is added to laundry soap. This compound absorbs solar radiation in the non-visible part of the spectrum and re-emits at longer wavelengths in the blue spectral region, which makes clothing appear 'whiter'. Another application of fluorescence encountered daily is cathode tube lighting. The internal walls of these tubes are covered with mineral salts (luminophores) that emit in the visible region due to excitation by electrons.

Phosphorescence corresponds to a different relaxation process. After the absorption phase, corresponding to the transfer of one electron into the S_1 level (singlet state), a spin inversion can occur if vibrational relaxation is slow, leading the electron to a T_1 state (triplet state) that is slightly more stable. Hence, return to the ground electronic state will be slower because it involves another spin inversion for this electron. For this reason, radiative lifetimes for phosphorescence can be up to 10^8 times greater than for fluorescence.

■ Fluorescence, which often occurs in cyclic, rigid molecules that contain π electrons, is enhanced by the presence of electron-donating groups and decreased by electron-withdrawing groups (Fig. 12.3). It also depends on pH and solvent. Non-rigid molecules, on the other hand, easily lose all of their absorbed energy by degradation and vibrational relaxation.

By analogy, this phenomenon can be compared to that encountered when a hammer hits a rubber block vs. an anvil. With the rubber block, energy will be dissipated in the mass (relaxation) and sound will not be emitted. However, on the anvil, part of the mechanical energy of the hammer will be retransmitted towards the outside (as a loud noise) and this can be compared to the phenomenon of fluorescence of non-rigid vs. rigid molecules.

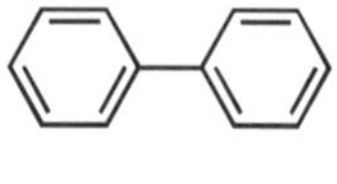

Biphényl (0,2)

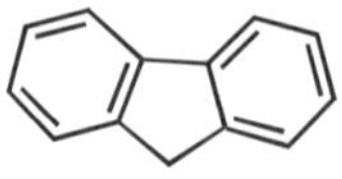

Fluorene (1)

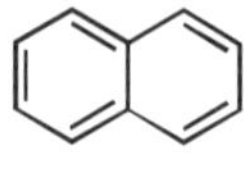

Naphthalene (0,55)

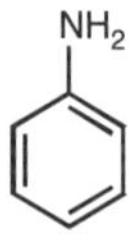

Aniline (0,6)

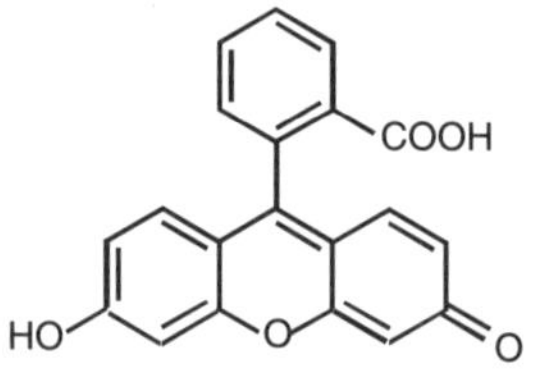

Fluoresceine (0,92 in NaOH 0,1 M and 0,65 to pH = 7)

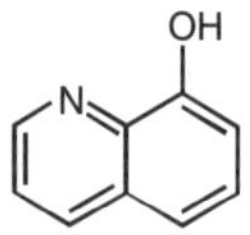

8-Hydroxyquinoline

Figure 12.3—*Fluorescent aromatic compounds*. The compound names are followed by their fluorescent quantum yields, whose values are obtained by comparison to compounds of known fluorescence. Measurements were made at 77 K. 8-hydroxyquinoline is representative of many molecules that form fluorescent chelates with certain metal ions.

12.3 Fluorescence intensity

The intensity of fluorescence depends on the position from where the emission occurs relative to the detector. Part of the exciting radiation is absorbed before it reaches the point of emission and part of the emitting radiation during fluorescence is trapped before it can exit the cell. The measured intensity of fluorescence is thus the result of emission from several small volumes in the space delimited by the entrance and exit slits (Fig. 12.4). Therefore, it is difficult to calculate the absolute intensity of fluorescence or emittance, I_f. The phenomenon of radiation damping – called internal filtering, due to the overlap of parts of the absorption and emission spectra (*colour quenching*) – is associated with energy transfer between excited species and other molecules or ions. These energy transfers occur by collisions or formation of complexes (*chemical quenching*). This is why the presence of oxygen can cause an underestimation of fluorescence.

In solution, the fluorescence quantum yield Φ_f (with a value of 0 to 1 inclusive), independent of the intensity emitted by the light source, is defined as the fraction of absorbed incident radiation that is re-emitted as fluorescence:

$$\Phi_f = \frac{\text{number of photons emitted}}{\text{number of photons absorbed}} = \frac{I_f}{I_a} \tag{12.2}$$

Knowing that $I_a = I_0 - I_t$ (I_t represents the transmitted light intensity), equation (12.2) relates I_f to the concentration C of the compound as follows:

$$I_f = \Phi_f(I_0 - I_t)$$

thus

$$I_f = \Phi_f \cdot I_0 \left(1 - \frac{I_t}{I_0}\right)$$

The absorbance A is defined in terms of the ratio of I_0 and I_t ($A = \log(I_0/I_t)$), thus:

$$\frac{I_f}{I_0} = 10^{-A}$$

which leads to the following equation:

$$I_f = \Phi_f \cdot I_0(1 - 10^{-A}) \tag{12.3}$$

In a dilute solution, A is close to 0. Therefore, the expression in parentheses in equation (12.3) can be simplified and becomes equivalent to $2.3A$. Equation (12.3) can be rewritten as:

$$I_f = 2.3\Phi_f \cdot I_0 \cdot A \qquad \text{thus} \qquad I_f = 2.3\Phi_f \cdot I_0 \cdot \varepsilon \cdot l \cdot C \tag{12.4}$$

where I_0 is the intensity of the excitation radiation, C is the molar concentration of the compound, l is the cell thickness, ε is the molar absorption coefficient of the compound and Φ_f is the quantum yield of fluorescence.

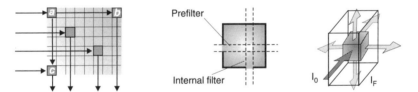

Figure 12.4—*Fluorescence intensity.* Depending on the point from which fluorescence is emitted in solution, a variable light intensity will reach the detector. By specific positioning of the excitation and emission windows, it is possible to estimate the re-absorption of fluorescence radiation (by comparison between sectors a and c in the figure) and the absorption of the incident radiation (by comparison between sectors a and b). In practice, fluorescence emitted from the central region of the cell is collected.

If all parameters due to the instrument and most due to the compound are factored into a constant K, the following equation can be used for weak concentrations:

$$I_f = K \cdot I_0 \cdot C \tag{12.5}$$

■ To obtain good results in fluorimetry, solutions must be dilute. Above a given limit, fluorescence is no longer proportional to concentration because of the non-linearity of Beer–Lambert's law. Excitation is proportionally weaker and association complexes are formed between excited and ground state molecules. This leads to the apparently paradoxical result that fluorescence can diminish as analyte concentration increases (Fig. 12.5).

Classically, quantitation by fluorimetry is done using a calibration curve obtained with reference solutions or by using a single point calibration. In the latter case, the unknown concentration is given by:

$$C_{unk} = C_{ref} \frac{I_{unk}}{I_{ref}} \tag{12.6}$$

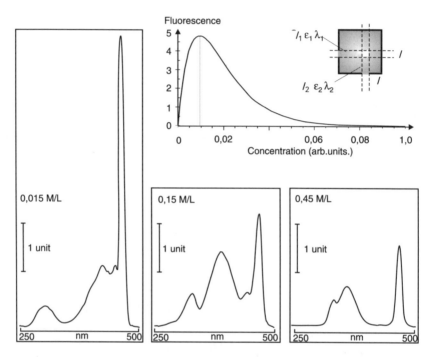

Figure 12.5—*Relationship between fluorescence intensity and concentration.* As concentration increases, the emission of fluorescence is no longer proportional to concentration. A maximum fluorescence intensity is observed, beyond which fluorescence decreases or tends to *roll over*. The example given is for biacetyl in CCl$_4$.

12.4   Rayleigh and Raman bands

When the excitation and emission wavelengths are close together, fluorescence of the sample must not be confused with two concurrent artefacts:

12.4.1   Rayleigh scattering

Rayleigh scattering is the re-emission by solvent molecules of a small fraction of the excitation radiation. This emission occurs at the same wavelength and is independent of the angle of observation (Fig. 12.6). The Rayleigh scattering intensity varies with the polarisability of solvent molecules.

12.4.2 Raman scattering

Raman scattering, which is 100 to 1 000 times weaker than Rayleigh scattering, is produced by the transfer of part of the excitation radiation to solvent molecules in the form of vibrational energy. This emission band is shifted towards the red. For each solvent, the difference in energy between the absorbed photon and the re-emitted photons is constant. It is thus possible, by modifying the excitation wavelength, to displace the shift in nanometres between the positions of Rayleigh and Raman emission. For water, this difference is $3\,380$ cm^{-1} (see Table 12.1 and Fig. 12.6).

Table 12.1—Position of the Raman scattering peak, calculated for four commonly used solvents and five excitation wavelengths from a mercury lamp.

Excitation (nm)	254	313	365	405	436	shift (cm^{-1})
water	278	350	416	469	511	3 380
ethanol	274	344	405	459	500	2 920
cyclohexane	274	344	408	458	499	2 880
chloroform	275	346	410	461	502	3 020

■ Raman scattering in water is used as a sensitivity test for fluorimeters. The test consists of measuring the signal to noise ratio of the Raman peak using a cell filled with water. For example, signal/noise will be measured at 397 nm (25 191 cm^{-1}) if the excitation energy used is 350 nm (28 571 cm^{-1}).

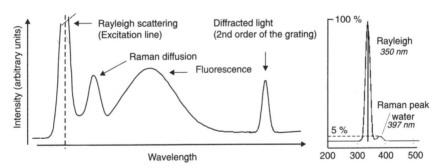

Figure 12.6—*The diverse components of a fluorescence spectrum.* The position of the Raman scattering band depends on the wavelength of excitation and the nature of the solvent.

12.5 Instrumentation

In fluorescence, the sample under analysis behaves like a light source, emitting in all directions. The emitted light is usually monitored perpendicular to the primary excitation source. For strongly absorbing solutions, measurements can be made on-axis with the incident radiation. For opaque or semi-opaque samples, a frontal measurement at various angles is recommended (Fig. 12.7). Commercial instruments use a xenon arc lamp with a power of 150 to 800 watts as an excitation source (see Fig. 12.8) and measurement of the light intensity is carried out using a photomultiplier tube or a photodiode.

Two categories of fluorescence instruments are available from manufacturers:

— fluorescence ratio fluorimeters
— spectrofluorimeters

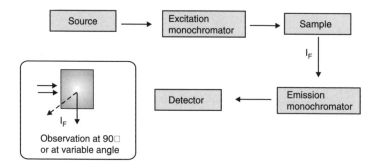

Figure 12.7—*The geometry of a fluorimeter and layout of the different instrument components.* Fluorescence is typically measured under steady-state conditions maintaining excitation, in contrast to studies on fluorescence dynamics.

Figure 12.8—*Xenon arc lamp used in fluorimetry.* The pressure of xenon in the lamp is in the order of 1 MPa. The lamp, made from a quartz envelope and without a filament, produces polychromatic (white) light whose power is wavelength dependent. (Reproduced by permission of Oriel.)

12.5.1   Fluorescence ratio fluorimeters (quantitation)

Light emitted from the source converges on the excitation monochromator, which allows a narrow band of wavelengths to be selected (15 nm) to induce fluorescence of the sample solution in the measurement cell. The emitted light, observed perpendicular to the direction of the incident beam, passes through the emission monochromator, allowing the selection of a narrow band of wavelengths for measurement (Fig. 12.9). The simplest instruments have a double compartment for measurement. This allows the sample solution and a standard fluorescent reference solution to be put into the optical path.

Standards used for this ratio method are generally quinine sulphate, rhodamine B or 2-aminopyridine solutions. In order to eliminate fluctuations due to the radiation source and other instrument parameters, comparative measurements are used. Instrument sensitivity is commonly expressed in terms of the signal to noise ratio of the Raman band of water (cf. 12.4). The nature of the solvent, the temperature,

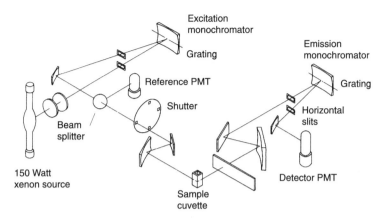

Figure 12.9—*Optical scheme of a spectrofluorimeter having two detectors, one of which is used to control the intensity of the light source.* A fraction of the incident beam is reflected by the beam splitter and monitored by a photodiode to control the intensity of the incident beam. Comparison of the signals obtained from both detectors allows the elimination of any drift in the light source. This procedure, for single beam instruments, gives approximately the same stability as with double beam instruments. (Model F4500 reproduced by permission of Shimadzu.)

the pH and the concentration are all parameters that affect the intensity of fluorescence.

12.5.2   Spectrofluorimeters

In contrast to the preceding set-up, spectrofluorimeters record an entire fluorescence spectrum. Each of the two motorised monochromators can scan a spectral band. It is possible to record the emission spectrum while maintaining a constant excitation wavelength or to record the excitation spectrum while maintaining a constant emission wavelength. Spectra often show small differences when they are obtained using different instruments.

■ Spectrofluorimeters have software that allows the best pair of excitation/ emission wavelengths to be chosen (Fig. 12.10). The process can be done manually using the following 'single factor method':

1. a UV spectrum of the compound is recorded using a spectrophotometer;
2. the spectrofluorimeter excitation is set at the maximum of the UV absorption spectrum;
3. the fluorescence spectrum is recorded;
4. the emission monochromator is set at the wavelength of maximum fluorescence and the excitation wavelength is varied. The excitation spectrum is obtained, which allows the best choice for the excitation wavelength (this can be different from the UV absorption maximum).

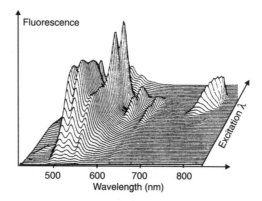

Figure 12.10—*Emission matrix—excitation of a mixture containing two fluorescent ions.* This topography of total fluorescence as a function of excitation and emission wavelengths constitutes a signature for each compound studied. This type of recording allows the optimum conditions to be determined.

12.6 Applications

Apart from 5 to 8% of all existing molecules that possess natural fluorescence, many others can be made to fluoresce by chemical modification or by association with a fluorescence molecule. By chemical reaction, a *fluorophore reagent* can be incorporated into the analyte (7-hydroxycoumarines can be used to this effect). This is called *fluorescence derivatisation*, a process analogous to that used in colorimetry. In HPLC, amines can be fluorescently labelled, permitting very low limits of detection to be achieved – in the order of attomoles (10^{-18} mol).

HPLC analysis of polycyclic aromatic hydrocarbons (PAH) in drinking water is one of the current and classical applications of fluorescence. In this case, the detector contains a fluorescence flow cell placed after the chromatographic column. This mode of detection is specifically adapted to obtain threshold measurements imposed by legislation. The same process allows the measurement of aflatoxins (Fig. 12.11) and many other organic compounds (such as adrenaline, quinine, steroids and vitamins).

For the determination of metal cations, chelation complexes formed with oxine (8-hydroxyquinolein), alizarine or benzoine are used. These complexes are extractable with organic solvents.

■ Although fluorescence lifetimes are very short, they can be measured using certain instruments. Many methods exist for making such measurements, such as the decrease of light intensity after excitation, or comparison of the fluorescence modulation as a function of rapid modulation of the excitation source. The first method requires the use of a pulsed laser and the second requires, a high frequency modulation of the light source. Analysis of the signal as a function of the excitation source frequency (modulation and phase) allows the fluorescence lifetime to be calculated.

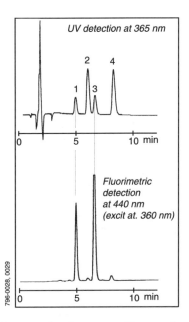

Conditions:
 RP-18 Column, 25 cm, 5 µm particles
 Mobile phase: water, acetonitrile, methanol
 (60:20:20)
 Injection of a mixture of 4 aflatoxins:

   1 - Aflat. G_2 (0.3 mg/L)
   2 - Aflat. G_1 (1 mg/L)
   3 - Aflat. B_2 (0.3 mg/L)
   4 - Aflat. B_1 (1 mg/L)

Chemical structures of the aflatoxins B & G

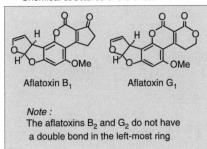

Aflatoxin B_1 Aflatoxin G_1

Note:
The aflatoxins B_2 and G_2 do not have
a double bond in the left-most ring

Figure 12.11—*Comparison of UV and fluorescence detection after chromatographic separation.* Aflatoxins, which are carcinogenic contaminants present in certain batches of grain cereals, are controlled by HPLC analysis. It can be seen that the peak intensities in UV detection vary with concentration whereas fluorescence detection is much more sensitive to aflatoxin G_2 and B_2. (Reproduced by permission of SUPELCO.)

12.7 Chemiluminescence

Some chemical reactions give rise to light emission. A well-known example is that of *luminol,* used for non-electric emergency lighting or the manufacture of luminous colour wands. The reaction that causes emission is an oxidation that takes place in the presence of a catalyst. Its quantum yield is close to 1.

Even though luminol can be used to analyse Fe(II) or hydrogen peroxide, there is no doubt that light emitted by oxidation reactions with ozone leads to the most numerous applications of chemiluminescence. This reaction, which offers high sensitivity, is used to measure nitrogen monoxide present in polluted atmospheres. Instruments used to make this measurement include an ozone generator (Fig. 12.12). The range of applications of this method can be extended to nitrogen dioxide, which is known to be transformed quantitatively into nitrogen monoxide, or to the determination of the total quantity of nitrogen present in a compound, which can be converted by combustion under certain conditions (Fig. 12.13).

The same principle can be applied to measuring sulphur that has been converted to hydrogen sulphide (H_2S) then oxidised by ozone.

However, of all the methods used to measure ozone, the chemiluminescence reaction with ethylene is the one most often used.

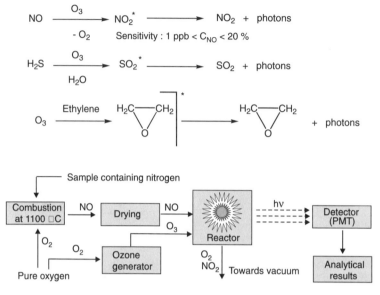

Figure 12.12—*Chemiluminescence reactions using ozone*. Schematic showing the principle of a nitrogen analyser based on the luminescence of nitrogen monoxide.

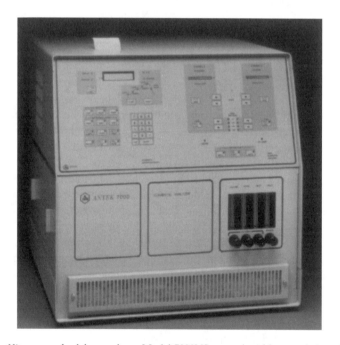

Figure 12.13—*Nitrogen and sulphur analyser*. Model 7000NS, reproduced by permission of Antek. This instrument includes two techniques of luminescence: nitrogen is measured by chemiluminescence (see Fig. 12.12) and sulphur by fluorescence of SO_2 (UV excitation).

Problems

12.1   Calculate the position of the radiation of Raman diffusion for a monochromator of excitation regulated to 400 nm; the solvent used is cyclohexane. (Recall that for cyclohexane, the Raman diffusion corresponds to a radiation displaced by $2\,880\,cm^{-1}$ with respect to the wavelength of excitation.)

12.2   A test for the sensitivity of a fluorimeter is to measure the intensity of fluorescence of the Raman diffusion peak of a cell filled with water and with an optical pathlength of 1 cm. If the wavelength of excitation is regulated to 250 nm, at which wavelength must the measurement be made (the Raman displacement of water is $3\,380\,cm^{-1}$)?

12.3   An unknown aqueous solution of 9-aminoacridine in water leads to an intensity of fluorescence which, measured at 456 nm, is of 60% with respect to an external reference. A standard solution of this compound, the concentration of which is 0.1 ppm in the same solvent, leads to a fluorescence of 40%, under the same conditions. Water alone, equally under the conditions of the experiment, presents a negligible fluorescence.
   1. Calculate the ppb concentration of 9-aminoacridine in the unknown sample.
   2. For quantitative analysis, how can one avoid confusing the Raman diffusion and the fluorescence of the compound?

12.4   3,4-benzopyrene is a dangerous aromatic hydrocarbon frequently present in polluted air measurable, in a solution of dilute sulphuric acid, by fluorimetry. The excitation wavelength is 520 nm and the measurement taken at 548 nm.
   10 l of contaminated air is bubbled in 10 ml of sulphuric acid. The fluorescence of 1 ml of this solution measured at 548 nm is 33.33 (arbitrary units). Two standard solutions, one containing 0.75 mg and the other 1.25 mg of 3,4-benzopyrene per ml of the same diluted sulphuric acid, led for extractions of 1 ml to values of 24.5 and 38.6 (same unit scale). A sample not containing the 3,4-benzopyrene was recorded as 3.5 (same unit scale) by the fluorimeter under the same conditions.
   1. Explain why benzopyrene is fluorescent.
   2. Calculate the mass of 3,4-benzopyrene per litre of air.
   3. Express the result of this measurement in ppm.

12.5   Ferrous iron catalyses the oxidation of luminol (see below) by hydrogen peroxide. The intensity of the chemiluminescence which follows increases linearly with the concentration of Fe(II) as it rises from $10^{-10}\,M$ to $10^{-8}\,M$. To measure a solution of unknown content in Fe(II), 2 ml of the solution was extracted and then introduced to 1 ml of water, followed by the addition of 2 ml of hydrogen peroxide and finally 1 ml of an alkali solution of luminol. The luminescence signal is emitted and integrated over a period of 10 seconds giving a value of 16.1 (arbitrary units). In a second attempt, 2 ml of the unknown Fe(II) solution is again extracted and on this occasion it was added to 1 ml of a solution. 5.15×10^{-5} M in Fe(II) and the same quantities of hydrogen peroxide and luminol were introduced as before. The signal, integrated over 10 seconds, was measured as 29.6.

Calculate the molar concentration of Fe(II) in the original solution.

luminol

12.6 For the determination of the zinc concentration in an unknown solution A, by the method of standard addition, a standard solution B is prepared whose pure zinc chloride concentration is 0.1 mmol/l ($M = 136.2$ g/mol, Zn = 65.38). 5.0 ml aliquots of the unknown solution A were introduced into four separatory funnels. 4.0 ml, 8.0 ml and 12.0 ml portions of the standard solution B were added to three funnels. After stirring and extracting (three times), with 5.0 ml of tetrachloromethane containing an excess of 8-hydroxyquinoline, the extracts were diluted to 25 ml. The following fluorescence readings were obtained:

ml of the solution B added	Fluorescence reading
0	7.65
4	13.95
8	19.6
12	25.8

1. Explain why the extraction with CCl_4 is performed in the presence of hydroxyquinoline.
2. From the data in the table and by the method of least squares derive an equation for the calibration curve.
3. Calculate the ppm concentration of zinc in the unknown solution A.

12.7 Fluorimetry is the method selected to measure the concentration of quinine in a commercially available drink. The excitation wavelength is 350 nm and the wavelength for taking the measurements is 450 nm. From solution A, of quinine, at 0.1 mg/l, a series of standards is prepared to establish a calibration curve.

Solution	Volume of A	Fluorescence (arb. units)	H_2SO_4 0.05 M
Standard 1	20	182	0
Standard 2	16	138.8	4
Standard 3	12	109.2	8
Standard 4	8	75.8	12
Standard 5	4	39.5	16
Blank	0	0	20

An extraction of 0.1 ml is taken from the original solution which is then diluted to 100 ml with H_2SO_4, 0.05 M. The value for the signal of this solution is 113.

1. Using least squares, equation from these data calculate the calibration.
2. What is the concentration weight/volume (w/v) of quinine (g/l), and in ppm?
3. Why in quantitative analysis is it preferred to take measurements using a fluorimeter rather than with a spectrofluorometer?

CHAPTER 13

X-ray fluorescence spectrometry

X-ray fluorescence is a spectroscopic technique of analysis, based on the fluorescence of atoms in the X-ray domain, to provide qualitative or quantitative information on the elemental composition of a sample. Excitation of the atoms is achieved by an X-ray beam or by bombardment with particles such as electrons. The universality of this phenomenon, the speed with which the measurements can be obtained and the potential to examine most materials without preparation all contribute to the success of this analytical method, which does not destroy the sample. However, the calibration procedure for X-ray fluorescence is a delicate operation.

This technique encompasses a large scope of instruments ranging from mobile spectrometers to high resolution spectrometers. On-line analysers and probes that can be fitted to scanning electron microscopes (SEMs) allow instant analyses to be performed (i.e. microanalysis by X-ray emission).

13.1 General principle

When a sample is irradiated by a source that emits photons of very high energy (from 5 to 60 keV), X-ray photoluminescence is observed and it is characteristic of the elements present in the sample.

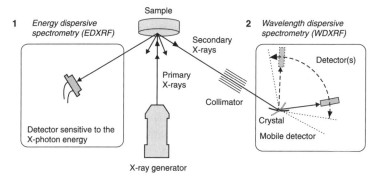

Figure 13.1—*X-ray fluorescence*. The sample, when excited by a primary X-ray source, emits fluorescence that can be detected according to two modes: 1) simultaneous detection by a cooled diode that detects the energy of a single photon; 2) sequential detection of the emitted wavelengths (using a 'θ, 2θ' goniometer assembly).

The fluorescence spectrum is made up of radiation with frequencies and intensities characteristic of the composition of the sample, allowing its qualitative or quantitative analysis.

■ X-ray fluorescence can be induced by other means of excitation besides photons, such as by bombarding the sample with fast electrons, protons or α-radiation. Although in the latter case it cannot properly be called fluorescence because excitation is not induced by photons, the X-ray emission produced yields an identical spectrum.

In practice, this non-destructive mode of analysis can be used for all elements starting from boron ($Z = 5$) in solid or liquid homogeneous samples without, in theory, any particular preparation. The emission spectrum depends very slightly on the chemical combination or chemical state of the elements in the sample.

Although qualitative analysis does not present any major difficulty because the spectra are easy to interpret, quantitative analysis is much more difficult: certain precautions must be taken during sample preparation. Ultimately, the results obtained can be as precise as those obtained by atomic absorption or emission if the influence of the matrix can be correctly discerned, especially when the sample is not homogeneous.

13.2   X-ray fluorescence spectrum

X-ray fluorescence of an element is a two-step process. For an isolated atom, the first step is excitation during which the interaction of the photon, given sufficient energy, leads to the stripping off of a lower shell electron such as a K electron. This *photoelectric effect* leads to internal ionisation and to the emission of a *photoelectron* (see Fig. 13.2).

■ Each photoelectron has an energy determined by the difference between the energy of the incident X-ray photon and that of the electronic level initially occupied by the ejected electron. The photoelectron energy spectrum is the basis of a method called ESCA (Electron spectroscopy for the chemical analysis).

The second step corresponds to the stabilisation of the ionised atom through reemission of part or all of its energy acquired during excitation. The energy gap caused by the preceding step is immediately followed by electron reorganisation that takes the atom back to its ground state (in 10^{-16} s). If E_1 is the energy level of the initial electron and E_2 is that of the electron that takes its place, a radiative transition of frequency ν with a probability of between 0 and 1 will occur:

$$h\nu = |E_2 - E_1| \tag{13.1}$$

Every atom, from $Z = 3$, leads to specific radiation that follows specific selection rules. For all the elements, fluorescence appears in the energy range of 40 eV to more than 100 keV (31 to 0.012 nm).

■ The interpretation given above is simplified, since fluorescence is not the only process that allows the atom to lose its excess energy. Other phenomena such as Rayleigh scattering (elastic scattering) and the Compton effect (inelastic scattering with release of Compton electrons) can complicate the X-ray emission spectrum.

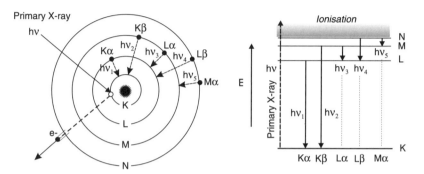

Figure 13.2—*Simplified schematic of an atom showing the origin, and the Siegbahn nomenclature, of some fluorescence radiation processes caused by impact of a photon having a high energy.* The position of the spectral line is not significantly influenced by the chemical combination in which the atom is found. For example, the $K_{\alpha 1}$ line from sulphur is observed at 0.5348 nm for S^{6+} and at 0.5350 nm for S^{0}, yielding a shift of 1 eV, which is comparable to the natural line width for X-rays.

Transitions are still designated according to Siegbahn nomenclature. Hence, for iron, the symbol $FeK\beta_2$ specifies the location of the gap (K shell), the distance that separates the two energy levels (initial and final states of the electron; α for 1, β for 2) and the relative intensity of the transition within the series (1 is more intense than 2). $K\beta$ transitions are approximately six times less intense than the corresponding $K\alpha$ transitions. Cascade electronic rearrangements are often observed for heavy elements (light elements cannot have L or M transitions). For example, carbon only yields a $K\alpha$ line at 4.47 nm (227 eV). H or He elements do not have X-ray fluorescence.

With the exception of level K, levels L, M, ... are multiples and thus Siegbahn's notation is not always sufficiently explicit. Therefore, transitions are more rigorously specified by quantum notation.

Because of the relation that exists between the energy of a photon and its wavelength, $E = hc/\lambda$, and of the several modes for recording spectra, radiation can be characterised either by its wavelength λ (nm or Å) or by its energy E (eV or keV). The two most commonly used equations of conversion are:

$$\lambda_{nm} = \frac{1240}{E_{(eV)}} \quad \text{or} \quad \lambda_{Å} = \frac{12.4}{E_{(keV)}} \tag{13.2}$$

13.3 Excitation modes of elements in X-ray fluorescence

It is necessary to use X-ray to induce X-ray fluorescence of elements in a sample source. These sources include X-ray tubes which have filters at the exit or which contain particular radioisotopes. If X-ray fluorescence is considered in its larger sense of X-ray emission, other excitation processes that employ particles can be used.

13.3.1 X-ray generators

A beam of electrons accelerated through a 50 kV potential strikes an anode (still called the *anticathode* in memory of Crooke's work) which is made of a metal with an atomic number between 25 and 75 (Fig. 13.3).

The electrons, which are slowed down as they hit the target, will emit a continuum spectrum with a limit towards the lower wavelength λ_0 that can be determined using equation (13.2). If the energy of the electrons is sufficient, the lines will appear as though they are superimposed on the preceding spectrum. These lines are characteristic of the metal that constitutes the anode. The higher the atomic number of the element, the more lines that appear, some of which will be highly energetic. Anodes, whose choice will depend on the application, are often made of rhodium (Kα line at $\lambda = 0.061$ nm or 20.3 keV) or tungsten (Kα line at $\lambda = 0.021$ nm or 59 keV). To reduce background signals, filters are used to bracket the primary X-ray radiation. Alternatively, a second target can be used, whose fluorescence will reflect that of the primary source and for which radiation due to deceleration will be absent.

■ In the past, spectrometers using direct excitation were designed using the sample as the anticathode. The power required to induce fluorescence through a mechanism involving electrons is lower than that required by photon impact. This technique, which has several constraints, can only be used for conducting samples.

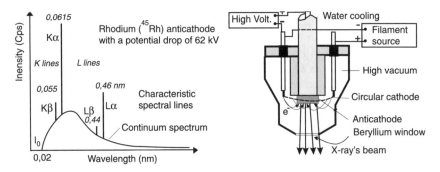

Figure 13.3—*X-ray emission spectrum produced by an anticathode (anode), measured with a spectrometer, and schematic of an X-ray tube.* The line spectrum can be observed as a superposition on the continuum spectrum. It is the continuum portion of this radiation that is solicited for applications that necessitate a high X-ray penetration power, such as for radiology. For analysis, the line spectrum is preferred. Water cooling is compulsory if the X-ray tube operates at high power (1–4 kW).

13.3.2 Radioisotope sources

Some radionuclides are transformed by internal electron capture (IEC), with the phenomenon summarised as follows:

$$\substack{A\\Z}X \xrightarrow{\text{IEC}} \substack{A\\Z-1}Y* \xrightarrow{h\nu} \substack{A\\Z-1}Y$$

It appears as if one of the level K electrons of the atom disappears into the nucleus. The void created induces X-ray fluorescence from the nucleus Y. There are several known radionuclides of this type, which have sufficiently long periods that they can be used as different energy sources (see Table 13.1). These are typically used in portable instruments. The activity of these isotopic sources is generally in the order of a few mCi and they can yield a flux of 10^6 to 10^8 photons/s/steradian. Because these sources require a permit for use as well as permanent protection because they emit continuously, their use is rapidly diminishing.

Table 13.1—X-ray sources using radio-isotopes

Transition	Period (years)	Emitted X-ray	λ (nm)	E (keV)
^{55}Fe $\rightarrow$ ^{55}Mn	2.7	MnKα	0.21	5.9
^{57}Co $\rightarrow$ ^{57}Fe	0.7	KeKα	0.19	6.4
^{109}Cd $\rightarrow$ ^{109}Ag	1.3	AgKα	0.056	22
^{242}Cm $\rightarrow$ ^{242}Pu	17.8	PuLα/Lβ	0.09–0.07	14–18

For X-ray generation, other radionucleides such as ^{241}Am ($\tau = 430$ years) can be used. This nucleus emits α radiation accompanied by δ photons of 60 keV energy. Finally, it is possible to mix a β-emitter and a second element, which is used as a target and plays the role of the anticathode in an X-ray tube. For example, the source ^{147}Pm/Al ($\tau = 2.6$ years) emits a decelerating radiation between 10 and 200 keV.

13.3.3 Other excitation techniques

Excitation of atoms by fast electrons can induce X-ray fluorescence. This is why analysis can be performed using scanning electron microscopes (SEM). When an object is 'illuminated' by highly energetic electrons, which are necessary to produce the SEM image, X-ray emission characteristic of the composition of that zone being irradiated is observed (Fig. 13.4). Many electron microscopes are equipped with an accessory that can collect a fraction of the X-ray fluorescence emitted by a sample. In this case, the sample plays the role of the anticathode in an X-ray tube. The energy of the electron beam is between 20 and 30 keV, for which there is a compromise that allows K or L lines to be emitted, leading to characterisation of most elements. This quasi-instantaneous analysis, called *X-ray microanalysis*, involves a sample volume of approximately $1 \, \mu m^3$ (see Fig. 13.9).

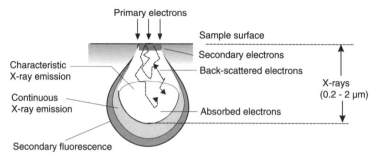

Figure 13.4—*Plume of interaction between an electron beam and a material.* Several phenomena occur within the material. A complex emission results in which it is possible to determine the origin of each of the emitted beams collected. If the energy is high, the X-rays will be formed at a greater depth within the material and it is thus more difficult to detect them.

13.4 X-ray absorption

In X-ray fluorescence, self-absorption due to optical *quenching* of the sample is always observed. For solid samples, radiation can be observed only at a depth of a few micrometres. The intensity P of radiation, after travelling a distance dx in a material which a *lineic absorption coefficient* μ (cm^{-1}), will decrease by dP for a penetration angle of 90°. The integrated form of the expression $dP = -\mu P\,dx$ is comparable to that described in colorimetry:

$$P = P_0 \exp(-\mu x) \tag{13.3}$$

For a given material of known composition, μ can be calculated using a table that gives the *mass absorption coefficients* of the elements that constitute that material. After calculating the weighted mass coefficient μ_M (cm^2/g) of the material, it is possible to obtain $\mu = \mu_M \cdot \rho$ where ρ is expressed in g/cm^3.

The linear and mass absorption coefficients decrease exponentially when the wavelength decreases. Using equation (13.3), it can be calculated that household aluminium foil with a thickness of 12 μm ($\rho = 2.7$) absorbs 57% of the TiKα line ($\mu_M = 264$) but only 1% of the AgKα line ($\mu_M = 2.54$).

13.5 Sample preparation

Liquid samples do not require preparation before analysis. However, it is preferable to transform solid samples, especially if the matrix is not well known. In fact, depending on the incident angle of the primary X-ray radiation, the sample thickness can be variable: from only a few angströms (for a grazing angle of radiation) to half a millimetre (see Fig. 13.4). Thus any superficial heterogeneity will translate as variations in the results.

■ Absorption by the matrix can cause either an underestimation of the results due to optical quenching, or an overestimation of the results when some of the X-ray fluorescence causes secondary excitation of other elements. For example, the presence of iron in aluminium causes an increase in the fluorescence of the latter because fluorescence emanating from iron excites aluminium.

The two techniques used for sample preparation are *mineralisation* and *pellet formation*. Mineralisation consists of mixing part of the sample in powder form with lithium borax ($Li_2B_4O_7$) and other additives. After fusion of these components, the glass bead obtained (called a *pearl*) consists mostly of light elements and is thus transparent. Pellet formation, using an hydraulic press, is an alternative to fusion. To ensure the cohesiveness of the pellet, a wax is added (made from an organic polymer formed from light elements). For solutions, a non-absorbing X-ray cell with a base of polypropylene or mylar (polyester) film is used.

13.6   Different types of instruments

13.6.1   Dispersive instruments – simultaneous system

Energy dispersive instruments are used for qualitative analysis and routine quantitation (Fig. 13.5) and represent the first category of instruments. They are generally equipped with a low power X-ray tube instead of a radioactive source in order to eliminate constraints caused by legislation.

Collection of a spectrum relies on the ability of the detector to recognise the energy of each of the emitted photons. This particular detector, installed close to the sample to collect part of the X-ray fluorescence, is either a proportional gas counter (neon)

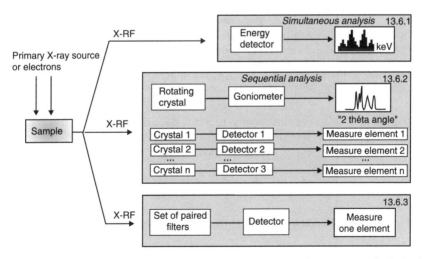

Figure 13.5—*Several approaches that allow recording of spectra or X-ray fluorescence results.* In the figure, references are indicated by their sections in the text.

or a Si-Li semiconductor (silicon diode doped with lithium). The latter is a crystal maintained at low temperature (cooled by liquid nitrogen or a Peltier device) and its surface is protected by a beryllium film (transparent for $Z < 11$) (Fig. 13.6). Each emitted X-ray photon is converted into an electrical charge proportional to its energy (1 electron for about 4 eV) and the signal is analysed by a multichannel detector.

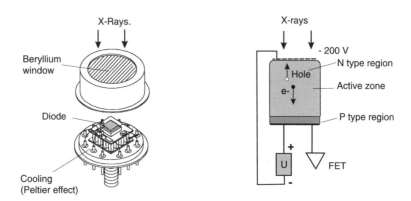

Figure 13.6—*Schematic of a cooled Si–Li detector*. The high quantum yield of this detector allows the use of primary X-ray sources of low power (a few watts or a radioisotope source).

For example, an analyser can have 2 048 channels that collect, during the period of measurement (a few minutes), the pulses caused by photons with energies in the range of 0 to 20 keV. The corresponding spectrum is a histogram (Fig. 13.7).

■ To classify the photons with better precision, their quantity is limited (to approx. 3 000/s) by reducing, if necessary, the source intensity. The instrument data system must allow unambiguous identification of the energy of the photons, which arrive as a pack. The resolution of an instrument is measured by the width at half-height of the Kα line of manganese, emitted by a [55]Fe radioactive source. For dispersive instruments, the resolution is in the order of 100 eV while the natural bandwidth varies between 2 and 100 eV.

It is possible to conduct simultaneous analyses over the whole spectral range. The quality of the data is related to the time t used to acquire it (sensitivity increases as $\sqrt{t}$). Correct positioning of the lines on the energy axis is often carried out, using the Kα line of cobalt as a reference.

13.6.2   Angular dispersion instruments – sequential and multichannel systems

In this second class of spectrometers, X-ray radiation emitted by the sample, after it has been filtered by a sheet collimator (Soller slits), impacts on a crystal analyser

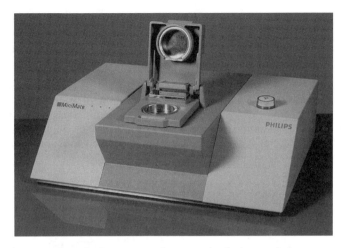

Figure 13.7—*An energy dispersive instrument and an example of a low resolution spectrum.* Spectrometer with a low-power X-ray tube, representative of a number of instruments of that type (model Minimate, reproduced by permission of Philips). This model is controlled by a microcomputer (not shown).

which is determined by the portion of the spectrum studied (Fig. 13.8). For low-energy lines, it is preferable to operate under a vacuum. The crystal behaves like a stack of mirrors that reflect light, in that it satisfies Bragg's law:

$$\lambda = 2d \sin \theta \tag{13.4}$$

Crystal	Plane (Miller index)	d(Å)
Topaze	303	1.356
Lithium fluoride	200	2.01
Silicon	111	3.14
Graphite	001	6.69
Oxalic acid	001	5.85
Mica	002	9.96
Lead stearate		51

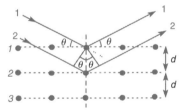

Figure 13.8—*Reflecting crystals used in goniometers of dispersive spectrometers and Bragg's relationship.* The higher the wavelength to be analysed, the higher the inter-reticular distance must be in the crystal.

The highest performance X-ray fluorescence spectrometers contain goniometers. Their resolution, expressed in eV, can reach a few tens of electron volts. These instruments can be classified as:

— *sequential analysers* (Fig. 13.9), where the detector and the crystal undergo simultaneous and synchronised rotations to ensure the detection of 2θ and θ (to $1/1\,000$th degree of an angle).

— *fixed channel* instruments, which have many detectors positioned around the same crystal. Several arrangements of the installed crystal detector allow simultaneous measurements of lines that are specific to the elements being studied. These types of instruments are adapted to scanning microscopes. Although they are of small dimension, these spectrometers have a good resolution (between 3 and 40 eV) and can be used for the analysis of elements from $Z = 7$ and higher.

■ **Compton effect.** A small fraction of the primary X-ray excitation beam is scattered in the form of radiation whose wavelength depends on the angle of observation. This radiation is superimposed on the X-ray fluorescence spectrum. The shift in angströms between the two wavelengths (excitation and Compton) is given by:

$$\Delta\lambda = 0.024 \, (1 - \cos\alpha) \tag{13.5}$$

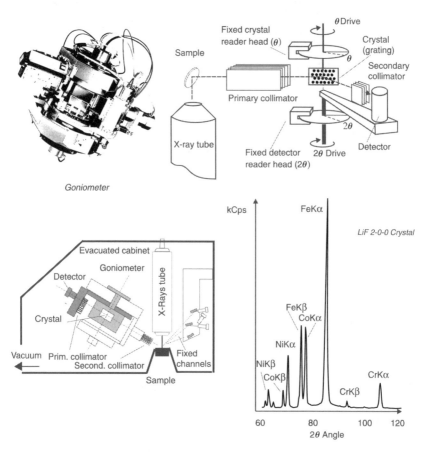

Figure 13.9—*Schematic of a sequential, crystal-based spectrometer and the spectrum obtained using the sequential method with an instrument having a goniometer.* The Soller slit collimator, made of metallic parallel sheets, collimates the primary X-ray beam emitted by a high power source (SRS 300 instrument, reproduced by permission of Siemens). A typical spectrum of an alloy, obtained by an instrument of this category, having an LiF crystal (200); with 2θ angle in degrees as the abscissa and intensity in Cps as the ordinate). Model Philips PW2400 Spectrum, reproduced with permission of VALDI-France.

13.6.3   Filter instruments – single channel systems

Instruments used for the continuous monitoring of industrial production constitute the third category of X-ray fluorescence instruments. In most of these cases, the object is to measure a single element present in a fabrication process.

One method of analysis of an element is based on the assumption that it can be measured using a single line, provided that line can be isolated from the rest of the spectrum. This is the basis of the method that uses equilibrated filters. The technique, used to relate the concentration of interest to the difference between two measurements, uses filters placed between the sample (or reference) and a counter of the Geiger–Müller type:

— one of the filters, the *transmission* filter, isolates radiation that is characteristic of the element.

— the other filter, the *absorption* filter, is opaque to this radiation.

Fluorescence originating from the filters themselves is a limiting factor in this method, which is used exclusively for routine analyses.

■ A filter set is formed by two sheets of pure metal of thicknesses such that the absorption curves differ only in the range of the peak to be measured. Each element possesses a pronounced absorption maximum, the *K corner*, that is close to the Kβ transition of that element (Fig. 13.10). For example, copper can be quantified by its Kα line using filters made out of Ni and Co.

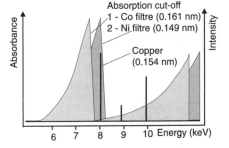

Element	K-Corner (A & keV)	Kα line
Cobalt	0.161 (7.7 keV)	0.179
Nickel	0.149 (8.3 keV)	0.166
Copper		0.154 (8.1 keV)

Figure 13.10—*Absorption cut-off of a few elements and their corresponding K lines.*

13.7   Quantitative analysis by X-ray fluorescence

The relationship between the weight concentration of an element and the intensity of one of its characteristic lines is complex. Several models have been developed to correlate fluorescence to weak, atomic concentrations. Many corrections have to be made due to inter-element interactions, preferential excitation, self-absorption, and fluorescence yield (heavy elements relax more quickly by internal conversion without emission of photons). All of these factors require the reference sample to be practically the same structure and atomic composition for all elements present as the sample under investigation. It is mostly because of these limitations that there are difficulties in obtaining quantitative results by X-ray fluorescence.

The intensity I_A of a given peak A (measured in number of counts) for a sample of narrow width, is related to the weight concentration C_A by Cliff–Lorimer's constant k_A as follows:

$$C_A = k_A \cdot I_A \qquad (13.6)$$

For solid samples, the surface must be perfectly clean, if not polished, because the analysis is conducted only a few μ_m below the surface.

13.8 X-ray fluorescence applications

One of the great strengths of X-ray fluorescence analysis comes from the fact that it is a non-destructive method that can be used on samples without the need for prior treatment.

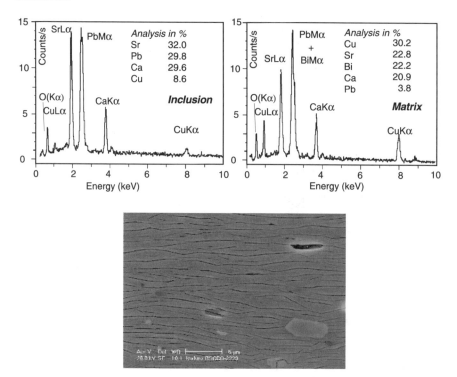

Figure 13.11—*Scanning electron microscopy (SEM) accompanied by X-ray fluorescence analysis.* Secondary electron image of a cross-section of a supraconducting polycrystalline ceramic with oriented grains of oxide $BiPb_2Sr_2Ca_2Cu_3O_{10}$ (Philips instrument, model XL30FEG). Energy emission spectra corresponding to the matrix and to a $5\,\mu m$-long inclusion (bottom). It should be noted that it is possible with this technique to obtain the composition at a precise point on the sample (Link-Oxford analyser) (study by V. Rouessac, reproduced by permission of CRISMAT, University of Caen).

■ This method is well adapted to qualitative analysis where automatic identification of the lines can be made by very sophisticated visual displays. For semi-quantitative analysis, software can be used to estimate a composition that is very close to that of the sample. However, for quantitative analysis, serious problems can be encountered and reference standards have to be analysed in matrices that are almost identical in composition to that of the sample.

X-ray fluorescence can be used to analyse all types of samples. Its applications are numerous, whether in research and development or in quality control of production. Initially, X-ray fluorescence was used in industries that treat metals of primary fusion or alloys and, more generally, in the mineral industry (for use one ceramics, cements, steel, glass, etc.). Because of the ease of use of common X-ray fluorescence instruments, its scope of application has expanded into other areas: the photographic industry and semi-conductors (for impurity control in silicon chips), the petroleum industry, geology, paper mills, gas analyses (such as nitrogen), toxicology and environmental applications (dust, fumes from combustion, heavy metals, and dangerous materials in waste such as Pb, As, Cr, Cd, etc.).

Finally, X-ray microanalysis allows the mapping of all the elements present in a heterogeneous sample, as observed with a scanning electron microscope, provided the sample is capable of conducting (Fig. 13.11).

Problems

13.1 A solution is prepared with 8 g of potassium iodide and 92 g of water.
1. Calculate the mass absorption coefficient (μ_m) of this solution measured with the Mo $K\alpha$ radiation of 17.4 keV. Consider that the solution (volume $\rho = 1.05$ kg/l) is an equilibrated mixture of these four elements (K, I, H and O).
2. What fraction of the Mo $k\alpha$ line remains after having crossed 1 cm of this solution?

Element	μ_M (cm²/g)	Atomic mass (g/mol)
K	16.2	39
I	36.3	127
O	1.2	16
H	0.4	1

13.2 Find the following classical conversion formulae:

$$\lambda_{(nm)} = \frac{1240}{E_{(eV)}} \qquad \lambda_{\overset{\circ}{A}} = \frac{12.4}{E_{(keV)}}$$

Recall: $h = 6.6260 \times 10^{-34}$ J s; $c = 2.998 \times 10^8$ m/s; 1 Å $= 10^{-10}$ m; 1 eV $= 1.602 \times 10^{-19}$ J.

13.3 An X-ray tube with a tungsten target and which serves as the source in an X-ray wavelength dispersive fluorescence spectrometer is equipped with a crystal of ethylenediamine tartrate. The plane of reflection in use corresponds to an inter-reticular distance of $d = 4.404$ Å.

1. Calculate the angle of deviation measured with respect to the direction of the incident radiation which collects the emission line $L\beta$ of bromine ($\lambda = 8.126$ Å) of a sample of sodium bromide (it may be considered that the observations recorded are of first order reflection).

2. Knowing that the wavelength of the tungsten line $K\alpha$ occurs at 0.209 Å, calculate the minimum voltage necessary across the X-ray tube, to excite this line.

13.4 Consider a wavelength dispersive instrument provided with a goniometer comprising a rotating topaz crystal of an appropriate size.

1. Knowing that the crystal spacing of the monochromator (303) has an interplanar distance of 0.1356 nm, calculate the limits of the wavelengths within which it is possible to record the spectrum if the angle of incidence varies through 10° to 75°.

2. Convert these two values into keV.

13.5 Sulphur presents two lines: $K\alpha_1 = 5.37216$ Å and $K\alpha_2 = 5.37496$ Å

1. What is the energy difference, expressed in eV, between $K\alpha_1$ and $K\alpha_2$?

2. If it is known that the width at the mid-height of these lines is 5 eV, what conclusion can be drawn?

3. If the position of the $K\alpha_2$ line increases by 0.002 Å when passing from S^{6+} to S°, show that this has no impact upon the location of this photoelectron in energy terms.

13.6 Is it possible to protect oneself from X-rays by being wrapped in aluminium foil of thickness 12 µm?

First calculate for the % transmission for the Ti $K\alpha$ line (4.51 keV, μ_m Al $= 264$ cm^2/g), then for the Ag $K\alpha$ line (22 keV, μ_m Al $= 2.54$ cm^2/g). The volumic mass of the aluminium is 2.66 g/cm^3.

Consider the same question but, in place of the aluminium, a film of latex is employed $(C_5H_8)_n$ of the same thickness (volumic mass $= 1$). The final calculation will be made with the transition $K\alpha$ line of titanium (μ_m C $= 25.6$ cm^2/g, μ_m H $= 0.43$ cm^2/g).

13.7 1. Why does the table of X-ray lines only begin with the third element of the periodic table?

2. Why, when measuring the elements whose energy of emitted radiation is inferior to 3 keV, is it necessary to replace the air in the apparatus by helium?

13.8 When establishing the Bragg equation the difference is calculated between the optical path of two rays incident and reflected, making the same angle, α. Why is this?

13.9 For determining the % mass of Mn present in a rock, the element Ba is used as an internal reference. Two solid standard solutions in the form of pearls of borax yielded the following results:

No of solution	% Mn by mass	Ratio Mn/Ba
1	0.250	0.811
2	0.350	0.963
unknown sol.	?	0.886

Evaluate the % mass for Mn in the original solution.

13.10 If a piece of nickel is suspended at a depth of 1 cm in an aqueous solution, what fraction of the intensity of the transition $K\alpha$, emitted when metal atoms are excited, will reach the surface? Mass attenuation coefficients for hydrogen and oxygen are H: 0.4 cm^2/g; O: 13.8 cm^2/g.

13.11 If the distance between the sample and the detector is 5 cm, calculate the % attenuation of a radiation possessing an energy of 2 keV produced by the presence of air over this distance. Take 80% nitrogen and 20% oxygen for the mass composition of air. The specific weight of air is 1.3 g/l. The mass attenuation coefficients of elemental nitrogen and oxygen are, for a radiation whose energy is 2 keV: $\mu_m = 494$ cm^2/g for N and 706 cm^2/g for O.

CHAPTER 14

Atomic absorption and flame emission spectroscopy

Atomic absorption and flame emission spectroscopy, also called flame photometry, are two methods of quantitative analysis that can be used to measure approximately 70 elements (metals and non-metals). Many models of these instruments allow measurements to be conducted by these two techniques, which rely on different principles. Their applications are numerous, as concentrations in the mg/l (ppm) region or lower can be accessed.

14.1 Principles common to the two methods

The principle behind these two methods of elemental analysis depends on measurements made on an analyte that is transformed into free atoms. To this end, for both techniques, the sample is heated in the instrument to a temperature of between 2 000 and 3 000 degrees to break chemical bonds, liberate the elements present and transform them into a gaseous atomic state. Thus, the total concentration of the element is measured without distinguishing the chemical structures present in the cold sample.

The thermal device used to elevate the temperature consists of a burner fed with a gaseous combustible mixture or, alternatively, in atomic absorption, by a small electric oven that contains a graphite rod resistor heated by the Joule effect. In the former, an aqueous solution of the sample is nebulised into the flame where atomisation takes place. In the latter, the sample is deposited on the graphite rod. In both methods, the atomic gas generated is located in the optical path of the instrument.

— In *atomic absorption spectroscopy* (AAS), the optical absorption of atoms in their ground state is measured when the sample is irradiated with the appropriate source.

— In *flame emission spectroscopy* (FES), the radiation intensity emitted by a small fraction of the atoms that have passed into the excited state by the elevated temperature is measured.

Measurement of the transmitted radiation is carried out at a wavelength specific for absorption (in AAS) or emission (in FES) of each element analysed.

■ In *flame emission spectroscopy*, light emission is caused by a thermal effect and not by a photon, as it is in *atomic fluorescence*. Flame emission, which is used solely for quantification, is distinguished from *atomic emission*, used for qualitative and quantitative analyses. This latter, more general term is reserved for a spectral method of analysis that uses high temperature thermal sources and a higher performance optical arrangement.

14.2   Interpretation of the phenomena involved

14.2.1   Kirchhoff's experiment

Although it is more than a century old, Kirchhoff's experiment demonstrating that incandescent gases can absorb the same radiation that they emit is a good illustration of atomic absorption and emission.

When white light impinging on a slit is dispersed by a prism, a continuous spectrum is obtained (Fig. 14.1a). If this radiation source is replaced by a Bunsen burner onto which sodium chloride is sprinkled, an emission of sodium is obtained consisting of a yellow doublet located at 589 nm among other lines (Figs. 14.1b illustrating *flame emission*, and 14.2). Finally, if the two preceding sources are placed in series in the optical path, a spectrum resembling that of Fig. 14.1a is obtained. However, this spectrum shows faded zones at the positions of the emission spectrum of sodium (Fig. 14.1c). This 'line reversal' is the result of the presence of a high proportion of ground state Na atoms in the flame that can absorb characteristic frequencies. This phenomenon is a manifestation of *atomic absorption*.

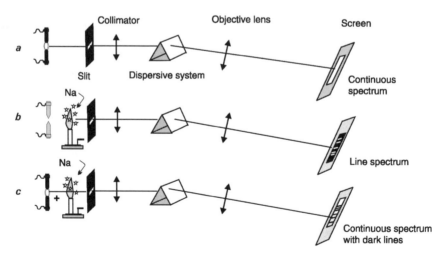

Figure 14.1—*Kirchhoff's 'line reversal' experiment.* See text for explanation. Schematics of the conventional optical set-up (collimator, objective) have been simplified for clarity.

■ Diverse instruments employing the principle of this experiment are used to measure mercury, a toxic and volatile element, present in many work areas. A device has been designed as a colorimeter dedicated for this single element. The source is a mercury vapour lamp and the cell is a transparent tube filled with the atmosphere to be monitored. If mercury vapours are present in the optical path, absorption of radiation emitted by the lamp will occur and this will lead to a decrease in the transmitted light intensity measured by the instrument.

14.2.2   Interpretation of Kirchhoff's experiment

An atom can only exist in potential energy states that are defined by its electronic configuration. When the element is brought to high temperature or irradiated with a light source, electrons from the outer shell are promoted, resulting in an excited state. This electron transfer corresponds to an absorption of energy. Conversely, when an atom returns to its ground state, it can re-emit the excess energy in the form of one or many photons. Thus, in the preceding experiments, the flame induces the most probable transitions in a sodium atom (Fig. 14.2).

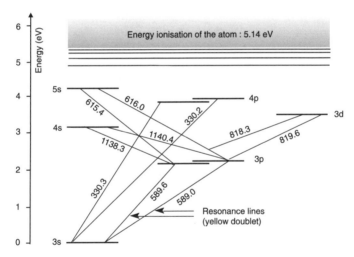

Figure 14.2—*Energy levels of the sodium atom.* Simplified representation of the excited states in the sodium atom. Emission lines are constructed according to selection rules. The values are indicated in nanometres.

14.3   Atomic absorption vs flame emission

The Maxwell–Boltzman distribution function permits calculation of the effect of temperature on each electronic transition. Designated by N_0 and N_e, the number of atoms in the ground and excited states, one obtains:

$$\frac{N_e}{N_0} = g \cdot \exp[-(\Delta E/kT)] \qquad (14.1)$$

where T is the absolute temperature, g is a small integer number (g depends on the quantum number of each element), ΔE is the difference in energy (in joules) between the ground (0) and excited (1) state populations and k is Boltzmann's constant ($k = R/N = 1.38 \times 10^{-23}$ J/K). If ΔE is expressed in eV instead of joules, then equation (14.1) becomes:

$$\frac{N_e}{N_0} = g \cdot \exp[-11\,600\,(\Delta E/T)] \qquad (14.2)$$

Current values of the various parameters expressed above are shown in Table 14.1 for certain elements. Examination of the data shows that the atoms are essentially in their ground state when ΔE is large and the temperature is low.

Table 14.1—Ratios of N_e/N_0 for certain elements at various temperatures.

Element	λ (nm)	E (eV)	g	2000 K	3000 K	4000 K
Na	589	2.1	2	1.03×10^{-5}	5.95×10^{-4}	4.53×10^{-3}
Ca	423	2.93	3	1.25×10^{-7}	3.60×10^{-5}	6.12×10^{-4}
Cu	325	3.82	2	4.77×10^{-10}	7.69×10^{-7}	3.09×10^{-5}
Zn	214	5.79	3	7.81×10^{-15}	5.68×10^{-10}	1.53×10^{-7}

It would appear in all cases that measurements should be based on atomic absorption instead of flame emission. Absorption spectra are simpler than emission spectra. In reality, however, the absorption measurements can be complicated by the presence of interferences, chemical interaction, instability of the energy level, and other phenomena that occur at elevated temperatures (Fig. 14.3).

Under the effect of temperature, each atomic transition leads to emission or absorption spread over a narrow interval of wavelengths. This uncertainty around the theoretical value constitutes the *natural bandwidth of the line* and leads to enlargement of the image of the line seen by a monochromator. The natural bandwidth ranges from 10^{-5} nm under ideal conditions to about 0.002 nm at 3 000 K.

■ The value of the ratio N_e/N_0 does not imply that all excited atoms return to their initial state as they emit a photon. As the temperature increases, the emission spectrum becomes more complex, particularly due to the emission of lines emanating from ionised atoms (see Fig. 14.3). It thus becomes necessary to have good quality optics in order to use this technique. The corresponding instruments are atomic emission spectrophotometers, which will be discussed in Chapter 15.

Previous experience indicates that flame emission is preferable for five or six of the elements. With present detectors, reliable measurements can be obtained as long as the ratio N_e/N_0 is superior to 10^{-7}. Therefore, elements such as the alkaline earths that give coloured flames are easily measured by emission (see Table 14.1).

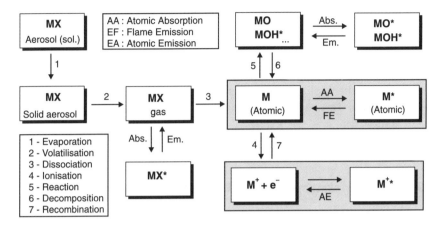

Figure 14.3—*Summary of the possible evolution of an aerosol in a flame.* Atomic absorption and emission occur in the shaded area of the drawing.

14.4 Measurements by absorption or flame emission

The quantification of elements becomes possible when their concentrations can be related to the intensity of the corresponding absorbance or emission.

14.4.1 Atomic absorption measurements

A signal measured by the instrument indicates absorbance of an element in the flame. This absorption depends on the number of ground state atoms N_0 in the optical path. This is where comparison with the Beer–Lambert law ends (ε is not calculated). Measurements are made by comparing the unknown to standard solutions.

$$A = k \cdot C \tag{14.3}$$

where A is the absorbance, C is the concentration of the element and k is a coefficient unique to each element at a given wavelength.

The instrument yields the absorbance by ratioing the transmitted intensities in the presence and absence of sample. Linearity is only observed for weak concentrations (typically below 3 ppm). The methods used, comparable to those used in molecular absorption spectrophotometry, involve classical protocols: methods using a calibration curve or standard additions, as long as the range of concentrations stays within the linear conditions of absorbance.

14.4.2 Flame emission measurements

For a population of n excited atoms, the emitted light intensity I_e depends on the number of atoms $\mathrm{d}n$ that return to the ground state during an interval of time dt (dn/d$t = k'n$). As n is proportional to the concentration of the element in the hot zone of the instrument, the emitted light intensity I_e, which varies as dn/dt, is also proportional to the concentration:

$$I_e = K \cdot C \tag{14.4}$$

This formula can be applied at low concentration and in the absence of self-absorption or ionisation. To conduct a measurement, the instrument is calibrated using a series of standards.

14.5 Basic instrumentation for atomic absorption

The optical scheme of an atomic absorption spectrophotometer is illustrated in Fig. 14.4, which shows a basic single beam instrument.

The light beam emitted by the source, which must be at the wavelength required for measurement, passes through the flame (or *graphite furnace*) in which the element is located in its atomic state. The beam is then focused on the entrance slit of the monochromator, located after the sample. The monochromator's role is to select a very narrow band of wavelengths. The optical path ends at the entrance slit of the photomultiplier tube.

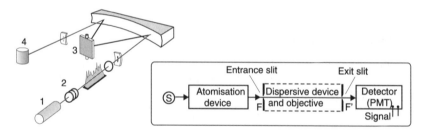

Figure 14.4—*The diverse components of a single beam atomic absorption spectrophotometer*. Model IL 157, built in the 1980s. 1, source; 2, burner; 3, monochromator; 4, detector (design according to Thermo Jarrell Ash Corp.).

■ In practice, a radiation source impinges on an entrance slit S (Fig. 14.5) located before the dispersive system. The exit slit is located after the dispersive system, close to the detector window, which selects a narrow bandwidth of the spectrum ($\Delta\lambda$ from 0.2 to 1 nm) and must not to be confused with the width of the exit slit or that of the image at the entrance slit.

In the absence of sample in the optical path, the detector perceives the full intensity I_0 of light emitted by the source for the wavelength interval selected by the entrance slit of the dispersive instrument. When a sample is present, the detector perceives a reduced intensity I (Fig. 14.5a). If the source emits a *continuum* of light, the ratio I_0/I will always be close to 1 because the absorption lines of elements are very fine (1×10^{-3} nm). On the other hand, if the source emits only at the wavelengths that a sample element is capable of absorbing (Fig. 14.5c), the ratio of intensities will be very small. With the high sensitivity of currently used photomultiplier tubes, the second situation is preferable; it is more efficient to measure a small change in light intensity against a dark background.

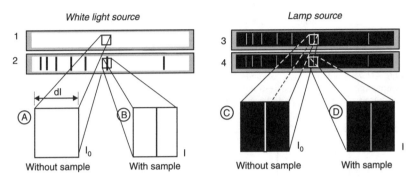

Figure 14.5—*Comparison of transmitted intensities in atomic absorption with a continuum source (a and b) and with a lamp that emits spectral lines (c and d)*. The square region shows the wavelength interval percieved by the photomultiplier tube (PMT). The PMT signal is proportional to the area of the white parts in the squares. In this way, the resolution is 'in the source', as expressed by Walsh, who is considered to be one of the pioneers of atomic absorption.

14.5.1 Hollow cathode lamps (HCL)

For the above reasons, atomic absorption instruments use discharge lamps that contain argon or neon, as the filling gas, at a pressure of hundreds of Pascal. The emission spectrum includes intense lines that depend on the element constituting the cathode in the lamp. Hence, to measure an element such as lead, the cathode must contain lead. There are approximately 100 different types of *hollow cathode lamps* made with pure elements, as well as lamps made of alloys and fritted powders for multi-element determination (Fig. 14.6). The anode is made of zirconium or tungsten and the windows of the lamp are borosilicate glass or fused silica (quartz), depending on the wavelength emitted by the cathode. The set-up shown in Fig. 14.6 cannot be used for sodium or mercury (their boiling points are too low), so classical lamps using metallic vapour are used instead.

When a 300 V potential is applied between the electrodes, the electrons ionise neon (or argon). Some of the ions possess enough kinetic energy to strip atoms away from the cathode, which becomes equivalent, at the surface, to an atomic gas. Using M(S) for the element M in its metallic state (cathode) and M(G) when it is in its atomic state, emission corresponds to a series of steps:

$$M(S) \xrightarrow{\text{Ne}^+} M(G)^* \longrightarrow M(G) + \text{photon}$$

The spectrum emitted by the lamp corresponds to the radiation emitted by the cathode superimposed on that emitted by the filling gas. The width of the emission

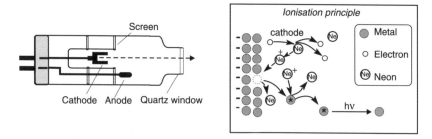

Figure 14.6—*Hollow cathode lamp.* Schematic of a typical lamp. The cathode is made from a hollow cylinder whose axis of revolution corresponds to the optical axis of a lamp. On the right is a diagram of the excitation of atoms in the cathode under impact with neon ions.

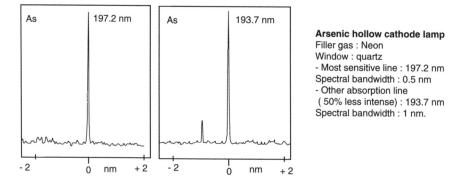

Figure 14.7—*Characteristics of a hollow cathode lamp (element: As).* Because the dynamic range in atomic absorption is narrow (two decades), the choice of a second wavelength of weaker absorption avoids the need to dilute a solution when an element is present in high concentration.

lines which depend on Doppler, Stark (ionisation) and Lorentz (pressure) effects is narrower than the corresponding absorption band. The monochromator allows elimination of a large part of the stray light due to the filling gas and selects the most intense spectral line to obtain maximum sensitivity (Fig. 14.7). This assumes that no interference is caused by other elements. Finally, electrodeless lamps with very intense emission use a radiofrequency emitter to excite the metallic vapour and are mostly used for elements such as As, Hg, Se and P.

14.5.2 Thermal devices for atomisation

— **Flame atomisation.** A burner is fed with a combustible gas mixture and constructed in a robust fashion to resist possible explosions of the gas (Fig. 14.8). The flame, which has a rectangular base of approximately 10 cm by 1 mm is aligned with the optical axis of the instrument. The sample is aspirated by the Venturi effect into the mixture of combustible gases feeding the flame.

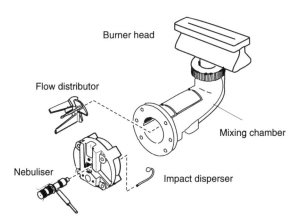

Figure 14.8—*Burner for an atomic absorption instrument.* This type of burner is used in models 3100–3300 from Perkin–Elmer. (Reproduced by permission of Perkin–Elmer.)

The flame is a complex medium in dynamic equilibrium that must be perfectly controlled. It is characterised by its chemical reactivity for a given maximum temperature (Table 14.2) and by its spectrum. Free radicals present in the flame have an emission and absorption spectrum in the near UV and this can sometimes interfere with the measurement of some elements. Thus, the observation height of the flame must be adjusted for some elements.

Table 14.2—Temperature limits for some gas mixtures.

Combustible mixture	Max. temperature (K)
butane/air	2 200
acetylene/air	2 600
hydrogen/oxygen	2 900
acetylene/oxygen	3 400
acetylene/nitrous oxide	3 100

— **Thermoelectric atomisation.** A flameless device without nebulisation, known as a *graphite furnace*, uses a graphite rod with a cavity that can hold a precise quantity of sample (a few mg or µl deposited using an automatic syringe). This rod, oriented parallel to the optical axis, is heated according to a four-step cycle (Fig. 14.9). The atomisation period is relatively short.

The graphite rod behaves like an ohmic resistor when it is subjected to a potential difference of a few volts. The rod is surrounded by a double sleeve containing an inert gas to protect it from oxidation and allow circulating water to cool the assembly. To avoid splashing, the temperature is gradually increased to first dry, then calcify and finally atomise the sample. The available thermal power is sufficient to reproducibly atomise the sample into the gas phase within three or four seconds (Fig. 14.9).

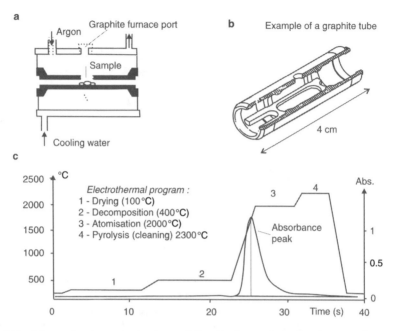

Figure 14.9—*Thermoelectric atomisation device.* a) Graphite furnace heated by the Joule effect; b) example of a graphite rod; c) temperature program as a function of time showing the absorption signal. The first two steps of this temperature program are conducted under an inert atmosphere (argon scan).

— **Chemical vaporisation.** Some elements (such as arsenic, bismuth, tin and selenium) are difficult to reduce in a flame when they are in higher oxidation states. For these atoms, the sample is reacted with a reducing agent prior to analysis (sodium borohydride or tin chloride in acidic media) in a separate vessel. The volatile hydride formed is carried by a make-up gas into a quartz cell placed in the flame (Fig. 14.10).

$$As^{+++} \xrightarrow{\text{NaBH}_4} AsH_3 \xrightarrow[\text{H}^+]{800^\circ C} As + 3/2\,H_2$$

Hydrides, which are easily thermalised at around $1\,000\,K$, liberate the atoms of an element. An electrodeless lamp is preferably used as a light source.

■ With mercury, Hg^0 is formed instead of the hydride. A special cell, which does not need to be put into the flame is used. This is called the 'cold vapour' method, and requires specialised instruments (reduction by $SnCl_2$).

14.6   Flame photometers

Photometric emission measurements are carried out using either atomic absorption spectrophotometers without the light source, or flame photometers. The latter are simple instruments whose price is ten times less that of atomic absorption spectro-photometers. Flame photometers are designed to make measurements on five or six elements. They include interchangeable coloured filters that can isolate a spectral band, including the chosen emission line (Fig. 14.11). Flame photometers are widely used for certain applications in quality control such as measurements of alkaline and alkaline earth metals in substances (e.g. calcium in beer or milk, potassium in cement, minerals, etc.). Some improved models possess two measuring cells that allow a direct reading of emission intensity ratios by comparison to an internal standard, allowing concentration determination. Response linearity is limited to concentrations between 10 and 100 ppm, so large dilutions of sample are typically required.

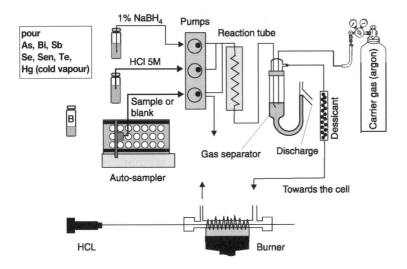

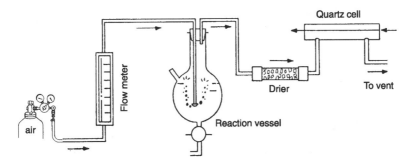

Figure 14.10—*Hydride reactor*. Used for some elements, this device includes a mixing tube where the hydride of a metal (or non-metal) is formed by a reaction with sodium borohydride. The flow of argon extracts the metal hydride that is formed (gas separator) and brings it into a quartz tube heated at 800 to 1 000 degrees in the flame.

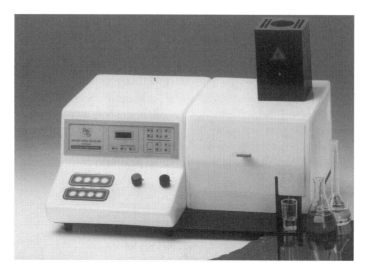

Figure 14.11—*Flame photometer*. (Reproduced by permission of ATS.)

14.7   Correction of interfering absorptions

Atomic absorption permits the measurement of some 70 elements at very low concentrations (Fig. 14.16). The scope of applications in atomic absorption is large. It is widely used because the method can accept samples of various forms. As in any other spectrophotometric method, it is necessary to carry out baseline correction to eliminate fluctuations coming from the lamp. Manufacturers have now designed double beam instruments in which the detector receives light that alternates between passage through the sample and passage through a reference (Fig. 14.12).

Instruments that have burners and require nebulisation of dilute aqueous sample solutions generally have low background noise in the signal. With graphite furnaces, incomplete atomisation of the solid sample at elevated temperatures can produce interfering absorptions. This matrix effect does not exist in an isolated state and thus cannot be eliminated by comparison with a reference beam. This is notably the case for solutions containing particles in suspension, ions that cannot be readily reduced and organic molecules, all of which create a constant absorbance in the interval covered by the monochromator.

14.7.1   Deuterium lamp background correction

Besides the double beam instrument that eliminates background due to light fluctuations from the source by measuring the background radiation from the flame, a second radiation source can be used to determine the absorption of the matrix.

For the selected wavelength, using the monochromator, the flame is swept alternately by light coming from the hollow cathode lamp and from the deuterium lamp, which constitutes a continuous source. When the deuterium lamp is selected, background absorption can be measured because the bandwidth of this lamp is hundreds of times larger than that of an absorption line. When the hollow cathode lamp is selected, total absorbance is measured (background absorption plus absorption from the element). The instrument then compares the transmitted intensities in both cases (a rotating mirror is used for a double beam).

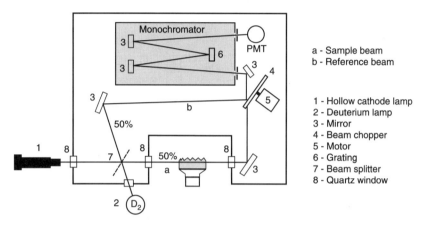

Figure 14.12—*Schematic of an instrument showing deuterium lamp background correction.* Perkin–Elmer, model 3300 with a Littrow-type monochromator. This double beam assembly includes a deuterium lamp whose continuum spectrum is superimposed, with the aid of semitransparent mirrors, on the lines emitted by the hollow cathode lamp. One beam path goes through the flame while the other is a reference path. The instrument measures the ratio of transmitted intensities from both beams. The correction domain is limited to the spectral range of the deuterium lamp, which is from 200–350 nm. (Reproduced by permission of Perkin–Elmer.)

14.7.2 Correction using the Zeeman effect

The influence of a magnetic field on gaseous atoms induces a splitting of each line into several polarised components. This phenomenon, which can be seen in the emission or absorption spectra of these atoms and is called the Zeeman effect, arises from perturbations in the energy states of electrons in the atom (Fig. 14.13). For example, the absorption wavelength of cadmium, situated at 228.8 nm, leads to three polarised absorption bands due to the Zeeman effect. One of these bands, the π component, retains the initial value of the wavelength whereas the other two, the σ components, are symmetrically shifted by a few picometres relative to the π component in a 1-tesla field. The direction of polarisation of the π and σ lines are perpendicular and the polarisation plane of the π component is parallel to the magnetic field (Fig. 14.14).

■ In the absence of a magnetic field B, the detector perceives a decrease in the light emitted from the source caused by background absorption plus the fraction absorbed by the element: these two absorptions are additive.

However, when a magnetic field B is applied in the region of the furnace, the selected absorption band separates into many lines. Those bands that have shifted in position can no longer interfere with the light emitted by the source. The line that persists at the initial wavelength (when $B = 0$), is polarised along the axis of the field. Therefore, it absorbs radiation from the source only in that direction. If a polariser is introduced between the oven and the detector, it may be possible to see the absorption of the element, depending on the polariser's orientation. For example, if the polariser is perpendicular to the direction of B, it will not detect the absorption; the element will have become transparent to the detector. In effect, the detector will no longer see the continuous absorption background.

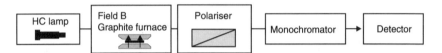

Figure 14.13—*Zeeman effect correction*. Instrument showing the principle used for correction of absorbance by the Zeeman effect. Two solutions are applicable: 1) magnetic field, B, switched alternately on and off and a fixed polariser; 2) fixed magnetic field, B, and a rotating polariser.

Application of this principle is used in two types of background absorption correction set-up. Single beam atomic absorption instruments have an electromagnet at the level of the graphite furnace (or flame) and a polariser in the optical path (Fig. 14.14). However, this accessory is quite expensive.

— In the first set-up, the field is kept at a fixed value and the polariser is rotated. The signal varies between two extreme values that correspond to the background absorption or to the background absorption increased by the π component of the element.

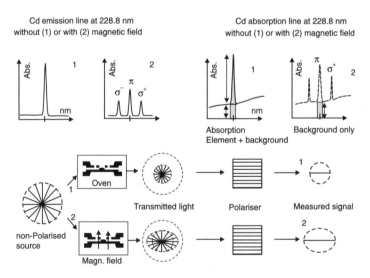

Figure 14.14—*Normal Zeeman effect*. Illustration of the oscillating field method (path 1 without field B: noise plus element, path 2 with field B: noise only).

— The second set-up uses an alternating on/off field and a fixed polariser with an orientation that suppresses detection of the π component. This system is equivalent to a double beam instrument with a common path. The atoms are affected by the Zeeman effect but particles in suspension are not (Fig. 14.14).

14.7.3   Background correction using a pulsed hollow cathode lamp

When the intensity of a hollow cathode lamp increases because of a reduction in the shunt resistance, the profile of the emission line changes. As the central part of the cathode becomes very hot, the line is broadened for several reasons. However, vaporised atoms emitted by the cathode will reabsorb in a colder part of the lamp in the form of a very fine line. The net result is that the emission curve dips in the middle because of self-absorption. This observation is the basis of the pulsed lamp technique for correction of background absorption (Fig. 14.15).

Under normal conditions (e.g. 10 mA) with the sample in the flame, a global measurement of background absorption and absorption by the element is obtained. However, under strained lamp conditions (e.g. 500 mA), only background absorption is measured. Comparison of both measurements allows the calculation of absorption due only to the analyte.

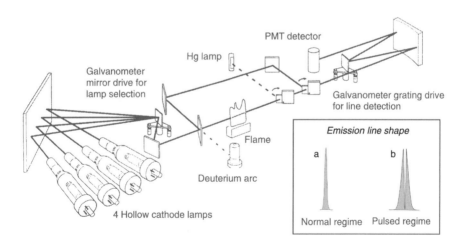

Figure 14.15—*Pulsed hollow cathode lamp background correction.* a) Shape of the emission line from a hollow cathode lamp under normal operating conditions, b) the 4000 Smith–Hieftje model from Thermo Jarrell Ash uses the principle of pulsed-source correction. The mercury source and the retractable mirrors are used for calibration of the monochromator. (Reproduced by permission of Thermo Jarrell Ash.)

14.8   Physical and chemical interferences

Whenever possible, an element is measured using its most intense absorption band corresponding to a resonance line. However, several factors can affect the position of absorption or emission lines and thus lead to incorrect measurements.

14.8.1   Spectral interferences

Each line has a finite width that depends on the instrument used. When using a graphite furnace, it is possible to observe interfering emissions from the walls of the graphite rod. It is also possible to have superimposed absorptions coming from molecules or background.

In atomic absorption, confusion seldom occurs, but the technique is still vulnerable to the superposition of two lines: the line of the chosen element and a secondary line of another element. This is why it is suggested that a second measurement is performed at another wavelength. In atomic emission, this problem is more frequent because spectra are more complex.

14.8.2   Superposition of emission and absorption of the same element

When a non-negligible fraction of atoms of an element emit while the rest absorbs at the same wavelength, the detector perceives a light intensity that is higher than it should be. In order to correct the signal, the emitted intensity has to be subtracted. Measurement with a pulsed lamp can be used for this: signal due to emission is constant and can be evaluated and subtracted from the modulated total intensity.

14.8.3   Chemical interactions

When trace analysis is performed, well-established protocols have to be followed to obtain reliable results. The production of free atoms in the flame or furnace can be modified according to the composition of the matrix in which the sample is found. Therefore, salts and mineral or organic reagents are often introduced in the solutions to be nebulised.

These *matrix modifiers* can act as 'liberating agents', R, with their action explained by the following equation:

$$R + M\text{-}X \rightarrow M + R\text{-}X$$

The element M will be more readily liberated if compound R-X is more stable. Thus, when calcium is to be measured in a matrix rich in phosphate ions or in refractory combinations containing aluminium, then sodium or lanthanum chloride is added. The desired effect is to liberate calcium and to increase the volatility of the matrix in order to ensure more efficient elimination during the decomposition step.

In thermoelectric instruments, the complexing agent ethylenediaminetetracetic acid (EDTA) can be added to increase the volatility of the analyte. Ammonium nitrate can also be added if the matrix contains a lot of sodium chloride.

Finally, the use of extremely hot flames with certain elements can cause ionisation of the latter, which decreases the concentration of free atoms in the flame. This effect can be corrected by adding an ionisation suppresser in the form of a cation whose ionisation potential is less than that of the analyte. A potassium salt at the level of 2 g/l is often chosen as an ionisation suppresser.

■ This variation in ionisation can spontaneously occur when the matrix naturally contains one or more alkaline elements. To avoid these random errors, a buffer of potassium or sodium salt is systematically added to the solutions. An alternative is to construct a calibration curve using a matrix that is very close to that of the analyte.

14.9 Sensitivity and detection limits in AAS

Sensitivity is defined as the concentration of the element, in $\mu g/ml$, which will lead to a 1% decrease ($A = 0.0044$) in the transmitted light intensity of an aqueous solution. Calibration curves should be in the range of 20 to 200 times this limit.

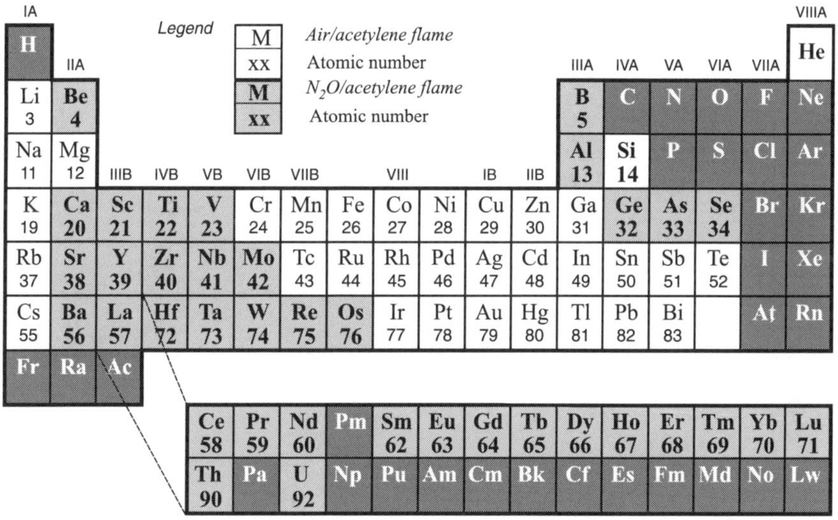

Figure 14.16—*Elements determined by AAS or FES.* Most elements can be determined by atomic absorption or flame emission using one of the available atomisation modes (burner, graphite furnace or hydride formation). Sensitivity varies enormously from one element to another. The representation above shows the elements in their periodic classification in order to show the wide use of these methods. Some of the lighter elements, C, N, O, F, etc. in the figure can be determined using a high temperature thermal source: a plasma torch, in association with a spectrophotometric device (ICP-AES) or a mass spectrometer (ICP-MS).

The detection limit is defined as the concentration of the element that will yield a signal whose intensity is equal to two times the standard deviation of a series of at least 10 measurements of the analytical blank or of a very dilute solution (confidence level 95%). In practice, concentrations should be at least 10 times higher than the detection limit to give reliable measurements (cf. 21.5.3).

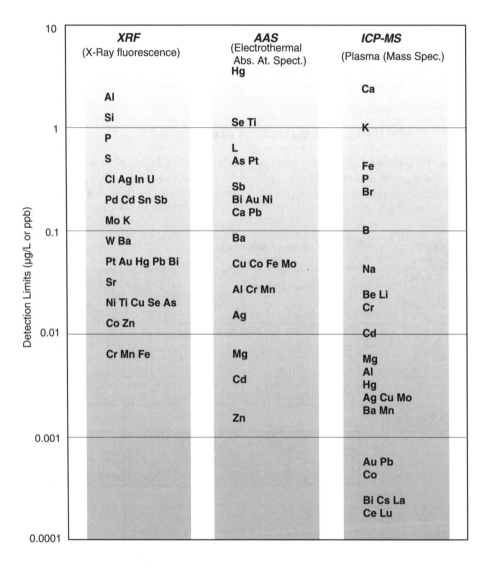

Figure 14.17—Relative detection limits of the main present methods in trace element analysis.

Problems

14.1   If the response of the detector of a flame photometer is proportional to the concentration of the element heated to a state of excitation, by what factor is the signal multiplied when the temperature passes from 2 000 K to 2 500 K?

(Establish the general expression and then apply it to the case of elemental sodium, whose resonance is promoted by a radiation of 589 nm wavelength.)

14.2   Why is it advisable to add ethylenediaminetetracetic acid (EDTA) to measure calcium when it is in the form of a phosphate?

14.3   Why is a potassium salt (such as KCl) frequently added when one wants to measure elemental sodium by flame photometer? Remember that the potential for the first ionisation of potassium is 419 KJ mol^{-1} while that for sodium is 496 KJ mol^{-1}.

14.4   The potassium concentration in a blood serum is to be analysed using a method of addition and flame emission. Two extractions of 0.5 ml of serum were taken to create two identical solutions and then both were diluted further with distilled water to a final volume of 5 ml. 10 µl of 0.2 M KCl is introduced to one of these. The values obtained from the apparatus were 32.1 and 58.6 arbitrary units. What is the potassium concentration of the serum?

14.5   An *AAS* method is employed for the determination of lead (Pb) in a sample of adulterated paprika by the introduction of lead oxide (of the same colour). An electrothermal atomic absorption instrument that provides a background correction based upon the Zeeman effect is used.

0.01 g of the paprika powder is placed in the tube of the graphite furnace. The determination of the area peak absorbance was made at $\lambda = 283.3$ nm first in the absence and later in the presence of a magnetic field. The value of the peak absorption following background correction was 1 220 (arbitrary units). Under the same conditions, 0.01 ml of a solution of 10 g/l Pb led to a value of 1 000 in the same units.

Calculate the % mass of lead in the sample of paprika under study.

14.6   A stock solution of calcium ions was prepared by dissolving 0.1834 g of $CaCl_2.2H_2O$ in 100 ml of distilled water and then further diluting by a factor of 10. From this new solution, three standard solutions were prepared by further dilutions of five, 10 and 20 times, respectively. The unknown sample is itself diluted 25 times. Sufficient strontium chloride was then introduced to eliminate any interference due to phosphate ions. An analytical blank containing the same concentration of strontium was the first solution to be examined by the air/acetylene flame. The results were as follows:

Blank	1.5	Ca = 40.1 g mol^{-1}
Standard 1:20	10.6	Cl = 35.5 g mol^{-1}
Standard 1:10	20.1	
Standard 1:5	38.5	
Sample	29.6	

What is the concentration (in ppm) of the calcium in the sample?

14.7 Among the numerous commercial derivatives of ethylenediaminetetracetic acid (EDTA), a mixed salt of Zn/Na containing one atom of zinc and two atoms of sodium per molecule of EDTA is found. This mixed salt presents itself in the form of a crystallised hydrate. An attempt is described to deduce the number of water molecules of this hydrate by measuring the presence of zinc by atomic absorption. The experimental approach was as follows:

A solution A, is prepared by dissolving 35.7 mg of the hydrate in 100 ml of water. 2 ml were then pipetted and diluted again with H_2O to a final volume of 100 ml. This constitutes the sample solution B. The standardisation is then performed with five solutions of known zinc concentration. (see table below).

1. To effect this measurement the slit bandwidth of the atomic absorption apparatus is 1 nm. What does this parameter represent?

2. Knowing that the calibration curve computed by the apparatus indicates that $A = 0.3692$, $C = 0.99\,mg/l$, calculate the number of molecules of water in the mixed salt hydrate of Zn/Na.

3. From the data and by the method of least squares derive an equation for the calibration curve. Calculate the zinc concentration in solution A.
   Recall: $H = 1.01$; $C = 12.01$; $N = 14.01$; $O = 16.0$; $Na = 22.99$ and $Zn = 65.39\,g/mol$ respectively.

4. Is the use of the standard calibration curve justified for atomic absorption?

Solution	Conc. Zn mg/l	Absorbance
Blank	0.00	0.0006
Standard 1	0.50	0.2094
Standard 2	0.75	0.2961
Standard 3	1.0	0.3674
Standard 4	1.25	0.4333
Standard 5	1.50	0.4817
Sample	C?	0.3692

CHAPTER 15

Atomic emission spectroscopy

Analysis by atomic (or optical) emission spectroscopy is based on the study of radiation emitted by atoms in their excited state, ionised by the effect of high temperature. All elements can be measured by this technique, in contrast to conventional flames that only allow the analysis of a limited number of elements. Emission spectra, which are obtained in an electron rich environment, are more complex than in flame emission. Therefore, the optical part of the spectrometer has to be of very high quality to resolve interferences and matrix effects.

Present atomic emission spectrophotometers allow multi-element analyses to be performed either simultaneously or sequentially. This ability is the result of progress made in optics, excitation sources, detectors and microcomputers. Atomic emission, which was initially used in the metallurgical industry, has now expanded into many areas and competes with atomic absorption.

15.1 Optical emission spectrophotometry (OES)

At a high enough temperature, any element can be characterised and quantified because it will begin to emit. Elemental analysis from atomic emission spectra is thus a versatile analytical method when high temperatures can be obtained by sparks, electrical arcs or inert-gas plasmas. The *optical emission* obtained from samples (solute plus matrix) is very complex. It contains spectral lines often accompanied by a continuum spectrum. Optical emission spectrophotometers contain three principal components: the device responsible for bringing the sample to a sufficient temperature; the optics including a mono- or polychromator that constitute the heart of these instruments; and a microcomputer that controls the instrument. The most striking feature of these instruments is their optical bench, which differentiates them from flame emission spectrophotometers which are more limited in performance. Because of their price, these instruments constitute a major investment for any analytical laboratory.

15.2   Excitation by inductively coupled plasma (ICP)

The *plasma torch* is a simple device that can reach temperatures of up to 8 000 K. Any element introduced into the torch will start to emit. By comparison, flames used in flame photometry are relatively cool.

> ■ A plasma is a globally neutral environment formed by atoms in equilibrium between their neutral and ionised (1 to 2%) state and by electrons ($10^{18}/cm^3$). Plasmas are considered the fourth state of matter. Essentially, plasmas that are inductively coupled are used in atomic emission analysis. The colour of the plasma depends on the gas used to form it.

A flow of argon in which ionisation has been initiated by a tesla discharge passes through an open quartz tube. A water-cooled copper tube is coiled around the quartz tube (Fig. 15.1). The copper tubing is connected to a radiofrequency generator (typically at 27 MHz) with a power of 1 to 2 kilowatts. The variable magnetic field that is created confines the ions and electrons to an annular path (with the appearance of an eddy current). As this environment becomes more and

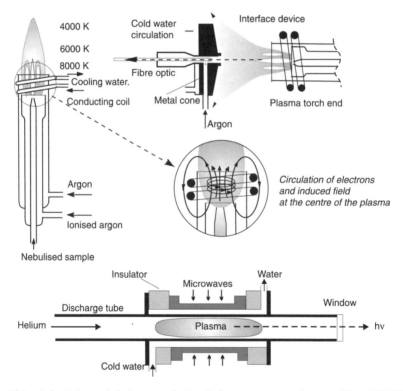

Figure 15.1—*Inductively coupled plasma torch.* A radiofrequency current (between 27 and 50 MHz) that induces circulation of the electrons in the inert gas drives the torch. The argon serves as an auxiliary gas, a cooling gas and the nebulisation gas. In the upper right is shown an optic device used to collect emitted light in the longitudinal axis of the plasma. Lower down, plasma generated by microwave.

more conductive, the temperature increases considerably by the Joule effect. The device behaves similarly to the secondary coil of a short-circuited transformer. The plasma is isolated from the torch wall by a gaseous sheath of flowing, non-ionised argon injected through an external tube concentric to the first one.

— If the sample is a solution, it is injected as an aerosol at the base of the torch via a third tube with a diameter 1 to 2 mm.

— If the sample is a solid, it is introduced inside the tube on the tip of a support mounted inside the torch.

Light collection from the plasma (in the transversal or longitudinal direction) is carried out as a function of the element depending on whether an ionic or atomic line is used for measurement. The temperature varies from 9 000 to 2 000 K between the bottom and the top of the torch. It is colder in the axis and at the solution nebulisation site.

15.3 Ionisation by arc, spark or electronic impact

Besides plasmas, which are at the forefront of thermal atomisation devices, other excitation processes can be used. These methods rely on sparks or electrical arcs. They are less sensitive and take longer to use than methods that operate with samples in solution. These excitation techniques, with low throughputs, are mostly used in semi-quantitative analysis in industry (Fig. 15.2). Compared to the plasma torch, thermal homogeneity in these techniques is more difficult to master.

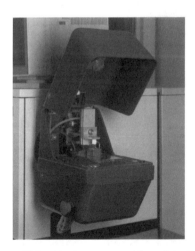 

Figure 15.2—*Two instruments with spark ionisation*. On the left, a Jobin-Yvon model JY-50E instrument. The opened spark chamber can be seen in the photograph. On the right, a model ARC Met-900 instrument from Metorex. The spark is produced by a gun connected by fibre optic to the spectrometer and situated in the console (reproduced by permission of American Stress Technologies).

■ The use of graphite electrodes is responsible for the observed characteristic emission lines of radicals and organic functional groups (e.g. cyanogen CN bands between 320 and 400 nm) and also for the background continuum they generate.

15.3.1   Constant current, low voltage arc

An electrical arc with a current intensity of several tens of amperes is created between two electrodes. The surrounding temperature stabilises between 3 000 and 6 000 K (Fig. 15.3). The sample is deposited on one of the electrodes, either in the form of metal fragments or as a powder mixed with graphite. Essentially, it is inserted into a cavity formed at the tip of the electrode. The element is thus vaporised by the arc.

15.3.2   Glow discharge sources

If the sample is a conductor, it is possible to use it as the cathode of a spectral lamp whose principle of operation is identical to that described for hollow cathode lamps (cf. section 14.5 and Fig. 15.3). The device must be sealed before it can be used, which represents a technical constraint. The advantage of this process, which is commonly used for surface analysis, is that it produces spectra with narrow emission lines because atomisation is made at a lower temperature than with the electrical arc method.

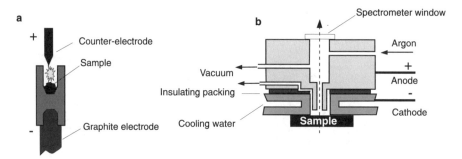

Figure 15.3—*Ionisation devices*. a) Continuous current arc (globular technique). The electrodes are inserted in a series circuit having a variable resistor, a cell and a continuous power source of a few tens of volts; b) Glow discharge device using argon (hollow cathode lamp type). Samples can be introduced as powders or non-conductive pellets.

15.3.3   High voltage AC spark

A small quantity of sample is vaporised as an aerosol when submitted to sparks generated under argon or helium between 20 and 50 keV. The high current density under an inert atmosphere creates an atomic gas that can be excited by collisions in the plasma. Spark-induced optical emission leads to ionic line spectra.

15.4 The principles of atomic emission analysis

In atomic emission analysis, one or several specific spectral lines are monitored for each analyte. It is technically difficult and requires a high performance instrument because emission of radiation does not only occur from the analyte but also from any additional material introduced in the high temperature thermal source (e.g. matrix, solution). Because emission can occur from either excited or ionised atoms, thousands of different spectral lines can be observed. Several of these lines are much more intense than those due to the analyte, which can be present at ultratrace levels.

> ■ Flame photometric analyses are much simpler. For example, when sodium is inserted in a flame at 2000 °C, the sodium atoms are the only ones emitting radiation. Light measurement can be conducted using a simple coloured filter that isolates the relatively large spectral band corresponding to the yellow colouration in the flame.

To isolate specific emissions of the analyte being analysed (i.e. optical transitions), a high quality optical set-up is required. Dispersive systems using *planar*, *concave* or *echelle* gratings are used in classical spectrophotometers or spectrographs (Figs. 11.10 and 15.6).

— In spectrographs, radiation lines collected as images of the source are equally spaced from one another. When a CCD is used as a detector (formerly a photographic plate) simultaneous recording is achieved.

— In a spectrophotometer, the fixed detector records the wavelength selected by a rotating grating located in the monochromator. This system allows sequential recording of line intensities.

The monochromator can be replaced by a polychromator. In this instance, a plate with multiple slits is placed at the focal point of the entrance slit to collect the selected lines of radiation. A detector is placed behind every slit (Fig. 15.5). This optical set-up is fixed since there is no rotation of the grating.

> ■ Originally, these spectra were recorded using spectragraphs with visual observation. Nowadays, spectrophotometers able to resolve numerous interferences and problems related to matrix effects are used. Each element, in its neutral or ionised state, is responsible for many lines with variable intensities. Thus the instrument is programmed to identify and quantify the elements in a sample.

15.5 Spectral lines

The active part of the optical set-up starts with the *entrance slit*, on which light produced by the sample (the light source) is focused. The slit width is not less than 10 µm because a minimum light intensity passing through is required. The source is thus transformed into a luminous object of linear form and the instrument's optical system will yield as many images of narrow lines as there are different wavelengths in the source. Hence, each line observed corresponds to a practically monochromatic image of the entrance slit of the spectrophotometer. Some elements can generate more than 2 000 lines.

■ In practice, the lines are distributed in the focal plane as a sequence of narrow parallel luminous segments. Two independent factors are responsible for the width of the spectral lines.

The first factor is technical, related to *instrumental parameters* such as the entrance slit width, the quality of the optics (e.g. focal distance), and diffraction through narrow orifices.

The second factor involves the theory that defines the *natural width* of the lines. Radiations emitted by atoms are not totally monochromatic. With plasmas in particular, where the collision frequency is high (this greatly reduces the lifetime of the excited states), Heisenberg's uncertainty principle is fully operational (see Fig. 15.4). Moreover, elevated temperatures increase the speed of the atoms, enlarging line widths by the Doppler effect. The natural width of spectral lines at 6000 K is in the order of several picometres.

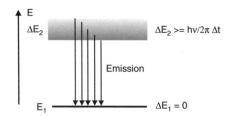

Figure 15.4—*Line width*. Transition between a stable ground state having an infinite lifetime and an excited state, which is short-lived. The uncertainty in ΔE_2 induces an imprecision on the corresponding wavelength.

15.6   Simultaneous and sequential instruments

As with any spectrophotometer instrument, the light intensity is determined at each wavelength using a photomultiplier tube. Emission measurements can be made over a wide dynamic range, which is an interesting feature because elements with widely different concentrations or sensitivity can be measured in a single sample solution.

Instruments can be classified into two major categories: those which measure several elements sequentially or simultaneously. Present trends favour systems with fixed optics delivering 'turnkey' solutions to defined applications.

15.6.1   Fixed optic, concave grating instruments (spectrograph type)

Predefined elements, characterised by precise spectral lines, are detected by as many photomultiplier tubes as there are secondary slits, each corresponding to an *analytical line* or measurement channel. If the instrument uses a concave grating, the entrance and exit slits are situated on the *Rowland circle*, which is at a tangent to the grating and has the same radius of curvature as the grating (Fig. 15.5).

■ Tens of elements can be measured in a single analysis. When a microcomputer is used, the results can be obtained automatically. It is possible to correct for background signals by simultaneously measuring several lines of one element. The linearity of the measurements covers a wider range than in atomic absorption. The optical part of these instruments is mounted on an ultra-stable metallic base.

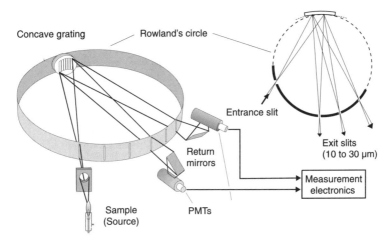

Figure 15.5—*Optical schematics of a polychromator with long focal distance and a concave grating.* Reflecting mirrors can be used when lines are too close to one another. Sometimes, several polychromators are combined. To the right, principle of the Rowland circle.

15.6.2   Fixed optic, echelle grating instruments

Another fixed optic arrangement uses an *echelle grating* in association with a focusing prism. This arrangement produces a double dispersion of the lines, in both the horizontal (due to the grating) and vertical (due to the prism) dimensions. This *order separating* device allows simultaneous detection over the whole spectral range (Fig. 15.6). The stationary optics are fixed during the instrument construction stage.

■ An echelle grating is characterised by a fewer number of grooves (about 100 per millimetre) than a traditional grating for the same spectral domain. As a consequence, for a given observation angle, many wavelengths are superimposed because there is a large separation between adjacent grooves. A prism which uses the second dimension of the focal plane to resolve the different images impinging on the entrance slit compensates for this multiplexing phenomenon (by the superposition of different orders *n* of the grating, where *n* is about 100). The spectrum is stationary relative to the optical assembly. The resolution achieved is approximately the same as that from an instrument with a 2 m focal length.

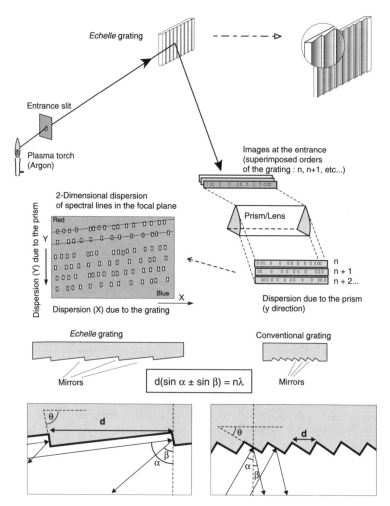

Figure 15.6—*Principle of dispersion in the focal plane using an arrangement comprising an echelle grating and prism.* For clarity, the associated optics (collimating and focusing lenses) are not shown in the top figure. In this set-up, the entrance slit is not very high.

There are also instruments with dispersion surfaces compatible with two-dimensional sensors. Their sensitivity and spectral response allow simultaneous measurement of thousands of lines (Fig. 15.7).

15.6.3 Wavelength scanning instruments (monochromator type)

In contrast to previously described instruments, the dispersive system in these instruments is movable in order to focus each wavelength on the fixed exit slit. A step motor allows a selection of wavelengths to be pre-programmed to sequentially

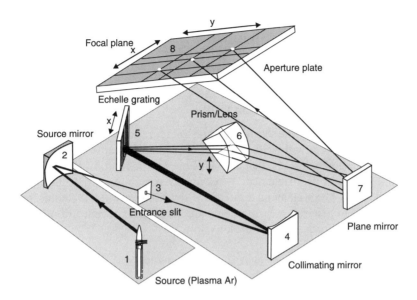

Figure 15.7—Optical diagrams for a spectrophotometer with an echelle grating. Model PU 7000 optical system (reproduced by permission of Philips). All spectral lines are captured, which allows a more complete study of the sample. The dynamic range of these instruments is still lower than that of a PMT.

examine the spectral lines. Light emitted from the source is analysed either by an Eberts arrangement (cf. 11.9.2) or by a dual monochromator of the Czerny–Turner type for high spectral resolution (a few nanometres), comparable to the line width at this temperature over a large spectral domain (160 to 850 nm). Several units of this type can be placed in series, constituting double or triple monochromators with very high performance. This set-up is possible without creating too cumbersome an installation.

The main feature of these instruments is their flexibility. A mercury vapour light source is used to internally calibrate the wavelength for rapid and precise positioning of the grating. The optical path lies in a rigid enclosure maintained under a vacuum. In this arrangement, the proximity of certain spectral lines necessitates a high degree of mechanical precision.

15.7 Performance

Its rapidity and detection limits, which are in the order of a few ppt (10^{-12}) for many elements, make atomic emission one of the best techniques currently available for elemental analysis. These sophisticated instruments, however, are not intended to replace the flame photometers that are still used for many simple measurements.

The detection limit values for different elements that can be found in comparative tables reflect the potential of these instruments. However, each application should be considered as an individual case. Certain instruments better tolerate dominating matrices such as soil or mud, where there are high concentrations of elements such as Si, Fe and Al.

15.7.1   Sensitivity and detection limits

Performance is defined by the sensitivity threshold, or the minimum concentration of element in solution that will yield an analytical signal with amplitude equal to twice that of the average background signal. This classical definition leads to optimistic values that can vary from element to element. The limit of detection represents the concentration of an element that can be detected with a 95% confidence limit (cf. chapter 21). In general, measurements are made in a concentration domain that corresponds to 50 times the limit of detection.

■ For a long time, the sensitivity of atomic emission was lower than that of atomic absorption for many elements. Improvements made to nebulisation systems have now lowered detection limits for elements such as As, Pb and Se.

15.7.2   Limit of resolution and resolving power

The *spectral resolution* can be measured directly from a spectrum. It depends on the emission peak width $\Delta\lambda$, at half height, measured in spectral units. Its value depends on the monochromator and its tuning. R defines the *resolving power* of a spectrometer: that is, its ability to separate two lines of very close wavelength:

$$R = \frac{\lambda}{\Delta\lambda} \tag{15.1}$$

Atomic emission instruments have resolving powers in the order of 30 000 to 100 000. The use of a wide exit slit (bandwidth perceived by the photomultiplier tube) degrades the resolving power of the spectrometer.

15.7.3   Linear dispersion and bandwidth

Several parameters, for example focal length, determine *linear dispersion*, which is expressed in millimetres per nanometre (or its inverse, which is called *reciprocal dispersion*). Linear dispersion represents the spread, in the focal plane, of two wavelengths differing by 1 nm. *Bandwidth*, which must not be confused with the width of the slit, is the interval of the spectrum that corresponds to the width in picometres exiting the slit. This width is generally greater than the natural width of the line being transmitted.

■ If the exit slit width is 20 μm and the dispersion is 2 mm/nm, the bandwidth will be equal to 10 pm ($1 \times 20 \times 10^{-3}/2 = 10^{-2}$ nm, or 10 pm).

15.8 GC coupled to atomic emission relatively

Flame photometric detection has been used for a relatively long time in conjunction with GC for determination of compounds containing phosphorus or sulphur. Based on this principle, it is possible to profit from the high temperatures of plasma emission analysis to obtain information on the elemental composition of eluting molecular compounds. For this purpose, an atomic emission spectrophotometer is placed at the outlet of a GC column (Fig. 15.8).

Temperatures reached are such that all elements can be detected by their emission lines. By choosing a line specific for one element, a selective chromatogram can be obtained for all compounds that contain this element. It is thus possible to search for classes of compounds corresponding to the association of many elements. This approach eliminates interferences due to sample matrices.

■ It is regrettable that this mode of identification of organic molecules implies their destruction. In this chemical butchery, structural information is lost and the only exportable result is the elemental composition.

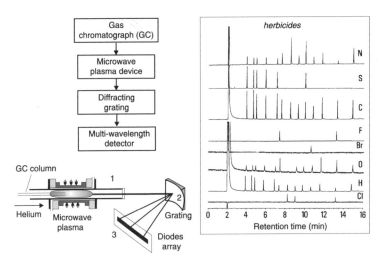

Figure 15.8—*Coupling of a gas chromatograph with an atomic emission spectrophotometer.* Effluents from the capillary column are injected into the plasma and decomposed into their elements. Each chromatogram corresponds to the compound containing the element of interest. For a given retention time, indication as to the elements included in a compound can be obtained. The plasma in this example is generated by heating the carrier gas (He) with a microwave generator confined in a cavity at the exit of the column. A diode array detector system can be used for simultaneous detection of many elements (chromatograms courtesy of a Hewlett–Packard document).

15.9 Applications of atomic emission spectrometry

Given the large number of elements that can be measured by atomic emission, this analysis method cannot be overlooked.

This technique can be used for the automatic determination of pre-chosen elements. Besides analytical applications in industry – for example, monitoring the wear of car and aircraft motors by measuring the metals present in the oil – environmental analysis is probably the area where best advantage of this method can be taken.

Samples for analysis can come from products of vegetable or animal origin (milk, meat), water, air (ashes emitted by an incinerator) or soils in which elements are present over a wide range of concentrations (from manure spreading on agricultural land to industrial sludge). This method also has applications in the area of forensic sciences and clinical medicine (tissue analysis or biological fluids).

An advantage of the atomic emission method is its linearity of response over a large range of concentrations, which permits treatment of complex matrices with a minimum of preparation. A single solution often allows the determination of many elements, from very high to trace concentrations, without the need for successive dilutions.

Problems

15.1 1. Why do the elements which are measured lead to atomic emissions of relatively low intensity?

2. Why does one generally observe ionic emissions?

15.2 When an atom emits radiation, the frequency of the corresponding photon is linked to the energy of the levels between which the transition occurs. According to the Heisenberg principle, if the excited state has a lifetime of Δt, then this results in a minimum uncertainty for the energy ΔE, such that: $\Delta E \Delta t \geq h/2\pi$.

From this equation calculate the natural width of the 589 nm emission line for sodium if $\Delta t = 1$ ns.

15.3 Five standard solutions were prepared for measuring the lead concentration in two solutions, A and B. The two solutions A and B contain the same concentration of magnesium used as an internal standard. The following data were obtained:

Conc. (mg/l)	Emission of signal (arbit. units)	Signal of Mg
0.10	13.86	11.88
0.20	23.49	11.76
0.30	33.81	12.24
0.40	44.50	12.00
0.50	53.63	12.12
sol A	15.50	11.80
sol B	42.60	12.40

From the data, calculate the lead concentration (mg/l) in the two sample solutions, A and B.

15.4 The resonance level of the sodium atom is sometimes said to be $16\,960$ cm^{-1} from the fundamental level. From this value calculate the corresponding wavelength and the energy of the transition associated with it.

15.5 Explain why it is easier to measure traces of radioactive, α-emitting isotopes of long half-life by ICP/MS rather than by radioactivity counting.

15.6 If the linear dispersion of a spectrograph used in atomic emission spectroscopy is 2 mm for a difference in wavelength of 1 nm and the size of the exit aperture is 20 μm, calculate the spectral interval of the emissions which reach the detector.

Anti-doping control

In 1992, the International Olympic Committee (IOC) produced a list of approximately 500 banned substances that can be classified into four categories of molecules: stimulants, analgesics-narcotics, anabolic steroids and beta-blockers. Several of these active molecules can only be detected by their metabolites in urine .

The IOC has accredited approximately 20 laboratories in the world, including the National Laboratory for Doping Control located at Chatenay-Malabry, to conduct these analyses. About 7000 analyses per year are made for France alone.

These analyses are conducted on the urine of athletes. Each sample is submitted to a complex analytical protocol that includes three levels:

1   Rapid screening.
Methods used at this stage range from colour tests to chromatography coupled to mass spectrometry (the detection of specific fragment ions appearing in mass chromatograms).

2   Identification/confirmation phase.
Each positive sample is then analysed for possible substances detected in the first stage. The analysis is conducted by GC- or HPLC-MS for positive identification and semi-quantification.

3   Quantification.
Following the results of phase 2, precise quantification may be necessary. This is usually conducted by conventional chromatographic or classical spectroscopic techniques (HPLC, UV absorption or fluorimetry).

The entire process is made even more complex by the duplicate analysis of all positive samples and by the use of security measures that are necessary at all stages in order to avoid errors, particularly in view of the impact of such results.

OTHER METHODS

CHAPTER 16

Mass spectrometry

Mass spectrometry (MS) is an analytical method based on the determination of atomic or molecular masses of individual species in a sample. Information acquired allows determination of the nature, composition, and even structure of the analyte. Mass spectrometers can be classified into categories based on the mass separation technique used. Some of the instruments date back to the beginning of the twentieth century and were used for the study of charged particles or ionised atoms using magnetic fields, while others of modest performance, such as bench-top models often used in conjunction with chromatography, rely on different principles for mass analysis. Continuous improvements to the instruments, miniaturisation and advances in new ionisation techniques have made MS one of the methods with the widest application range because of its flexibility and extreme sensitivity.

This technique is used in several diverse areas: organic and inorganic chemistry, biochemistry, clinical chemistry, environmental sciences, agriculture, geochemistry as well as many industrial control processes and for regulation compliance.

16.1 Principles

A minute quantity of sample in the gas phase (or suitable form) is ionised and the resulting charged species are analysed by a magnetic and/or electric field, depending on the type of instrument. The study of ion trajectories in the analyser tube, maintained under vacuum (10^{-4} Pa), allows the determination of the *mass/charge ratio* and, eventually, the nature of the ions. This destructive method is extremely sensitive.

■ If a sample of methanol (CH_3OH) in the gas phase is bombarded by electrons, a small fraction of the molecules will be transformed into charged species, some of which are positive ions ($CH_3OH^{+\cdot}$). These ions, formed in an excited state, have a surplus of internal energies that will immediately induce fragmentation of a great number of these ions. However, not all the ions will dissociate by the same reaction pathways, thus leading to different fragment ions whose mass is less than that of the molecular ion. All fragment ions, formed by simple bond cleavage or rearrangement reactions that follow cleavage, contain information on the structure of the initial molecule (see Fig. 16.1a).

During mass spectral analysis, the sample undergoes many successive processes that can be described in the following manner:

1. *ionisation*: the sample is vaporised and ionised in the *ion source* of the instrument by one of many possible processes. At this stage, molecules of the compound will lead to a statistical distribution of fragment ions.
2. *acceleration*: immediately after their formation, ions are readily extracted from the ion source, focused by a series of electronic lenses and accelerated to increase their kinetic energy.
3. *separation*: the *mass analyser* will filter ions according to their mass/charge ratio. Some instruments combine many mass analysers in series.
4. *detection*: after separation, ions terminate their path and strike the *detector*, which measures electrical charge and amplifies the weak ionic current.
5. treatment of the resulting detector signals leads to *display of the mass spectrum*.

The mass spectrum of the analyte represents the statistical abundance of each type of ion formed, as a function of its mass/charge ratio in increasing order (Fig. 16.1). Under the same operating conditions, the spectrum is reproducible and character-istic of the sample.

■ This method only yields the mass/charge ratio of ions, m/q. Logically, to calculate m, one must know q. However, the ions carry a net charge of type $q = ze$ (where e represents the elementary charge of an electron and z is a small integer), which leads to determining m to the nearest factor of z. This is why the m/z scale is used on the abscissa in mass spectra (m being in atomic mass units). For small and medium-sized molecules (where $M < 1000$), which typically generate singly charged ions ($z = 1$), the increasing order in mass is the same as the ratio m/z. This assumes that the spectra are graduated in atomic units (u) or Daltons (Da). Generally, m/z is expressed in Thomsons (Th), a unit rarely indicated on spectra.

Ion abundances can be recorded in two different ways:

— the *continuous spectrum* of a selected mass or interval of masses. In this mode, ion signals appear as peaks of different widths, depending on the instrument (Fig. 16.1). Higher performance instruments can determine the ion masses with a precision better than 10 parts in a million (10^{-5} Da). The present mass limits of the instrument, which is constantly being improved, can reach 10^6 Da.

— the *fragmentation spectrum* (*bar spectrum* or stick diagram). The bar spec-trum is produced by summing the intensities of ions at the nominal masses closest to their exact mass, and intensities are usually expressed as a % of the most intense peak, typically called the *base peak*. This graphical representation of masses, whose heights are proportional to their abundance, is essentially a histogram (Fig. 16.1). Bar spectra are easily archived and can be used for further comparisons. This nor-malised representation, sometimes presented in tabular form, has a limitation in that several ions with different elemental compositions (different exact masses) can occur at the same nominal mass.

■ The expression *mass spectrum* should not cause confusion. The instrument does not lead to a spectrum in the traditional sense like that encountered in methods measuring the interaction of a sample with electromagnetic radiation. The expression *mass spectrum* originates from the resemblance of recordings obtained on the first instruments using photographic plates with those of optical spectrographs from the same era. Mass spectrometry does not belong to optical methods of spectroscopy.

Mass spectrometry has gradually become an irreplaceable method for the investigation of compound structures. It is also applied to the determination of the elemental composition of inorganic samples (ICP/MS). When combined with chromatography for the separation of compounds, MS can be used to study mixtures of molecular species. The *on-line* coupling of these techniques, GC/MS, constitutes one of the best methods of analysis for mixtures and samples containing trace amounts of analyte.

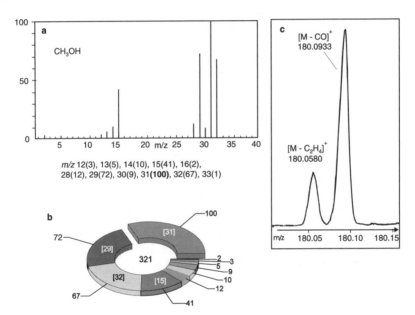

Figure 16.1—*Bar (fragmentation) spectrum and mass spectrum presented in graphical and tabular form.* a) Bar spectrum of methanol; b) non-conventional representation of the same spectrum in the form of a circular diagram: for each 321 ions formed, there are statistically 100 ions of mass 31 u (Da), 72 of mass 29 u, etc. The various ions constitute different populations; c) part of a high-resolution recording of compound M with two ions with very close *m/z* ratios (one due to loss of CO and the other due to loss of C_2H_4).

■ Compound identification by mass spectrometry can be achieved using either of the following methods:

— The structure of the initial compound can be reconstructed from the ion fragments such as in a puzzle. The level of difficulty increases as the mass of the compound increases because successive fragmentation reactions of the ions are complex. The task can be simplified if several spectra are recorded under different conditions and if the appropriate computer software is available.

— Identification can also be achieved using a library of mass spectra. A library containing a large number of spectra will lead, in favourable cases, to successfully finding the mass spectrum of the molecule in question.

16.2   The Bainbridge spectrometer

The use of a magnetic field to determine the m/z ratio of ions is the basis of an instrument built by Bainbridge in 1933 to study isotopes (Fig. 16.2). A number of currently used mass spectrometers are based on the use of the same classical equations.

After their creation, positive ions are accelerated through a voltage difference V. They thus acquire a velocity v that depends on their mass m. Following acceleration, the ions enter a transversal magnetic field of intensity B. The orientation of this field does not modify the ions' velocity but forces them on a circular trajectory that is a function of their m/z ratio. The fundamental relationship of dynamics $\vec{F} = m\vec{a}$ (a designates acceleration), applied to ions of mass m on which a Lorentz force $\vec{F} = q\vec{v} \wedge \vec{B}$ is exerted leads to the following relationship:

$$\vec{a} = \frac{q}{m} \vec{v} \wedge \vec{B} \tag{16.1}$$

Only the centripetal component of the acceleration vector is active. The ion trajectory lies in a plane perpendicular to $\vec{B}$ and containing $\vec{v}$. Since $a = v^2/R$, replacing the acceleration a by its form in equation (16.1) (where $q = z \cdot e$ is in coulombs, v in m/s, B in tesla and m in kg), gives:

$$R = \frac{mv}{zeB} \tag{16.2}$$

The above equation demonstrates that separation occurs as a function of the *moment* of the ions (their *quantity of movement*). Therefore, the m/z ratio of ions (equation (16.3)) can only be obtained if the ions' speed v is known:

$$\frac{m}{z} = \frac{RBe}{v} \tag{16.3}$$

In older instruments, a *velocity filter* placed before the magnetic field was used to eliminate v. This was done by combining two opposing forces created by an electric and magnetic field, which would only permit ions remaining on a central trajectory to exit the filter.

Because the net resulting force must be 0, then: $qE = q \cdot v \cdot B$ so $v = E/B$.

■ **Coulombic force and the Lorentz equation** – When a potential difference of V volts is applied between two parallel plates separated by a distance d, an electric field will be generated: $\vec{E}(V \cdot m^{-1})$. If the environment is homogeneous, the field will be uniform and oriented towards the lower potential, thus $\vec{E} = \vec{V}/d$. The field $\vec{E}$ will determine the equivalent force $\vec{F}$ acting on an ion of charge q in the field, and this will be independent of its mass ($\vec{F} = q \cdot \vec{E}$ if the ion carries a single charge,

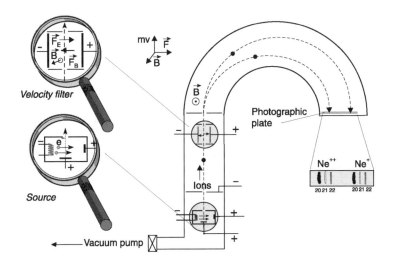

Figure 16.2—*Bainbridge spectrograph with a magnetic deflection of 180° and a velocity filter.* The filter eliminates the necessity to have a heterogeneous monokinetic beam. A photographic recording of the spectrum of neon is shown. The two series of bars are due to the stripping of one or two electrons from each isotope of neon (m/z: $^{20}\text{Ne}^{+\cdot}$, $^{21}\text{Ne}^{+\cdot}$, $^{22}\text{Ne}^{+\cdot}$, $^{20}\text{Ne}^{++}$, $^{21}\text{Ne}^{++}$, $^{22}\text{Ne}^{++}$).[1] The father of these instruments is J-J. Thomson who, in 1913, showed the existence of neon isotopes. Later Aston showed the sulphur and chlorine existence of isotopes.

$q = e = 1.6 \times 10^{-19}$ C). The coulombic force exerted is independent of the velocity of the ion.

— The force exerted on an ion carrying a charge q with a velocity $\vec{v}$ and subjected to a magnetic field of intensity $\vec{B}$ is given by the Lorentz equation:

$$\vec{F} = q\vec{v} \wedge \vec{B} \qquad (16.4)$$

This equation can be deduced from the general Laplace formula that expresses the force exerted on a conductor of length dl, through which a current I passes, in a magnetic field of intensity $\vec{B}$. The orientation of the Lorentz force ($\vec{F} = I \cdot dl \wedge \vec{B}$) can be found by different approaches such as the right-handed three-finger rule or using the orientations of a direct trihedron.

16.3 Magnetic analysers (EB type)

Currently used magnetic analyser mass spectrometers constitute a logical evolution of the previously described instrument. They provide very accurate m/z values, but are limited for high mass analyses (problems produced in the magnetic sector). They also include an electrostatic sector E placed after ion acceleration and before the magnetic field, B (Fig. 16.3).

---

[1] The symbol $+\cdot$ signifies that the ion is both a radical (i.e. has an unpaired number of electrons) and a cation.

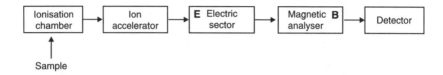

Sample

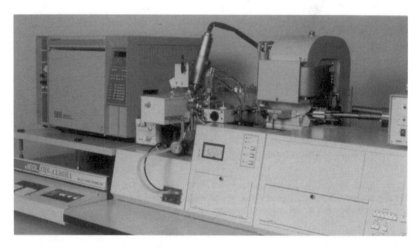

Figure 16.3—*Magnetic analyser-based instrument.* A model JMS AX505 from Jeol is shown. The characteristic shape of the electromagnet can be seen on this photograph (magnetic sector). The detector lies at the extreme right of the photograph and the ion source appears at the centre of the picture. It should be noted in the figure that the instrument is also coupled to a GC. (Reproduced by permission of Jeol, Japan.)

■ Known as **electromagnetic separators**, the use of preparative scale mass spectrometers (calutrons) allowed the United States to isolate a few kilograms of ^{235}U in 1943 to produce the first atomic bomb (the Manhattan project). This process, which is relatively inefficient under a pressure of 10^{-3} Pa, is seldom used today.

16.3.1   Ion acceleration

Approximately 5% of the positive ions formed either before or in the ionisation chamber will reach the detector. The ions must first be accelerated through a potential difference, V, which can be as high as 10 kV (Fig. 16.4). The process must be conducted under a good vacuum ($P < 10^{-4}$ Pa) to avoid electrical arcing and to minimise collisions. Under these conditions, all ions with the charge $q = ze$ will acquire the same kinetic energy $E_{kin(1)} = zeV$ (equation (16.5)). Thus, their velocity after acceleration will be inversely proportional to the square root of their mass m_i (equation (16.6)).

$$E_{kin(1)} = zeV = \frac{1}{2}m_i v_i^2 \tag{16.5}$$

$$v_i = \frac{\sqrt{2zeV}}{m_i} \tag{16.6}$$

However, because the ions have a velocity before acceleration, their total kinetic energy must take into account the initial kinetic energy $E_{\text{kin}(0)}$, which is small, such that:

$$E_{\text{kin(tot)}} = E_{\text{kin}(0)} + E_{\text{kin}(1)} \tag{16.7}$$

Equation (16.7) is critical because the mass accuracy will depend on the velocity of individual ions with the same mass. The use of a high accelerating potential V will restrain the range of velocities for ions of the same mass (Fig. 16.4).

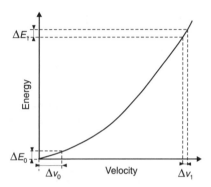

Figure 16.4—*Influence of the accelerating voltage on the velocity distribution of ions with the same mass. Assuming that Δv_0 is the velocity distribution before acceleration, it can be observed that the distribution Δv_1 after acceleration is much smaller ($\Delta E_1 = \Delta E_0$ but $\Delta v_1 \ll \Delta v_0$).*

16.3.2 Electric sector

Using an electric sector made of two cylindrical electrodes (Fig. 16.5) can increase precision in mass measurement. The electric sector will compensate for the non-homogeneous energy of the ion beam. The force exerted on the ions by the electric sector does not modify their kinetic energy, since the force is perpendicular to the central trajectory which has a radius of curvature R'. Its intensity is equal to $F = mv^2/R'$. Replacing the term mv^2 by its equivalent in terms of acceleration ($mv^2 = 2zeV$), the transmission conditions of the ions through this filter can be obtained. When the electric field $\vec{E}$ and V follow equation (16.8), the filter will transmit only ions which have acquired the proper energy under acceleration due to the potential drop V. The field $\vec{E}$ can be calculated using the voltage difference U across the sector plates and the distance d between them ($\vec{E} = U/d$). Thus the sector will act as an *energy filter*. Moreover, the electric sector has *direction-focusing properties*. It can bring into focus ions that have been emitted at different angles before entering the sector (Fig. 16.5).

$$\vec{E} = \frac{2U}{R'} \tag{16.8}$$

16.3.3 Magnetic analysers

In the EB configuration, a magnetic analyser follows the electric sector. Instruments of this type can be classified into two main categories: the Nier–Johnson geometry in which the radius of curvature in the magnetic field is the same direction as that in the electric field and the Mattauch–Herzog geometry, where the curvature in each field is in opposite directions (see Fig. 16.5). In the latter arrangement, if the magnetic ana-

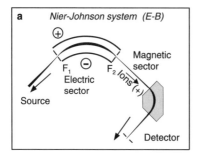 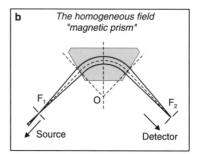

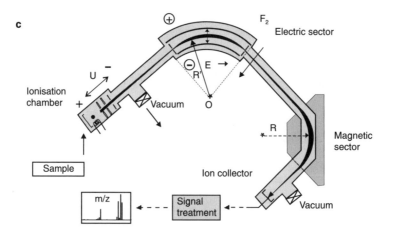

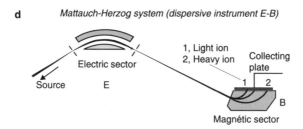

Figure 16.5—*Magnetic analyser mass spectrometer.* a) Nier–Johnson system, b) directional focusing properties of the magnetic field, c) principle of a double focusing mass spectrometer d) Mattauch–Herzog system.

lyser has the proper geometry, all masses are simultaneously focused in the beam, which is a necessary condition for spectrometers that measure isotope ratios, as described further on. Eliminating the velocity v by combining equations (16.2) and (16.5) allows the *deflection (focusing) equation* for a magnetic sector to be written as:

$$\frac{m}{z} = \frac{R^2 B^2 e}{2V} \tag{16.9}$$

For a Nier–Johnson geometry, ions following a trajectory with radius R, imposed by the analyser tube, will be detected. Hence, to obtain a spectrum within a given mass range, the field B has to be progressively modified (or scanned). In the scanning mode, all ions, irrespective of their m/z value, will successively follow the unique path that leads to the detector.

> ■ The relationship between m and V is the basis of a method for accurate mass measurement called *peak matching*. In this technique, compound X whose mass is to be measured, is injected into the mass spectrometer along with a mass standard (usually a polyfluorated compound whose fragment ion masses are known with high precision). For the measurement, a reference peak close to the mass of interest is chosen. For example, if the mass to be determined is in the order of 200 Da, the C_4F_8 ion of perfluorokerosene (PFK) of mass 199.9872 will be used. In practice, a dual channel oscilloscope is used with a decade box (made of a series of precision resistors) and both the unknown and the reference peaks are forced to overlap as an indirect method for modifying V.
>
> $$\frac{m_x}{m_{ref}} = \frac{V_{ref}}{V_x} \tag{16.10}$$

The magnetic sector, also called a *magnetic prism*, has the same directional focusing properties as the electric sector: a beam of ions of the same mass and kinetic energy entering the magnetic sector F_1 with a small angular dispersion α will be refocused in F_2, which is the image of F_1. At a given moment during the scan, only ions with a defined mass can follow the radius of curvature R (Fig. 16.5). This two-sector geometry allows simultaneous energy and angular focusing of the ions (*directional and energy focusing*). In magnetic sector instruments, the weight and the size of the magnet limit the mass range.

16.4 Time of flight (TOF) analysers

The principle underlying time of flight (TOF) mass spectrometers is based on the relationship that exists between mass and velocity at a given kinetic energy. The instrument, which uses pulsed ionisation, measures the time taken by each mass to travel the length L of a field-free analyser tube. The basic equation (16.11) used in linear TOF analysers is obtained by eliminating the velocity v from equation (16.5) in conjunction with the relationship $L = vt$:

$$\frac{m}{z} = \left(\frac{2eV}{L^2}\right) t^2 \tag{16.11}$$

In practice, TOF analysers can measure masses in excess of 300 kDa. The resolution on a TOF mass spectrometer is related to the duration of the ionisation pulse, to the initial energy distribution of the ions and to the precision with which the time of flight can be measured. Resolution can be increased using almost instantaneous ionisation methods (a few nanoseconds with a laser pulse) and the process can be repeated several hundred times per second. A *reflectron* (Fig. 16.6) can also be used in the analyser to compensate for kinetic energy distributions. In this electrostatic mirror (the reflectron), ions with higher velocities (i.e. higher energies) will penetrate deeper into the zone of the field than low energy ions, which allows ion focusing at the collector. Thus, ions of the same mass with different energy distributions will be collected at the same time.

■ **Ion mobility spectrometer (IMS).** Portable spectrometers used for the detection or determination of target compounds are also based on the flight time of ions reaching the detector. However, the flight tube is operated at atmospheric pressure (Fig. 16.7). This technique is very different to TOF mass spectrometry, with which it is often compared; instead it is close to a gas-phase electrophoresis process. In the environment of the ion source, which contains a film of ^{63}Ni (β^- emitter), the presence of air leads to ionic clusters $[(N_2)_x(H_2O)_yH]^+$ and $[(N_2)_x(H_2O)_yO_2]^-$ that will react with the compounds present depending on their electron affinity. This chemical ionisation-type process forms an aerosol that will migrate under the effect of the electric field to the detector in a few milliseconds. The mobility of ions depends not only on their mass, charge and size but also on air pressure and temperature. Limits of detection of a few ppb can be obtained with this technique.

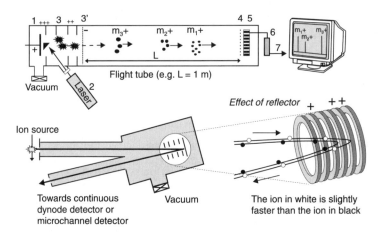

Figure 16.6—*Linear time of flight (TOF) and principle of the reflectron.* 1) Sample and sample holder; 2) MALDI ionisation device; 3 and 3′) extraction and acceleration grid (5 000 V potential drop); 4) control grid; 5) multichannel collector plate; 6) electron multiplier; 7) signal output. The bottom figure shows a *reflectron*, which is essentially an electrostatic mirror that is used to time-focus ions of the same mass, but which have different initial energies. This device increases resolution, which can attain several thousand.

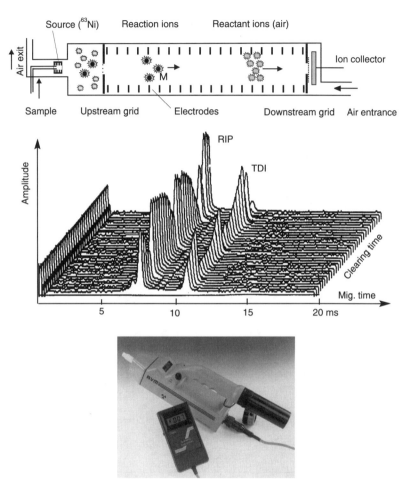

Figure 16.7—*Ion mobility spectrometer*. Ions enter the analyser tube by control of the polarity of the acceleration grid. An example of a recording in the repetitive mode (RIP: Reactant Ion Peak) for tolyl isocyanate (TDI), the compound analysed, is shown. A commercial model of an Environmental Vapour Monitor is also shown (reproduced by permission of Grasby Electronics, UK). This instrument uses a GC column to improve compound identification.

16.5 Ion cyclotron resonance

This type of mass spectrometer, which is not widely used, allows mass determination with a high precision. An ion cyclotron resonance spectrometer is basically an ion trap: ions formed by electron impact, for example, are subjected to the orthogonal magnetic field B, which induces cyclotronic movement in the xy plane (Fig. 16.8). The radius of the circular movement, which depends on kinetic energy, is given by equation (16.2). If the velocity v is small and the magnetic field B is intense, the radius of the trajectory will be small and the ions will be trapped in the ionisation chamber. This assumes that conditions of high vacuum are present and that the walls repel the ions. Using the classical equations $v = \omega R$ and $\omega = 2\pi\nu$, equation (16.12) can be obtained:

$$\frac{m}{z} = \frac{eB}{2\pi\nu} \qquad (16.12)$$

The rotation frequency ν of each of the ions depends on its mass and not on its velocity. Hence, at a given time there will be as many frequencies as there are ions of different m/z ratio in the trap. All of this occurs in a very small rectangular cell that is inserted into an intense magnetic field (3 to 8 teslas) generated by a superconductor (Fig. 16.8). Mass analysis is achieved by measuring precisely the different frequencies of the ions.

At the beginning of the analysis, the ensemble of frequencies is incoherent and has no measurable macroscopic effect on the cell. Ions with the same m/z ratio must be made coherent to conduct a frequency analysis. This is achieved by irradiating the cell with a short radiofrequency pulse (*ca.* 1 ms lifetime) that includes all the frequencies to be determined. During the irradiation pulse, ions will increase their

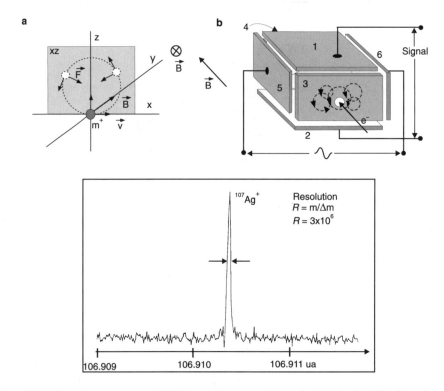

Figure 16.8—*Ion cyclotron resonance (ICR) mass spectrometer.* Ion trajectories in the ICR cell are shown. Plates 5 and 6 are used for excitation, plates 3 and 4 are used to trap ions and plates 1 and 2 are used as the detection system. Ions can be formed inside or outside the ICR cell. Example of the high resolution that can be obtained with this type of spectrometer ($R = 3 \times 10^6$), cf. 16.8.3.

energy: the radii of their trajectories will increase. Moreover, ions of the same m/z ratio will become coherent (i.e. come into cyclotronic resonance) and will pass close to the wall at the same instant. The net result will be a detectable signal formed by the superposition of the current induced by each population of m/z. The intensity of this interferogram [$I = \mathrm{f}(t)$] will decrease as the excitation is stopped and contains information on each of the frequencies that form it. A Fourier transform analysis can be made on the data to convert $I = \mathrm{f}(t)$ into $I = \mathrm{f}(\nu)$, which leads to the generation of a mass spectrum $I = \mathrm{f}(m/z)$.

16.6 Quadrupole analysers

Many applications of mass spectrometry do not require high resolution. Thus, other types of smaller, less expensive and lower performance mass spectrometers have been developed. Quadrupole mass filters that use an electric field to separate ions are in this category. These devices are used as mass detectors (GC/MS, LC/MS and ICP/MS) and in a variety of industrial processes for gas and residual gas analyses.

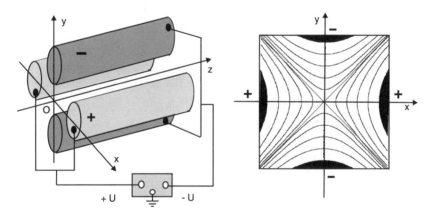

Figure 16.9—*Representation of a quadrupole*. Notice the pairing of oppositely charged electrodes. This experimental design requires high-precision machining of the hyperbolic electrodes. To the right a series of equipotential hyperbolic lines in the central part of the quadrupole is shown.

16.6.1 Potential and electric fields in a quadrupole

An ideal quadrupole field can be generated using four parallel electrodes ($L = 5$ to 20 cm) which have a hyperbolic cross-sectional field at their interior (Fig. 16.9). The electrodes are coupled in pairs and a potential difference U is applied across the pairs. If the distance between two opposite electrodes is $2 \cdot r_0$, then the potential Φ within the xy plane of the quadrupole will be given by:

$$\Phi = U \cdot \frac{x^2 - y^2}{r_0^2} \qquad (16.13)$$

Under homogeneous conditions, the potential Φ is 0 along the optical O-z axis.

Equation (16.13) implies that all the points in the x-O-y plane, for which $y = \pm x$ define the asymptotes of the equilateral hyperbola, are at the same potential. The field lines are orthogonal to equipotential curves, which implies that for each point M_{xyz} inside the quadrupole, the value of the electrical field will be:

$$\vec{E} = -\text{grad } \Phi$$

The voltage at the surface of each electrode inside the quadrupole is given by:

x electrode $\Phi = +U \Rightarrow y^2 = x^2 - r_0^2$

y electrode $\Phi = -U \Rightarrow y^2 = x^2 + r_0^2$

When a positive ion enters the filter, maintained under vacuum, at the origin O, its velocity vector in xyz space will determine its trajectory. The central portion of the quadrupole behaves like a tunnel in the O-z axis. The walls of the tunnel can either attract or repel an ion depending on the ion's position. The two positively charged electrodes will focus the ion in the O-z axis, corresponding to a potential valley (stability zone) while the two negatively charged electrodes will have a defocusing effect (potential maximum, unstable y-O-z plane).

■ The image of a marble that rolls and comes to rest at the bottom of a horizontally oriented drainpipe resembles an ion in a potential valley. A potential maximum is analogous to trying to place the marble on the outside of the drainpipe; it always rolls off, representing an unstable position.

16.6.2   Use of a quadrupole as a mass filter

An AC voltage V_{RF} of radiofrequency ν and maximum amplitude V_M is superimposed on a DC potential U (ν is in the order of 2 to 6 MHz and V_M/U is in the order of 6 MHz). The alternating applied potentials of the two sets of electrodes are out of phase by π (Fig. 16.10).

$$V_{RF} = V_M \cos(2\pi\nu t) \tag{16.14}$$

Thus, the potential at each point in the quadrupole filter as a function of time will be the sum of expressions (16.13) and (16.14):

$$\Phi = [U + V_M \cos(2\pi\nu t)]\frac{x^2 - y^2}{r_0^2} \tag{16.15}$$

The resulting field within the quadrupole is made up partially of a fixed voltage and partially of a variable voltage. Under these conditions, ions entering the quadrupole at O are subjected to a force that is variable both in intensity and in direction. As a result of the effect of the field, the ions follow complex three-dimensional trajectories that are generally unstable, leading to their contact with one of the electrodes.

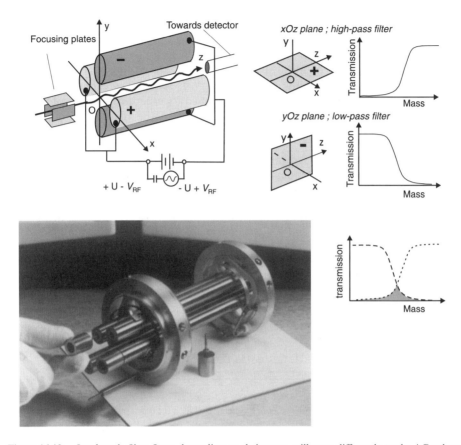

Figure 16.10—*Quadrupole filter*. Ions, depending on their mass, will react differently to the AC voltage. A model of a quadrupole filter with a pre-filter is shown (Fisons).

To calculate trajectories, the forces F_X, F_Y and F_Z exerted on the ion must be known (Mathieu equations). Because this operation involves complex computation, only two particular cases will be discussed.

— Assume an ion has a zero velocity component in the O-y axis: its trajectory will be in the x-O-z plane. Because the electrodes in that plane will alternately be positive or negative ($V_{RF} > U$), heavy ions with a high inertia will not substantially respond to the variation of the field: on average, they will sense the positive potential U and be focused on the central axis. However, ions with lower mass, having a lower inertia, will oscillate with the change in voltage and as they gain amplitude will be discharged and thus lost on one of the electrodes. Thus the two positive electrodes constitute a *high-pass filter*.

— Consider now an ion that has a zero velocity component in the O-x axis: it will remain in the y-O-z plane. Under these circumstances, heavy ions with a high inertia will eventually be attracted by the negative electrodes. Lighter ions that can rapidly react to the influence of the radiofrequency field will be transmitted. Thus, the z-y plane constitutes a *low-pass filter*.

There are two ways to operate a quadrupole filter: the voltages U and V_{RF} are maintained constant while ν, the radiofrequency, is scanned, or alternatively, ν is maintained constant and both voltages are scanned together keeping their ratio at a fixed value. It is thus possible to sequentially obtain the mass spectrum of the compound.

■ The resolution R of a quadrupole mass filter (see equation (16.17) is related to many parameters. The instrument can be operated with a constant Δm, in which case resolution increases with mass (Fig. 16.11).

The maximum resolution is given by equation (16.16) where e represents a unit charge, V_z the acceleration potential, L the length of the quadrupole mass filter, r_0 the radius of the filter and V_{max} the maximum voltage of the radiofrequency.

$$R = \text{Constant} \cdot \frac{L^2 V_{max}}{r_0 e V_z} \tag{16.16}$$

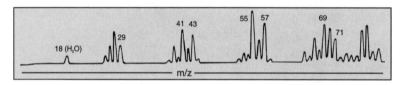

Figure 16.11—*Partial spectrum of a hydrocarbon obtained with a gas analyser operating under the principle of a quadrupole filter. In this case, the instrument has been used with the constant Δm over the mass range.*

16.7   Ion traps

Three-dimensional ion traps that operate on the principle of the quadrupole are another type of mass analyser (with or without a DC component). In ion traps, the ions are confined between the electrodes which have a particular shape that resembles the set-up of a quadrupole. Although they are physically simple devices, the fundamental principle of ion traps is very complex and they are more sensitive yet less expensive than quadrupoles. The volume determined by the so-called annular, superior and inferior electrodes is simultaneously the ion source and the *mass filter* (see Fig. 16.12). Ion traps are usually coupled to separation techniques (GC/MS, LC/MS).

The mode of operation of an ion trap can be described in the following way: the ions are generated in the central part of the filter by electron ionisation using a short electron pulse. A radiofrequency voltage is then applied to the annular electrode, which confines the ions in the source where they follow complex trajectories in the presence of a low helium pressure of about 0.01 Pa. The mass spectrum is obtained by increasing the radiofrequency amplitude, which destabilises ions of increasing mass. The increase in voltage causes the ions to increase the amplitude of their

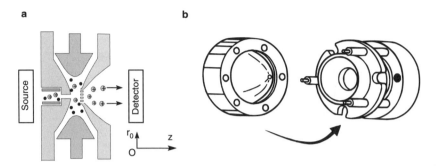

Figure 16.12—*Ion trap*. a) Design of the electrodes on an ion trap; b) the electrodes used in the ion trap.

oscillations in the axial direction and they are eventually ejected through one of the end-cap electrodes behind which stands an electron multiplier. Chemical ionisation can also be used if a reagent gas is introduced concurrently with the sample.

16.8 Performance of mass spectrometers

16.8.1 Upper mass limit

Every mass spectrometer has an upper mass limit for the m/z ratio it can measure. Thus the mass in Daltons corresponding to a particular m/z will depend on the number of charges z that the ion carries. For example, with an upper m/z value of 2 000, the instrument can detect an ion of mass 80 000 Da if that ion has a charge of $q = 40e$.

16.8.2 Sensitivity

The sensitivity of a mass spectrometer is given in terms of the amount of sample consumed per second (e.g. a few pg/s or fmol/s) to give a signal with normalised intensity. Since the ion beam persists for some time, spectra are obtained by repetitive scanning.

16.8.3 Resolution power

In a spectrum, the width of every mass peak is related to the performance of the instrument. An important property of the instrument is its ability to separate masses close to one another, which is called the *resolving power R (resolution)*. R is calculated by dividing the m/z ratio of the chosen peak by its width $\Delta(m/z)$ (see Fig. 16.13) which is usually measured at half-height (FWHM, full width at half maximum). For magnetic sector instruments, width is traditionally measured at 10% of the peak height.

$$R = \frac{m/z}{\Delta(m/z)} \qquad \text{or} \qquad R = M/\Delta m \qquad (16.17)$$

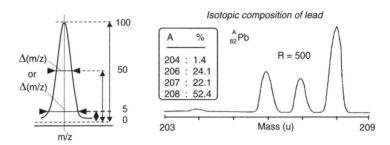

Figure 16.13—*Resolution.* The figure on the left defines the parameters used to calculate resolution. To the right a low-resolution spectrum of a sample of lead is shown. The highest mass resolutions are achieved with cyclotron resonance instruments (see Fig. 16.8). The resolution greatly depends on the compound chosen for the calculation. For instruments in which Δm is a constant, the upper mass limit theoretically corresponds to the maximum resolution, i.e. this the value at which masses m and $m + 1$ can no longer be distinguished from one another.

16.9   Sample introduction

Given the variety of forms and the nature of possible samples, there are numerous methods of introducing a sample to a mass spectrometer. The sample can be ionised either before or in the ion source of the spectrometer.

16.9.1   Direct sample introduction

For gases or volatile liquids, a small quantity of sample is injected with a micro-syringe into a reservoir that is connected to the ionisation chamber via a minute orifice. Because the reservoir is maintained under vacuum and can also be heated, the sample is vaporised. This procedure is called a *molecular leak* or *molecular pumping*.

For continuous monitoring of industrial processes, the sample can be diffused through a porous membrane into the mass spectrometer.

Solid samples that have a sufficient vapour pressure at 300 °C are deposited on the tip of a heated metal probe which is then inserted into the instrument through a vacuum lock. With some ionisation methods, the solid sample is mixed with a liquid matrix (e.g. glycerol or benzoic acid).

16.9.2   Sample introduction in hyphenated techniques

Complex mixtures are often analysed by hyphenated techniques, in which a separation method is interfaced with a mass spectrometer. Coupling of a gas chromatograph to a mass spectrometer (GC/MS) is simple: the chromatographic capillary column is directly inserted into the ionisation chamber. Compounds eluting from the column in a gaseous state are introduced into the ion source and analysed in the order in which they exit the CG column. The vacuum system of the mass

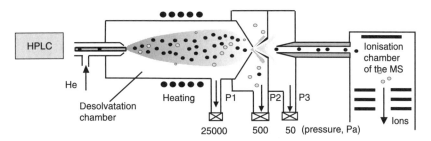

Figure 16.14—*HPLC/MS interface using a particle beam interface.* The angular diffusion of heavy molecules under vacuum is narrower that that of smaller molecules of solvent, thus they have a higher probability of being transported to the mass spectrometer along the central axis of the system.

spectrometer usually has sufficient pumping capacity to handle column flows of 1 to 2 ml/min.

Interfacing HPLC or HPCE (capillary electrophoresis) to mass spectrometry is technically more complex than with a GC because these techniques use a solvent that is often aqueous (water is a poison to mass spectrometers). The use of micro-columns in HPLC is desirable for coupling to MS because micro-columns operate at very low flow rates. They are also compatible with different ionisation techniques for the analysis of high molecular-weight species.

The *particle beam interface* is an example of an HPLC/MS interface (see Fig. 16.14). In this interface, the mobile phase is vaporised in the form of a spray into a desolvation chamber before entering the area where the sample is concentrated by evaporation. Because solvent molecules are lighter, their angular dispersion is wider than that of the analyte, which can be transferred into the transfer capillary. This type of interface is slowly being abandoned in favour of others that have a greater sensitivity.

16.10   Major ionisation techniques (under vacuum)

16.10.1   Electron ionisation (EI)

Electron ionisation is still the most widely used technique for the analysis of volatile molecules. It is considered to be a 'hard ionisation' process, which leads to repro-ducible spectra that can be compared to a library of mass spectra for compound identification. In this technique, ionisation occurs in the ion source by the collision of the sample molecules with electrons that are emitted from a filament by a thermo-ionic process (Fig. 16.15).

In EI, it is customary to use an ionisation energy of 70 eV. This is achieved by accelerating the electron produced by the filament through a potential drop of 70 V, applied between the filament and the chamber. Ionisation efficiency in EI is in the order of one ion produced for every 10 000 molecules. In some cases, reducing

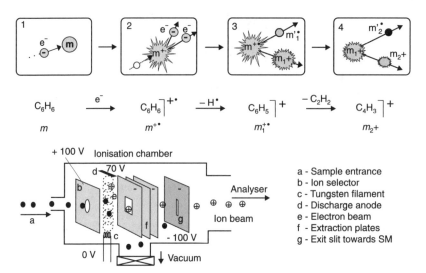

Figure 16.15—*Electron ionisation (EI).* The collision of an electron with a sample molecule m produces ionisation that leads to formation of a parent ion and fragment ions. Ions that result from the reaction m_1^+ and m_2^+ are also called secondary or daughter ions. Since they carry no charge, neutral fragments produced during decomposition, $(m, m_1'$ and $m_2')$, are not detected. An illustration of electron ionisation of benzene is shown. Also shown is a schematic of the ionisation chamber (ion source). Using a parallel magnetic field can increase the effective path of an electron in the ion source, which increases ionisation efficiency.

electron energy to about 15 eV can increase the relative intensity of the molecular ion $M^{+\bullet}$ (Fig. 16.16).

■ In electron ionisation, the yield of positive ions is much greater than that of negative ions, Thus, the technique is usually used in the positive ion mode. Nonetheless, the negative ions produced can be studied by inverting the polarity of the accelerating voltage and that of the magnet.
A relatively explicit nomenclature is used to characterise the type of ions formed: singly charged, multiply charged, monoatomic, polyatomic, molecular, pseudo-molecular, parent, daughter, metastable, secondary fragments, etc.

16.10.2   Chemical ionisation (CI)

Chemical ionisation results from the gas-phase collision between the analyte and species formed from the reagent gas introduced concomitantly in the ion source and bombarded by electrons. Methane, ammonium or isobutane are often used as reagent gases (Fig. 16.17). The reagent gas is introduced into the ion source at a pressure of a few hundred pascals, which reduces the mean free path and favours collision. Chemical ionisation produces positively and negatively charged species.

The pseudomolecular ion $MH^{]+}$ (which is not a radical ion) is usually more stable than the $M^{]+\bullet}$ ion produced by electron ionisation. Hence, fewer fragment ions are observed in chemical ionisation and so it is called a 'soft ionisation' technique.

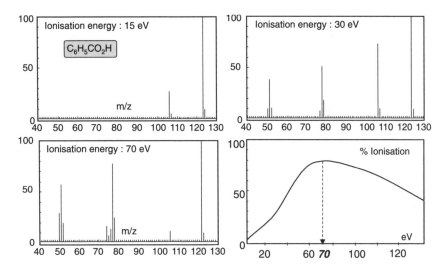

Figure 16.16—*Influence of electron energy on fragmentation.* An example with benzoic acid is shown here. An ionisation efficiency curve as a function of electron energy is presented. It should be noted that the spectra are normalised; that is, the most intense peak has a value of 100. Some claim that this is an application of Procruste's method (according to Greek legend, the bandit Procruste would force travellers to lie on his bed and, depending on their size, he would either cut their feet or stretch their bodies so they would fit the dimensions of the bed).

Primary reactions : CH_4 $+ e^-$ $\longrightarrow$ $CH_4\rceil^{+\bullet} + 2\,e^-$ *(Ionisation)*

Secondary reactions : $CH_4\rceil^{+\bullet}$ $\longrightarrow$ $CH_3\rceil^+ + H^\bullet$

$CH_4\rceil^{+\bullet} + CH_4 \longrightarrow$ $CH_5\rceil^+ + CH_3^\bullet$ *(Autoprotonation)*

$CH_3\rceil^+ + CH_4 \longrightarrow$ $C_2H_5\rceil^+ + H_2$

Collisions with M : $CH_5\rceil^+ + M$ $\longrightarrow$ $MH\rceil^+ + CH_4$ (ion at M+1)

$C_2H_5\rceil^+ + M$ $\longrightarrow$ $MC_2H_5\rceil^+$ (ion at M+29)

If M is of the type RH : $CH_5\rceil^+ + RH \longrightarrow$ $R^+ + CH_4 + H_2$ (ion at M-1)

If it is a tert-butyl cation : $(CH_3)_3C\rceil^+ + M \longrightarrow$ $MH\rceil^+ + CH_2{=}C(CH_3)_2$

Figure 16.17—*Chemical ionisation.* This figure shows reactions occurring when methane is used as a reagent gas. The last equation represents the reaction that occurs when isobutane is used. Because of the high pressure used, the intensity of ions from the reagent gas is high, thus the mass spectrum is not scanned below 50 Da.

The protonation reaction leading to $MH^{]+}$ depends on the proton affinity of M. Because hydrocarbons of type RH have a low proton affinity, the R^+ ion is often seen in the spectrum. Since proton transfer in chemical ionisation can lead to either $M + H^{]+}$ or $M - H^{]+}$, uncertainties can exist as to the true mass of the molecular ion.

In ammonia chemical ionisation, $NH_4^{]+}$ behaves in a similar way to $CH_5^{]+}$. In isobutane, the main reacting ion is $(CH_3)_3C^{]+}$, which is stable and leads to formation of the ion $(M + 1)$ by a reaction leading to formation of isobutene (see Fig. 16.17).

16.10.3   Fast atom bombardment (FAB)

This ionisation technique and others described in the following sections are used for non-volatile compounds. In FAB, a beam of fast atoms (high-energy argon or xenon) is used to induce ionisation. The fast atoms are produced by first ionising argon or xenon gas in a cell, accelerating these primary ions formed, then by colliding these ions with slow-moving neutrals of the same gas (Fig. 16.18). In these collisions, the charge of the fast-moving ion is transferred to the slow-moving neutral species resulting in a fast neutral atom, which is then directed at the sample.

In FAB, the sample is usually dispersed in a non-volatile liquid matrix, such as glycerol or diethanolamine, and deposited at the end of a sample probe that can be inserted into the ion source. The sample on the probe is ionised when bombarded by the fast atom beam. However, ionisation of the matrix also occurs, leading to a very large background signal. The technique is thus limited for the analysis of small molecules. Fast-moving ions (Cs^+ or Ar^+) can be used instead of fast-moving atoms, which is the basis of a technique called liquid secondary ion mass spectrometry (LSIMS).

16.10.4   Matrix assisted laser desorption ionisation (MALDI)

This desorption ionisation technique leads to weak fragmentation. The analyte is incorporated into a solid organic matrix (such as hydroxybenzoic acid) and the mixture is placed on a sample holder that is irradiated with UV laser pulses (e.g. N_2 laser, $\lambda = 337\,nm$, pulse width $= 5\,ns$). The laser energy is absorbed by the matrix and transferred to the analyte, which becomes desorbed and ionised (Fig. 16.18c). Although MALDI is considered to be a soft ionisation technique, a substantial amount of energy is involved. Because the technique involves pulsed ionisation, it is well suited for time-of-flight mass analysis of biomolecules. The analysis of small molecules ($M < 500\,Da$) is limited because the matrix decomposes upon absorption of the laser radiation. However, solid supports such as silicone can be used as the matrix to overcome this disadvantage.

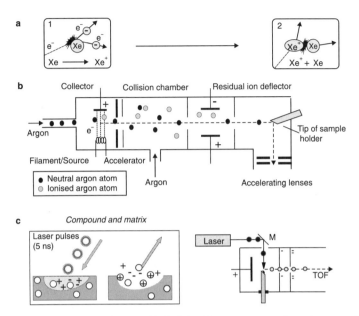

Figure 16.18—*FAB and MALDI techniques.* a) Principle of fast-atom generation using xenon; b) formation of a fast-atom beam of argon in a collision chamber and bombardment of the sample (using a FAB gun); c) MALDI. The impact of a photon leads to a similar result as with a fast atom. The mechanism for desorption ionisation is not entirely known. These ionisation modes are particularly well suited for the study of medium to high molecular weight species. They are mostly used in biomedical sciences but not for routine determinations.

16.11 Atmospheric pressure ionisation (API)

Several of the more recent applications of mass spectrometry involve the determination of elements and of unstable, polar biomolecules ($M > 2\,000$ Da). These applications have become possible with the advent of atmospheric pressure ionisation (API) techniques that permit the analysis of solutions.

16.11.1 Inorganic analysis using argon plasma ionisation

Argon plasmas are used in optical emission spectrometry to atomise and ionise elements leading to the emission of characteristic spectral lines. Hence, a plasma torch ($7–8\,000$ K) can be used for ionisation in mass spectrometry. Ions produced in the plasma are introduced into the mass analyser through a small orifice (called a skimmer) placed in the axial direction. Because the mass spectrometer is operated under a vacuum, the ions are sucked into the mass analyser through the skimmer. An aqueous solution of the sample can be aspirated into the plasma or, alternatively, the plasma can be placed at the exit of a gas chromatograph (e.g. *speciation* of organometallic compounds by GC/ICP-MS). Since all chemical bonds are broken in the plasma, the only accessible information is that concerning the concentration of monoatomic ions (Fig. 16.19).

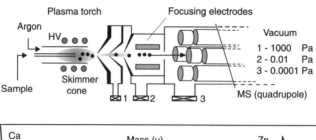

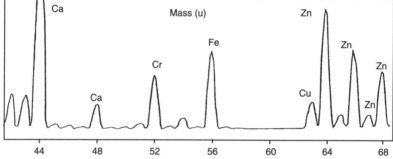

Figure 16.19—*ICP – MS recording plasma torch ionisation for elemental analysis.* A spectrum of a sample containing metals in the mass range of 42 to 68 Da is shown. The sample contains a high concentration of calcium that constitutes a major peak at 40 Da (not shown on the figure). The resolution in this recording is not sufficient to separate isobaric ions, thus peaks in the spectrum are the sum of different species with the same nominal mass.

16.11.2   Spray ionisation of biomolecules

These ionisation devices are usually placed at the exit of micro-HPLC or CE columns where the mobile phase is nebulised into the mass spectrometer. Ions are generated either by adding an electrolyte to the mobile phase, such as NH_4^+, Na^+, K^+ or H^+ or by using an electric field (*electrospray*, ESI, or *ion-spray*) (Fig. 16.21b). Droplets in the spray undergo heating or collision with a dry gas, which causes the evaporation or desolvation of solvent molecules. This process eventually leads to the formation of single (non-fragmented) ions that are protonated or 'cationised' and carry a variable number of charges (approximately one charge per 1 000 Da). The mass of the compound can usually be calculated from the molecular ion cluster (Fig. 16.16).

Atmospheric pressure chemical ionisation (APCI, Fig. 16.21c) is a high sensitivity process that can also lead to the presence of multiply charged species $(M + nH)^{n+}$.

One of the main advantages of these soft ionisation techniques is that they lead to the formation of multiply charged, pseudomolecular ions (z can be greater than 30). Hence the mass range of the spectrometer can be extended to over 10^5 Da (to include proteins, polysaccharides and other polymers) (Fig. 16.20). These ionisation devices are often coupled to the mass spectrometer through a heated capillary transmitting the ions.

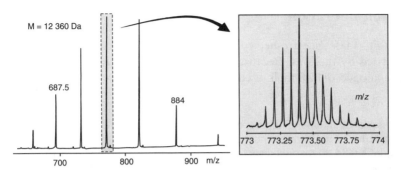

Figure 16.20—*Multiply charged molecular ions.* An electrospray spectrum of horse cytochrome c, a protein of molecular weight 12 360 Da is shown. Between two consecutive peaks in the molecular ion cluster, the charge state varies by one unit. The second spectrum corresponds to a high-resolution spectrum in the 772–774 *m/z* range. In this isotopic cluster, all ions carry the same number of charges. It is possible from either of these spectra to calculate the approximate molecular weight and the number of charges carried by the ions (spectra reprinted with permission from F. W. McLafferty *et al., Anal. Chem.*, 1995, **67**, 3802–5. Copright 1995 American Chemical Society).

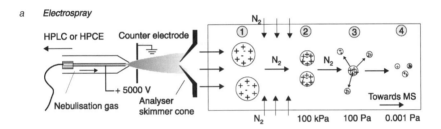

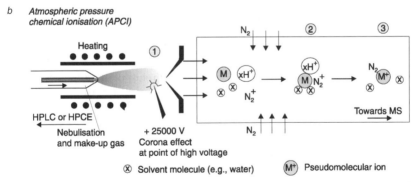

Figure 16.21—*Atmospheric pressure ionisation electrospray.* The end of the capillary, which is held at a high potential, leads to formation of charged droplets (1). As the droplets evaporate, the charge density at their surface increases leading to coulombic explosion. The process eventually leads to the formation of multiply charged molecular ions (3). A sheath gas of nitrogen is used to accelerate the evaporation process (4); *atmospheric pressure chemical ionisation.* The solution is sprayed into a heated zone (1) where a needle held at high potential is placed (corona discharge). Molecules present in the liquid phase (2) are mostly ionised and form clusters with the analytes (3). The clusters are broken down under a stream of nitrogen leading to charged analyte molecules. Thermospray is an ionisation process which uses both heating and an electrical field.

16.12   Ion detection

Detection in mass spectrometry is based on the measurement of electrical charge carried by the ions. Because many ions belonging to the same species are present, analogue detection is used. However, in some cases single ions can be measured.

■ It is essential that the number of ions detected reflects the number of ions formed, irrespective of their mass. The mass spectra are generally obtained by scanning the magnetic field. In magnetic sector instruments the m/z ratio varies with the square of the magnetic field, therefore the spacing between ions of different m/z ratio is not constant.

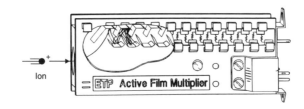

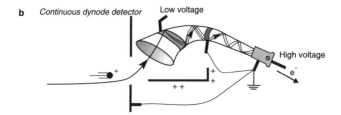

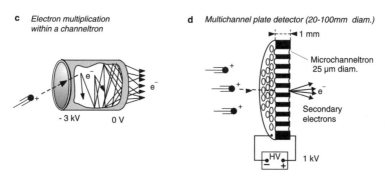

Figure 16.22—*MS detectors.* a) Multiple stage electron multipliers (reproduced by permission of ETP Scientific Inc.); b) channeltron®: the conical shape of the cathode allows the detection of ions with slightly different trajectories; c) electron multiplication within a channeltron; d) entrance of a multichannel plate detector (microchanneltron).

Types of detection system:

— *multistage electron multipliers*. In this device, positive ions hit a conversion cathode and liberate electrons. The electrons produced are then accelerated via a series of dynodes, producing an electron cascade (see Fig. 16.22). This type of detector is extremely sensitive, having a gain of up to 10^8. The conventional Cu/Be alloys used, which are sensitive only at low vacuum conditions, have been replaced by aluminium-based dynodes that have increased the performance of these detectors.

— *continuous dynode multipliers (channeltron$^{\circledR}$)*. Ions are deflected towards the collector entrance, which has a conical shape and is made of lead-doped glass. The ejected electrons produce a cascade (i.e. amplification) upon collision with the walls as they move towards the positive electrode (Fig. 16.22). This type of detector is usually mounted off-axis to avoid signals due to neutral species or photons (photons are emitted by the filament).

— *microchannel plate detectors*. These detectors consist of a large number of microchanneltrons arranged in an array. This resembles an electronic version of a photographic plate. Each of the microchannels (which are 25 μm in diameter) is coated with a semiconducting material and acts as a continuous dynode (Fig. 16.22). The electron cascade is collected at the anode. This system allows simultaneous detection of ions of different masses.

■ Photographic plates used in early instruments have now been abandoned because they are slow, non-reproducible and their response is nonlinear with ion intensity (low dynamic range). However, the principle of simultaneous detection is very attractive. Spark source Mattauch–Herzog spectrographs have long used this detection system.

16.13 Tandem mass spectrometry

Instruments that incorporate two or three mass analysers in a series have been developed to study ion fragmentation. Several of the same type of mass analyser can constitute a tandem mass spectrometer, or they can be constructed using different mass analysers (hybrids). Hybrid spectrometers include the combination of magnetic sector followed by quadrupole, multiple quadrupole, quadrupole TOF, etc. In these instruments, a collision cell is placed between each analyser (Fig. 16.23). Tandem instruments have different scanning modes.

— *Daughter ion scan*. In this mode, an ion is selected with the first mass filter and enters the collision cell where it collides with a target gas (argon or N_2). The fragments formed in the collision cell are then analysed by scanning the second mass analyser. In this way, the mass spectrum of a selected ion is obtained.

— *Parent ion scan*. In this mode, the second mass analyser is set to only allow passage of a certain m/z ratio ion. While the second analyser is fixed, the first mass analyser is scanned over the mass range. In this way, all parent ions leading to a common fragment can be identified.

— *Constant neutral loss*. In this mode, both mass analysers are scanned simultaneously. However, the mass scale of the two analysers is offset by a given mass, which constitutes a neutral fragment (e.g. CO or C_2H_2). Thus, all ions present in the ion source leading to the loss of the same neutral mass are detected.

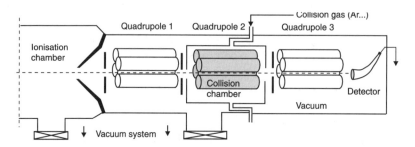

Figure 16.23—*Triple quadrupole MS – MS instrument*. In the triple quadrupole arrangement, the middle quadrupole is used as a collision chamber. It is operated in the radiofrequency voltage mode only, where it will transmit all masses. A gas pressure introduced in the second quadrupole is responsible for collision activation. Triple quadrupole instruments can conduct all three types of MS – MS analysis described above.

■ After having described the five types of mass analysers most commonly used, it is reasonable to question whether all types of mass analyser yield the same mass spectrum. No type of mass analyser is universal and each has its own characteristics. For example, the sampling method of each mass analyser is different. Mass analysers such as magnetic sector and time of flight have very fast sampling times (nano- to microseconds) while quadrupole analysers sample on a millisecond time scale and ion traps can have sampling times as high as 1 s. Thus the mass spectrum of a compound will vary depending on the analyser used for its recording. Mass spectrometers are very different to optical spectroscopy methods, hence recording conditions must be standardised before comparisons are made.

APPENDIX

Applications in mass spectrometry

Mass spectrometry is presently one of the fastest moving areas in the analytical instruments industry. It is used mainly in three areas: organic analysis, elemental analysis and isotopic analysis. This section is focused on the structural determination of organic compounds by mass spectrometry.

16.14   Determination of empirical formulae

16.14.1   Methods based on isotopic abundance

Most elements possess stable isotopes. A compound analysed by mass spectrometry will lead to an isotope cluster whose relative intensities can be calculated.

■ Sample: benzene, C_6H_6 ($M = 78$). Since approximately 1% of all carbon atoms is ^{13}C, it can be estimated that 6% of the benzene molecules will include a ^{13}C in their skeleton and will have a mass of 79 (the contribution of deuterium, ^2H, is negligible, being in the order of 0.01%). This explains the presence of a small peak at $m/z = 79$ in the mass spectrum of benzene. Similarly, a peak will be observed at $m/z = 80$ due to the presence of molecules that incorporate either two ^{13}C atoms, two deuterium (D) atoms, or one ^{13}C and one D atom.

Equations (16.18) and (16.19) give the relative intensity in percent of the $M + 1$ and $M + 2$ peaks for molecules containing C, H, N, O, P (where nC, nN, nO, etc. represent the number of carbon, nitrogen and oxygen atoms respectively in the molecule) suppose $M = 100\%$:

$$(M + 1)\% = 1.11n\text{C} + 0.36n\text{N} \tag{16.18}$$

$$(M + 2)\% = \frac{(1.11n\text{C})^2}{200} + 0.2n\text{O} \tag{16.19}$$

The relative intensity of the $M + 1$ and $M + 2$ peaks can be used to deduce the empirical formula of the compound. Software packages can be used to perform this operation, based on the relative abundance analysis. This is a commonly used method for low-resolution mass spectra, but it has limitations.

Table 16.1—Atomic mass of the isotopes of elements commonly encountered in organic compounds.

Element	Atomic mass (g.mol^{-1})	Nucleide (%)	Mass (Da)
Hydrogen	1.00794	^{1}H (99.985)	1.007825
		^{2}H (0.015)	2.014050
Carbon	12.01115	^{12}C (98.90)	12.000000
		^{13}C (1.10)	13.003355
Nitrogen	14.00674	^{14}N (99.63)	14.003074
		^{15}N (0.37)	15.000108
Oxygen	15.99940	^{16}O (99.76)	15.994915
		^{17}O (0.04)	16.999311
		^{18}O (0.20)	17.999160
Fluorine	18.99840	^{19}F (100)	18.998403
Sulphur	32.066	^{32}S (95.02)	31.972070
		^{33}S (0.75)	32.971456
		^{34}S (4.21)	33.967866
Chlorine	35.45274	^{35}Cl (75.77)	34.968852
		^{37}Cl (24.23)	36.965903

http://www.cchem.berkeley.edu/Table/

16.14.2 High resolution accurate mass determination

Accurate mass determination using techniques such as peak matching is a reliable way to obtain the empirical formula. Knowledge of the accurate mass to four or five decimal places (see Table 16.1) allows the elemental composition of the compound to be obtained. Before the use of computer algorithms, empirical formula determination was made with the help of Beynon tables. Currently, data systems propose the most probable empirical composition for a given mass. The types of elements suspected to be in the compound can be input into the algorithm to reduce calculation time.

16.15 Determination of isotope ratios for an element

Determination of isotopic abundance by mass spectrometry is a universally employed method. For example, it is possible to determine the geographic source of plant-based organic compounds or mineral compounds, or even to improve the precision of radiocarbon dating using the 'isotopic signature' of a species.

■ Isotopic distribution within an element will vary between living organisms depending on the biosynthetic pathways that lead to its formation. Furthermore, the rate at which a molecule crosses cellular membranes will depend on the molecule's isotopic distribution. Hence, detectable differences in isotopic composition can be observed in the products formed. Detection of adulterated vegetable oils, flavourings and fruit juices, as well as the study of metabolism in plants and numerous biomedical applications, use isotopic abundance as a tool. For example, the

$^{13}C/^{12}C$ ratio in natural vanillin is smaller than that observed in synthetic vanillin. The same effect applies to glucose, whose isotopic carbon distribution varies depending on the biological cycle of the plant.

Variations in the ratio $^{13}C/^{12}C$ can be determined relative to an adopted universal standard. This reference standard is calcium carbonate from Pee Dee (USA), which has an elevated abundance in ^{13}C (where $^{13}C/^{12}C = 1.12372 \times 10^{-2}$). In practice, the determination is made by measuring the peak intensities of $^{13}CO_2$ (45) and $^{12}CO_2$ (44) obtained by combustion of the sample. The relative deviation, in thousandths, δ, of the compound can thus be obtained. The value of δ is usually negative.

$$\delta = 1000 \left[\frac{\left[\dfrac{^{13}CO_2}{^{12}CO_2} \right]_{sample}}{\left[\dfrac{^{13}CO_2}{^{12}CO_2} \right]_{reference}} - 1 \right] \qquad (16.20)$$

■ Since the procedure involves combustion, measurement of the isotopic distribution at specific sites in the molecule is not possible. When this information is required, NMR is used to determine the D/H ratio at each site.

In the presence of a mixture of two compounds A and B, with isotopic ratio distributions δA and δB respectively, the total isotopic variation δM in the sample, determined experimentally, will be a weighted combination of δA and δB. If the fraction of B is represented by x and that of A by $1 - x$, this leads to the following equation:

$$\delta_M = (1 - x)\delta_A + x\delta_B$$

$$x = \frac{\delta_M - \delta_A}{\delta_B - \delta_A} \qquad (16.21)$$

The determination of deviations in isotopic ratios requires very precise measurements. The combustion stage involved is usually carried out immediately before injection into the mass spectrometer. Some instruments have been developed that include a gas chromatograph in line with a tubular combustion oven (containing copper oxide at 800 °C) and a low-resolution magnetic sector instrument. The mass spectrometer is equipped with a multicollector that allows recording at each individual mass (see Fig. 16.24).

■ **Acceleration mass spectrometry (AMS)** – The precise measurement of isotopic ratios for very low abundance isotopes is beyond the capability of conventional mass spectrometers. In these cases of isotopes at minute trace levels, some 50 mass spectrometers exist worldwide. The *tendetrons* used for these types of analyses are derived from Van de Graaff-type particle accelerators. These instruments are based on tandem mass spectrometry.

The sample, introduced in a solid form, is bombarded by caesium ions. The first mass analyser is used to extract negative ions at selected masses. The ions are then accelerated through a potential difference of several megavolts to eliminate isobaric

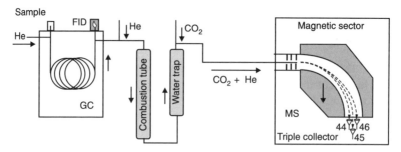

Figure 16.24—*Instrument for isotope ratio measurements.* The high sensitivity required by these types of analyses is achieved by placing several Faraday-type detectors after the magnet, each recording the current at a single mass. A reference standard is injected with the sample.

interferences. During their flight, the ions collide with a solid or gaseous target, which results in the formation of multiply charged positive ions. Abundance measurements are made by alternately recording isotopic signals from the sample, standard and analytical blank.

The application of this technique for dating objects using ^{14}C has led to a method that is more sensitive and precise than the classical Libby method of radiocarbon dating.

16.16   Identification using spectral libraries

Spectral searches using a library of reference spectra can be a useful tool in identification. Search algorithms have improved over the years and now use the concept of artificial intelligence. Several software packages can be used to conduct searches in spectral libraries in which the main peaks of known compounds are encoded. The compounds offering the best matches are retained as potential candidates. Library searches involve three stages:

— *data reduction.* The spectrum of the compound is reduced to a maximum number of peaks (*ca.* 16) preferentially keeping high-mass ions that have more information content than low-mass ions. Similarly, spectra contained in the library are coded for a maximum number of peaks, usually eight.

— *pre-search.* Reduced spectra in the library that have peaks in the same position as that of the compound are selected, although the intensity of the ions can vary.

— *main search.* After pre-selection, a refined algorithm using criteria such as peak intensity, rarity of mass, constant neutral loss, etc., is used to calculate a similarity index for each spectra. Then the potential candidates retained are ranked in order of decreasing similarity. The selection criteria can be modified to further refine the search procedure. This technique is now commonly used for the analysis of compounds that constitute spectral libraries. The NIST (National Institute of Standards and Technology) library has over 250 000 entries.

16.17   Fragmentation of organic ions

Manual interpretation of a mass spectrum from basic principles is always important for identification. Although several specialised works have been published on the topic, this type of analysis still requires a lot of experience. Organic chemists are usually familiar with these methods of interpretation since more reaction mechanisms lead to decomposition than those observed in the condensed state. However, because of the very short period of time between ion formation and ion detection (a few microseconds), species that are unstable under normal conditions can be observed.

16.17.1   Basic rules

When an organic molecule M is ionised by electrons, the primary process that occurs is the stripping off of an electron in a bonding or non-bonding (n) orbital. For compounds incorporating heteroatoms, ionisation will preferentially occur at the heteroatom. In this primary process, the molecule M is ionised but not fragmented:

$$M + e^- \rightarrow M^{1+\cdot} + 2e^-$$

The resulting radical cation (usually symbolised as $M^{1+\cdot}$) will eventually fragment. Since a mass spectrum results from the fragmentation of billions of molecules, it will show all possible fragmentations. However, the various fragmentation products will have different intensities that reflect the relative probabilities of fragmentation.

Many factors are involved in the fragmentation of $M^{1+\cdot}$ into daughter ions and neutrals (radicals or molecules):

— weak bonds will fragment more readily

— fragmentation leading to stable ionic or neutral species will be favoured

— rearrangement reactions will be favoured when they involve a six-membered ring leading to a hydrogen transfer at the radical-cation site.

16.17.2   Fragmentation of an ionised σ bond

Ionisation of a C–C σ bond in a *hydrocarbon* such as propane will lead to the formation of an ethyl cation and a methyl radical:

$$CH_3CH_2CH_3^{1+\cdot} \rightarrow CH_3CH_2^+ + CH_3^{\cdot}$$

Similarly, the ionisation of isobutane will more often lead to the formation of an isopropyl cation by elimination of a methyl radical because the isopropyl cation is thermodynamically more stable than the methyl cation.

16.17.3 α fragmentation

In ketones, ionisation of the keto group by ejection of one electron of an oxygen atom will lead to homolytic cleavage of the σ (C–C) bond that is α to the site of ionisation. Fragmentation of the butanone molecular ions leads to formation of an acetyl ion $CH_3CO^{]+}$ ($m/z = 43$) and an ion at $m/z = 57$, $CH_3CH_2CO^{]+}$, that is 10 times less intense (see Fig. 16.25). Elimination of the larger alkyl chain as a radical is favoured. In the general case of a ketone RCOR', four fragment ions, $RCO^{]+}$, $R'CO^{]+}$, $R^{]+}$ and $R'^{]+}$ are observed.

Rupture of the C–C bond adjacent to oxygen in ether ROR' leads to the formation of oxonium ions, which are stabilised by resonance (Fig. 16.26). A similar situation is encountered in amines (formation of the iminium ion at mass 30 for primary amines $CH_2 = \overset{+}{N}H_2$). Cleavage of the C–O bond in ethers is also observed and leads to the cations $R^{]+}$ and $R'^{]+}$.

α Fragmentation

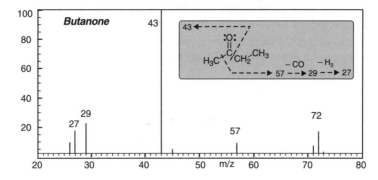

16.17.4   Rearrangement reactions

Besides fragmentation that occurs by simple bond cleavage, ions can be formed through molecular rearrangement. These rearrangements can be simple, as in the

Figure 16.25—*Electron ionisation mass spectrum of butanone.*

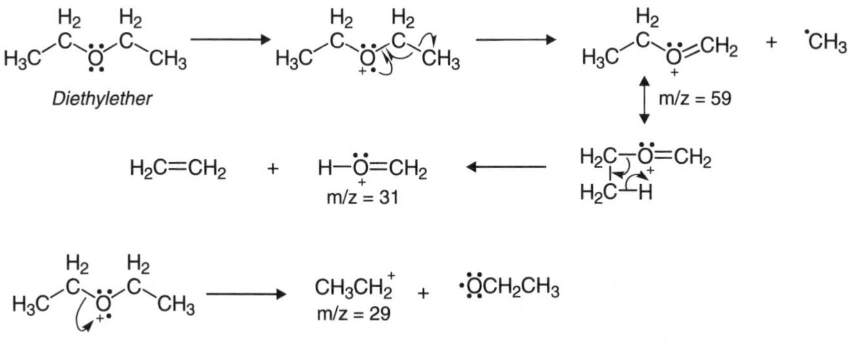

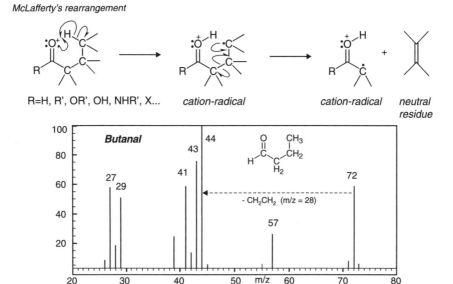

Figure 16.26—*Fragmentation pathway of diethylether in an ether.*

McLafferty's rearrangement

R=H, R', OR', OH, NHR', X... cation-radical cation-radical neutral residue

Figure 16.27—*McLafferty's rearrangement, in the case of butanal, C_4H_8O.*

elimination of H_2, or more complex. Rearrangement ions can often be observed (Fig. 16.27).

A common and well-known rearrangement reaction is the McLafferty reaction. Rearrangement is observed in compounds containing a carbonyl C=O or a double bond C=C. The reaction involves a hydrogen transfer to the ionisation site (oxygen atom) through a six-membered intermediate ring. A neutral molecule, ethylene, is eliminated in the process (see below). The fragment ion formed by this reaction is a new radical cation that has an even-numbered mass if the compound does not contain nitrogen.

The study of fragmentation processes has led to semi-empirical rules used for compound identification. Interpretation from first principles is also used to validate results obtained by spectral library searches.

16.17.5 Metastable ions

The lifetime of every ion will depend on its internal energy. Ion decomposition is a statistical phenomenon that, therefore, depends on internal energy. Thus, metastable ion decomposition can assist in spectral interpretation.

Some ions formed in the ion source are sufficiently stable (with low internal energy) to reach the detector and appear at their normal position in the mass spectrum (Fig. 16.28). However, if the lifetime of the ion, governed by its internal energy, is in the order of a few milliseconds, then the ion can decompose during its flight time. This unimolecular decomposition during its flight will lead to the following reaction:

$$M^{+\cdot} \rightarrow m^+ + m'^\cdot$$

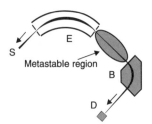
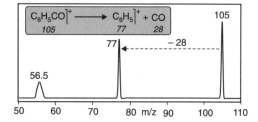

Figure 16.28—*Metastable peaks*. The figure shows the three masses involved in a metastable transition. The relationship between these three masses is given in the above text.

A daughter ion m^+ that is formed before the electric analyser in a double focusing instrument of EB geometry will not be detected. The ion will not have sufficient kinetic energy to be transmitted by the sector and will become lost. However, if dissociation occurs after the electric sector and before the magnetic analyser, the ion with momentum mv will be transmitted by the magnetic sector. Because its energy is less than that of a normal ion, the ion m^+ will be observed at an apparent mass m^*, which does not correspond to its real mass. The metastable peak, m^*, observed in the spectrum as a broader peak, is related to the mass of both the parent ion and daughter ion by the following equation:

$$m^* = \frac{m^2}{M}$$

Because of the mathematical relationship involved, the mass of a *metastable* peak m^*, called the *diffuse peak*, will seldom be at a nominal mass. It will usually appear at fractional values (Fig. 16.29).

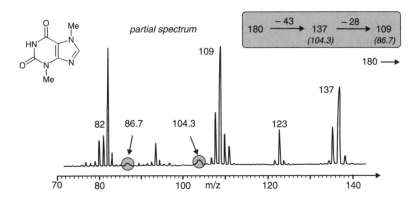

Figure 16.29—*Metastable peaks observed in the mass spectrum of theobromine.* The molecular ion (180 Da) loses the CONH' radial (43 Da) giving rise to an ion at 137 Da. This fragmentation is accompanied by a metastable peak at 104.3 Da. The fragment at 137 Da further decomposes by loss of CO (28 Da) to yield a second metastable transition observed at a mass of 86.7 (Reproduced by permission of Kratos.)

Identification of the different types of ions observed in a mass spectrum through *peak-matching* and metastable ion analysis allows the determination of molecular structure. Several newer mass spectrometric techniques Mass analysed ion kinetic energy (MIKE) or reversed Nier–Johnson geometry) can also be used in spectral interpretation. These techniques are described in specialised monographs.

Problems

16.1 The isotopic factor δ of carbon, measured from the carbon dioxide resulting from the combustion of a natural vanillin, is $\delta = -20$. A value arising from a synthetic vanillin is $\delta = -30$. Calculate the percentage of these two species within a sample of mixed vanillin, knowing that a previously measured value $\delta = -23.5$ has been obtained.

16.2 To evaluate the volume of a large tank of complex shape the method of isotopic dilution is to be used, in association with a soluble salt of lutetium. This element, which has mass $M = 174.97\,g/mol^{-1}$, has two stable isotopes ^{175}Lu (97.4%) and ^{176}Lu (2.6%). The procedure is as follows:

After filling the tank with water, 2 g of the trichloride of lutetium hexahydrate $LuCl_3.6H_2O$, mass $389.42\,g/mol^{-1}$, is added.

1. What is the mass X of lutetium, introduced into the tank?

After stirring until the lutetium salt is well mixed, the extraction of a sample of 1 litre of solution is taken from the tank. To this volume of solution is added 20 µg of lutetium hexahydrate trichloride prepared from the pure isotope ^{176}Lu ($^{176}LuCl_3.6H_2O$ has for mass $390.4\,g\,mol^{-1}$; $^{176}Lu = 175.94\,g\,mol^{-1}$) is added to this solution.

2. Calculate the mass of ^{176}Lu, expressed in µg, added to the sample.

3. By a classical method it is found that the ratio of the two isotopes is now $^{175}Lu = 90.0\%$ and $^{176}Lu = 10.0\%$. What type of apparatus would seem to be the best adapted to measure the ratios of isotopes?

4. Discuss the choice of lutetium for use in the experiment.

5. Finally, calculate the volume of the tank.

16.3 A magnetic sector mass spectrometer-type apparatus houses a circular trajectory of radius 25 cm which is imposed upon the ions. The accelerator voltage is raised to 5 000 V. A mass spectrum is recorded between 20–200 Da.

If it is assumed that each ion carries a single charge ($e = 1.6 \times 10^{-19}$ C, 1 Da $= 1.66 \times 10^{-27}$ kg):

1. What range of magnetic fields would be required if the accelerating voltage is held constant?

2. Why, in general, are recordings not made by a sweeping variation of voltage?

16.4 Calculate the ratio of the $(M + 1)^+$ to M^+ peak heights for footballene (formula: C_{60}), knowing that carbon has two isotopes ^{12}C: 12 amu (98.9%) and ^{13}C: 13 amu (1.1%).

16.5 Vanadium (symbol V) has two isotopes whose relative abundances are $^{51}V = 99.75\%$ and $^{50}V = 0.25\%$.

To determine the vanadium concentration in a sample of steel, 2 g of the steel is dissolved in an acid medium and 1 μg of ^{50}V is added to the resulting solution.

After stirring, an analysis by ICP – MS is performed which obtained a mass spectrum with two peaks centred upon the masses 50 and 51, possessing the same area size.

1. What is the % content of each isotope of vanadium in the unknown steel if the ratio of the areas of the two peaks is the same as that of the masses of the two isotopes?

2. Give a more thorough answer knowing that $^{50}V = 49.947$ g/mol and $^{51}V = 50.944$ g/mol.

3. Explain why this problem would become more complicated if the steel contained either titanium or chrome.

16.6 The method of isotopic labelling is employed to determine the presence of traces of copper with precision. An ICP/MS installation was employed. The results of the experiment rely upon the ratio of the intensities of the peaks corresponding to the masses 63 and 65 of the two isotopes ^{63}Cu and ^{65}Cu of elemental copper. The experiment is described below. First, an injection of a small quantity of the unknown solution was injected. The intensities of ^{63}Cu and ^{65}Cu are 82 908 and 37 092 respectively (arbitrary units).

1. Calculate the percentage of copper in the unknown sample. Recall that $^{63}Cu = 62.9296$ g mol^{-1}, $^{65}Cu = 64.9278$ g mol^{-1}.

2. The above experiment was repeated with a solution of ^{65}Cu which was used as an isotopic marker. For measurements a 250 μl aliquot of the solution containing ^{65}Cu, whose concentration in the metal is 16 mg/l, is added to an

aliquot of 250 µl (approximately 250 mg), of the unknown sample solution. After stirring well, the intensities of the ^{63}Cu and ^{65}Cu peaks were determined and were 31 775 and 79 325 respectively (arbitrary units).

From the above data, calculate the mass concentration of elemental copper in the 250 µl aliquot of the sample solution.

Deduce the ppm copper concentration in the unknown sample solution.

16.7 The dibenzosuberone 1 is a ketone whose structure is given below and the spectrum in Figure 16.1, p. 291.
1. Calculate the exact mass from the largest molecular peak, and write the isotopic compositions of the different species constituting peak $M + 1$.
2. In the mass spectrum of this compound, among other fragments, one can see two with the same nominal mass which could represent the loss of either CO or of C_2H_4.

    Explain the loss of CO from the parent ion and indicate if, by loss of C_2H_4, a positive ion, a radical or a cation-radical must result.
3. Indicate for the two peaks the corresponding molecular formula and precise molecular weight.
4. Knowing that the resolution factor of the mass spectometer is 15 000, is it possible to distinguish the different species constituting the peak $M + 1$?

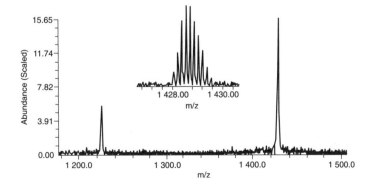

16.8 The study of macromolecules by mass spectrometry presents differences with respect to the studies of molecules of medium or more modest size.
1. What are these principal differences?
2. Calculate, using two different methods and the information available from the recording below, an approximating molecular mass of ubiquitin (a protein found in baker's yeast).

CHAPTER 17

Labelling methods

Several labelling techniques are used for the quantification of analytes present in complex mixtures. In these methods, the sample is spiked with a known quantity of a labelled form of the analyte. After mixing the labelled analyte with the sample, a small portion of the sample is withdrawn and measurement of the ratio of labelled to unlabelled analyte is used to determine the initial quantity of analyte in the sample.

Two types of label are most frequently used. The first consists of using an isotope (stable or radioactive) followed by recovery of part of the compound to be analysed using physico-chemical processes (recrystallisation, chromatography, etc.). The second consists of using an enzyme to measure extremely small amounts of organic compound. In the latter, the compound can be isolated using an immunochemical reaction. This type of approach led to the development of radio-immunological assays (RIA), which are now slowly disappearing, and immunoenzymological assays (IEA), of which ELISA (Enzyme Linked Immuno Sorbent Assay) is a particular type. These methods have been used in numerous applications in many fields of science.

Neutron activation, where the sample is bombarded by neutrons, is used for elemental analysis and can be included in the same category as labelling techniques. While it is a method of high sensitivity, its disadvantages include the need to render the sample radioactive.

17.1 Principle of labelling methods

A trivial but utopian approach that could be used to quantify a compound in a mixture would be to totally separate the compound from the mixture and weigh it. Unfortunately, this approach can very seldom be used because the extraction of extremely small quantities of a compound is neither precise nor quantitative. Analytical chromatography is a technique that can be used to isolate a compound from a mixture, but it is an indirect way to achieve extraction. The group of methods described below stem from an entirely different principle.

In the general approach, a known quantity of a labelled compound, which is a compound with a *tag* that allows it to be distinguished from the unlabelled form and other compounds, is added to the sample. After thorough mixing of the labelled compound with the sample, a fraction of the sample is isolated and the ratio of both forms of the compound is used to obtain the amount initially present in the sample. It is important that labelling of the compound does not affect its isolation behaviour. This is not a classical method of standard additions.

> ■ Labelling techniques are not only found in chemical analysis. For example, to estimate the number of salmon in a pond, a few individuals are captured, tagged (i.e. labelled) and reinserted into the pond. After a few days, a number of fish (i.e. the sample) are removed from the pond, and using the fraction of tagged salmon in this sample, the entire population can be deduced. For instance, if the number of salmon initially labelled was 500 and 10 of them are found in a group of 200 fish removed a few days later, then the total number of salmon (x) will be: $10/200 = 500/x$, so $x = 10\,000$).

The same principle is used in several analytical methods where molecules or elements are labelled:

— the isotopic ratio of an atom present in the analyte can be modified using either a stable isotope or radioisotope whose activity is measured in becquerels. Mass spectrometry or NMR can be used to measure stable isotope ratios. This is called *isotopic analysis*.

— an enzyme or a fluorescent derivative can be used as a labelling reagent. These are the principles used in *immunoenzymatic* or *immunochemical* analyses.

17.2 Isotopic dilution with a radioactive label

In these types of isotopic analysis, the same compound as that to be measured is used (element or molecule) where one of the atoms in it has been replaced by a radio-isotope to allow radioactivity measurements. A small, precisely known quantity of the labelled compound, called the tracer, is added to the sample and, after homogenisation, an aliquot of the spiked sample is isolated by a fractionation technique such as recrystallisation or chromatography. The specific activity of the tracer is measured before and after fractionation.

Assume that compound X is to be measured in a sample. A quantity m_S^* of the labelled compound X (the reference tracer) is added to a precisely known amount of sample. If A_S represents the specific activity of tracer and A_X represents the specific activity of compound X *after* fractionation from the sample, then, because the total quantity of tracer must be conserved, one obtains:

$$A_S m_S^* = A_X(m_S^* + m_X^*) \tag{17.1}$$

The quantity m_X of analyte in the sample is obtained by rearranging equation (17.1):

$$m_X = m_S^*\left(\frac{A_S}{A_X} - 1\right) \tag{17.2}$$

If A_X is much smaller than A_S, equation (17.2) can be approximated by the following equation:

$$m_X \cong m_S^* \frac{A_S}{A_X} \tag{17.3}$$

Having determined m_x, and knowing the quantity of sample used, the concentration of X in the initial sample can be easily obtained.

In such a procedure, the tracer must be uniformly mixed with the sample and the aliquot taken must be sufficiently large so that weighing errors are avoided. If the concentration of the compound to be measured is extremely small, its isolation in the pure or even partially pure state can be difficult: the lack of precision in this method will be due to the fractionation procedure and not to the measurement of activity. This type of approach can be adapted to HPLC measurements with the proper detector when the amounts to be measured are extremely small (such as in radiochromatography).

■ When the compound to be measured is already radioactive, the preceding method is called *reverse isotopic dilution*. In this case, the procedure is the same; however, the radioisotope is diluted with a stable isotope.

17.3 Measurement of radioisotope activity

The preceding approach relies on the measurement of two radioactivities. Thus, it requires a proper environment and in most cases the proper authorisation to manipulate radioactive substances.

17.3.1 Radioactive decay

Any radionuclide is characterised by its half-life τ whose value is independent of the type of decay products that are created. Half-life is defined as the time required (from initial time $t = 0$) for the decomposition of half the atoms in the sample. The law of radioactive decay allows calculation of the number of atoms N left at time t in a population with N_0 atoms initially. The integrated form of this law is given by the following equation:

$$N = N_0 \exp[-\lambda t] \qquad \text{with} \qquad \tau = \ln \frac{2}{\lambda} \tag{17.4}$$

where λ (in units of time^{-1}) in equation (17.4) is the radioactive constant of the radionuclide being considered. In practice, it is not N, the number of atoms remaining, that is known but rather the activity $A (A = dN/dt)$, expressed in becquerels (1 Bq = 1 disintegration per second) and the curie, Ci, (which) has a value of 3.7×10^{10} Bq). The activity, which can be measured with the proper detector, is directly related to the concentration of the radionuclide ($A = \lambda N$). A is related to A_0, the initial activity, by an equation analogous to the law of radioactive decay:

$$A = A_0 \exp[-\lambda t] \qquad (17.5)$$

17.3.2 Radioactively labelled molecules

This method can only be used for compounds containing elements that have radio-isotopes. A relatively large number of organic molecules are available with radio-active labels. In fact, when each molecule of a compound contains a ^{14}C atom at a given site, activity can attain 60 mCi/mol (Fig. 17.1).

In general, ^{14}C is preferred to ^3H (tritium), which can be readily exchanged and for which autoradiolysis can create problems in conserving tritiated molecules.

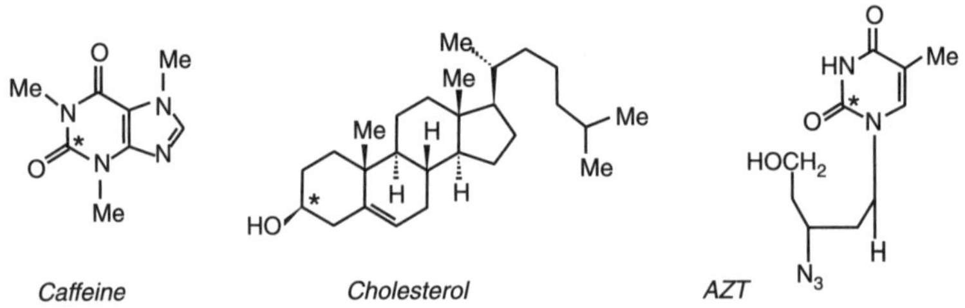

Caffeine   *Cholesterol*   *AZT*

Figure 17.1—*Three molecules labelled at a single site with ^{14}C.* This radioisotope has a sufficiently long half-life that labelled compounds can easily be stored, and corrections to compensate for natural decay are not necessary.

Table 17.1—Characteristics of commonly used radioisotopes.

Isotope	Half-life	Type of emission	Energy (MeV)
^3H	12.26 years	β^-	0.02
^{14}C	5730 years	β^-	0.156
^{32}P	14.3 days	β^-	1.7
^{35}S	88 days	β^-	0.167
^{125}I	60 days	I.E.C.*	0.149

* Internal electron capture, cf. 13.3.2.

■ The isotope ^{14}C used for labelling organic compounds is prepared in a nuclear reactor. In the reactor, low-energy neutrons called *thermal neutrons* originating from the controlled fission of ^{235}U bombard solid targets containing nitrogen atoms (aluminium or beryllium nitrite). The radioactive material formed is isolated from the target sample by oxidation to $Ba^{14}CO_3$, which is the form in which ^{14}C is sold. Many organic reactions involving $^{14}CO_2$ can be used to synthetically incorporate the radioisotope into a given molecule.

17.3.3 Scintillation counters

The emission from the radioisotopes is often insufficient to penetrate the window of a Geiger–Müller counter. Therefore, the compound whose activity is to be measured is often mixed in solution with a scintillator, called a *fluor*, which transforms β rays into luminescence proportional to the number of β particles emitted. The sample is dissolved in a solvent (toluene, xylene or dioxane, the latter being used for water-soluble compounds) that acts as a relay to transfer the energy to the scintillator. The scintillation mixture contains PPO (2,5-diphenyloxazole), which emits in the UV and POPOP, which emits in the visible and is well adapted to detection with photomultiplier tubes (Fig. 17.2). The quantum yield of emission will depend on the energy of the emitted particles.

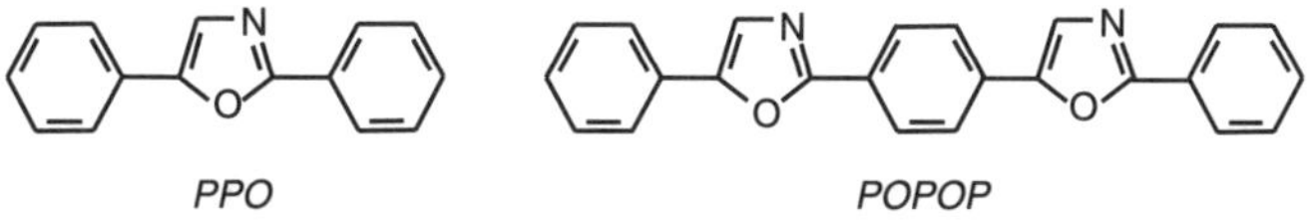

PPO   POPOP

Figure 17.2—*Two examples of commonly used scintillators: PPO and POPOP.*

■ The counting device must be calibrated. Part of the energy dissipated as β rays is absorbed without being transformed into photons (chemical quenching). Moreover, part of the luminescent radiation emitted by the scintillator can be re-absorbed by the solution if the fluorescence spectrum overlaps the absorption spectrum (quenching, cf. Chapter 13). The photons, after passing through a band-width filter, are integrated over a given period of time by one or two photomultiplier tubes operating together (Fig. 17.3). When using HPLC, the scintillator liquid is introduced into the measurement cell at the same time as the mobile phase. Alternatively, the mobile phase is passed through a capillary made of scintillating glass.

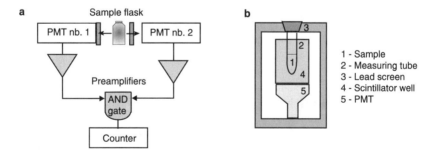

Figure 17.3—*Counting system.* a) Device used to measure the activity of a low-energy radioisotope using the method of two coincident detectors. A single β emission can produce hundreds of photons. It is thus possible to measure photons in opposite directions using two photomultiplier tubes (PMT). Counting only occurs if both PMTs produce a signal that is not offset by more than a few nanoseconds; b) device involving a PMT in a counting well used to measure luminescence produced by a sample that has been mixed with a scintillation cocktail (in aqueous or non-aqueous media).

■ Radioactive measurements involve risk to personnel and the environment and therefore their use is subject to very strict rules. In order to obtain a permit to use radioactive chemicals, a laboratory must be equipped for special waste disposal. To avoid external escape of radiation, standard precautions are used. These include the use of gloves, protective screens, low amounts of activity and low-energy emitting radiochemicals. Of the most commonly used radioisotopes, ^{35}S is the only one that emits high energy γ radiation. A 1-cm thick Plexiglas window doped with lead is usually sufficient to protect the user from low levels of activity (lower than 1 MBq). However, care must be taken to avoid internal contamination. Even a low energy radioisotope can be dangerous if it is introduced into an organism.

There is no simple and direct relationship between the intensity of radiation (Bq) and irradiation dose (Sievert) because the latter is a function of the total energy per unit mass of the irradiated substance.

17.4   Choice of a sub-stoichiometric method

Sometimes the analyte is in such low concentration that it is impossible to isolate. It can be noted from equations (17.1) and (17.3) that it is not necessary to know A_x and A_s individually if their ratio can be determined. To achieve this, a reproducible reaction can be conducted on the labelled standard (analytical blank) and, in an identical fashion, on the sample in order to obtain the same quantity of *derivatised* compound. Thus the sub-stoichiometric method is similar to the immunochemical method for trace analysis.

■ For example, assume that a sulphate has to be measured at the level of a few percent in an aqueous solution. After the addition of a known quantity of $^{35}SO_4^{2-}$, a barium salt of a lower quantity than that necessary to precipitate all sulphate ions is added (formation of insoluble $BaSO_4$). By comparing the activity of the recovered precipitate with that obtained for the same quantity of labelled sulphate alone, under the same experimental conditions, it is possible to obtain the activity ratio and thus deduce the quantity of sulphate in the original sample (this is known as inverse isotopic dilution).

17.5   Labelling with stable isotopes

The isotopic dilution method can be extended to non-radioactive tracers by using mass spectrometry or NMR to determine the variation in isotopic ratios. This method can be used for the measurement of molecules or elemental species (about 60 elements have stable isotopes). This approach allows ultra-trace analysis because, contrary to radioactive labelling where the measurement relies on detecting atoms that decompose during the period of measurement, *all* labelled atoms are measured. *Isotopic mass spectrometers* are well suited for these measurements.

■ Let us assume that trace levels of caffeine in a solution have to be determined. This molecule of mass 194 has an $N-CH_3$ group in the 5-membered ring (Fig. 17.4). By synthesis, it is possible to replace the three H atoms on the methyl group with three deuterium atoms (isotope of hydrogen, written ^2H or D). Caffeine labelled with $N-CD_3$ has a mass of 197. The determination of this species can be conducted by GC/MS or HPLC/MS. A known concentration of caffeine-d3 is added to a known volume of the sample. During the chromatographic co-elution of both forms of caffeine (which have the same retention time) the mass spectrometer will alternately measure the intensities of the peaks at $m/z = 194$ and $m/z = 197$. The mass chromatograms of the 194-peak and are the 197-peak is plotted and the areas under the respective peaks are determined. A calibration curve can be used to obtain the concentration of caffeine (Fig. 17.4).

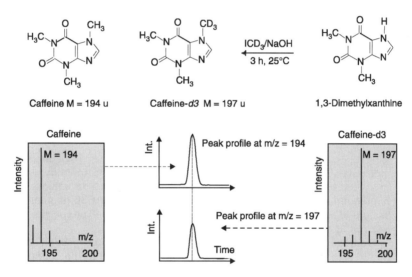

Figure 17.4—*Measurement of caffeine by HPLC/MS with isotopic dilution.* The stable isotope used in this determination is deuterium D. Caffeine-d3 is obtained by methylation of 1,3-dimethylxanthine.

17.6   Immunoenzymatic (IEA) methods of measurement

The sensitivity and simplicity of an analytical method always improves when selective detection can be used. Because of detection selectivity, a sample in a complex matrix can be analysed with minimum sample preparation. Wouldn't it be ideal to have an ultra-specific reagent that would allow the determination of a compound in an environment containing thousands of substances?

This has become a reality with the adaptation of immunoanalysis to small molecules. This approach has seen a drastic expansion in biomedical sciences because of developments in genetic engineering and biochemistry. The same methods have been common in environmental sciences for some years now.

17.6.1 Basic principles

It is necessary to give some background information to understand the principle underlying immunochemical tests such as ELISA (*Enzyme Linked Immuno Sorbent Assay*). Such information is not always available to chemists, since chemistry and immunology have traditionally been separate disciplines.

Introduction into the body of an alien substance (an *antigen*) of sufficient molecular weight induces the production of biomolecules called *antibodies*. Antibodies are glycoproteins, also referred to as immunoglobulins (principally IgG).

In the presence of each other, antigens and antibodies agglutinate (bind). The forces that bind them are principally due to hydrogen bonding and hydrophobic interactions. The hydrostatic pressure exerted by surrounding water molecules also ensures cohesion.

Antigens not only *induce* an immune response, but will also *react* with antibodies existing in a living organism. These two properties are different. Small molecules like pesticides are not able to induce the production of antibodies. They can only react with antibodies already present. They are called *haptens*. To generate antibodies specific to a hapten, haptens must be covalently bonded to a protein (e.g. BSA *Bovine Serum Albumin*) or polysaccharide. Therefore, the hapten must have reactive sites. On the other hand, this modification in the structure must not influence the specificity of the antibodies that will be produced.

■ The generation of specific antibodies is a difficult task. For example, if a single molecule of a hapten is bound to a protein of 66 kDa, the antibody will be directed to several sites on the protein (called *epitopes*). It is thus desirable to have several hapten molecules bound to the same protein (i.e. about 50 haptens) at different sites. Under these conditions, the antibodies produced will be specific to the hapten and not to the protein.

Antibodies are typically produced by the subcutaneous injection of the hapten-bound protein into five or six mice, rats or rabbits. This multi-point primary injection corresponds to approximately 100 μg of pesticide. The animals do not all react in the same fashion even though they have the same genetic background. Two weeks later, the same process is repeated with five times less immunogen (hapten-bound protein). Two weeks after that, blood samples are taken from the animals to verify whether the antibodies have appeared. The animals are re-injected a further two or three times to create a hyper-immunisation. The serum of these animals is now a source of the desired antibody, which is of the polyclonal type.

17.6.2 Competitive ELISA assay

The following steps are involved in a heterogeneous-phase competitive ELISA assay (Fig. 17.5):

1. A standard solution of the compound to be measured or the sample is put into a tube (or the well of a microtitre plate) whose walls have been sensitised by coating them with the adapted antibodies. The bottom of the tube is transparent for detection purposes.

2. A known volume of solution containing the compound covalently linked to an enzyme is added (the *enzyme conjugate*, which is often horseradish peroxidase). During an incubation period, both forms of the compound (that to be measured and the labelled form) are in competition for binding with the antibodies present on the container walls.

3. After the incubation period (e.g. 30 min), the tubes are washed several times with water. Only bound molecules will remain in the tubes. The amount of enzyme-conjugated compound bound to the wall (i.e. to the antibody) will increase accordingly with decreasing amounts of free, unlabelled compound in the sample.

4. Substrate *S* specific to the enzyme is then added along with a chromogen *C* (e.g. tetramethylbenzidine) that is designed to react with the product of the enzymatic reaction of *S*.

5. The fixed enzyme will transform molecule *S* into species *P*, which will react with *C* to give a coloured product. The amplification factor imparted by the enzyme is responsible for the high sensitivity of this assay. As the amount of compound present in the sample decreases, the quantity of fixed enzyme increases and so, therefore, does the colour intensity.

6. Finally, the reaction is quenched after a short period of time by addition of a strong acid, which destroys the enzyme. The intensity of the colour in the tube is then read using a colorimeter.

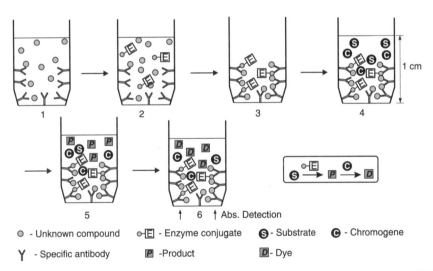

Figure 17.5—*The steps in a competitive ELISA assay.* Several assays of this type are used in clinical analysis.

■ Horseradish peroxidase (M = 44 kDa) is a stable and reactive enzyme. It possesses four lysine residues that are not involved in its activity but are used for linking. According to its name, hydrogen peroxide is a substrate for this enzyme. As it decomposes, hydrogen peroxide generates oxygen that acts on the chromogen, TMB (tetramethylbenzidine).

This methodology traditionally used by biologists is now applied in the areas of environment and food analysis (with pesticides, aflatoxins, anabolic steroids, PAHs). However, it is only possible to measure a compound if the adapted antibodies and enzyme conjugates are available.

17.6.3   The concentration/absorbance relationship

The number of antibody binding sites on the wall is relatively small compared to the number of molecules in solution. Labelled and non-labelled molecules will react with the antibodies in proportion to their relative concentrations (Fig. 17.6).

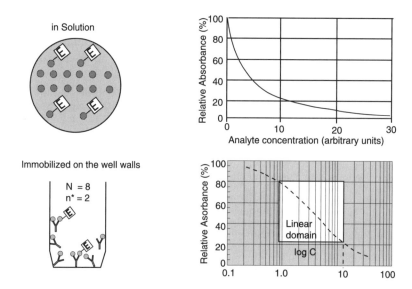

Figure 17.6—*Relationship between concentration and absorbance for ELISA assays.* The ratio of both types of molecules at the wall is the same as that in solution (3 in the example shown). Linearity, plotted on a semi-logarithmic scale, is only observed for a very narrow range of concentrations.

If N is the total number of binding sites on the wall, n^* the number of labelled molecules adsorbed, C^* the concentration of labelled species in solution and C the concentration of the non-labelled unknown, we can write:

$$\frac{n^*}{N - n^*} = \frac{C^*}{C} = R \tag{17.6}$$

The absorbance A is proportional to n^* by the constant k (i.e. $A = kn^*$) thus:

$$A = k \cdot N \cdot \frac{C^*}{C + C^*} = k \cdot N \cdot \frac{R}{R + 1} \tag{17.7}$$

A typical feature of the experimental curve in such assays is its non-linearity. The percent absorbance of the solution relative to the reference, which is obtained with the blank (enzyme conjugate absorbed on all antibody sites), is plotted as the ordinate. Concentration is plotted on the abscissa. Equation (17.7) shows that absorbance does not vary linearly with concentration. However, if the concentrations are plotted on a logarithmic scale, part of the graph will be linear to a first approximation.

> ■ The absorbance measurement conducted in these assays can be replaced by fluorescence detection (fluorescence ELISA). In this case, the enzymatic conjugate chosen is an alkaline phosphatase that converts the specific substrate into a fluorescent compound.

17.7 Other immunoenzymatic techniques

There are several ways to conduct competitive ELISA assays in heterogeneous-phase media. Ohmicron has developed an approach where the antibodies are covalently bound to magnetic particles of 1-μm diameter instead of to the container wall (Fig. 17.7). This approach has two distinct advantages:

— the quantity of antibody that can be used is not limited, in contrast to using antibody-coated walls

— the antibody antigen reaction is favoured because both elements can be well mixed: this technique is midway between homogeneous phase and heterogeneous phase.

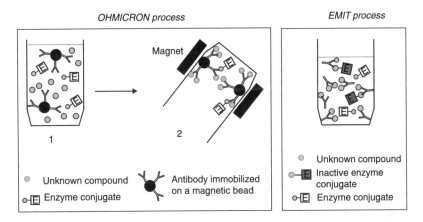

Figure 17.7—*The Ohmicron and EMIT processes, which are modified ELISA assays.*

The assay using magnetic particles is essentially the same as traditional ELISA and is conducted in a similar way. However, after incubation and before rinsing, a magnetic field is applied to hold the antibodies against the container walls.

The technique known as EMIT (*Enzyme Multiplied Immunoassay Test*, Syva-Biomerieux) in homogeneous phase is well known for the determination of drugs and hormones. In EMIT, the free antibodies are placed in a tube into which are added the sample and a known quantity of the antigen with its associated enzyme (the enzyme conjugate). Competition occurs in solution; however, the conjugated antigen that is bonded to the antibody loses its enzymatic properties. Thus, there is no need for rinsing and the substrate can be directly added. The intensity of the colour in this assay will be due to the fraction of the non-fixed enzyme. If the concentration of the compound is small, much of the enzyme will be denatured and the colour will be weak.

17.8   Advantages and limitations of the ELISA test in chemistry

ELISA assays yield reliable semi-quantitative results more rapidly than chromatography. They are mostly used to eliminate negative samples that may not be useful for submission to the subsequent steps of rigorous chromatographic analysis. However, ELISA assays have certain limitations:

1. The possibility of recognising several molecules with different sensitivities causes dispersion in the results. How does one know if a cross-reaction (a false positive) hasn't occurred?
2. The measurement range is relatively narrow. When concentration increases, the measurement is less precise (an effect of using logarithmic scale). A good dilution factor must be found.
3. The wells in microtitre plates have to contain the same quantity of antibody with the same reactivity, which is technically difficult. Immunisation of a laboratory animal is a one-off event, which might explain why results differ from one ELISA kit to another.
4. The kits must be kept in a cold environment. Therefore, the method is limited by kit lifetime and is less reliable for field analyses.
5. The cost of these analyses is not as low as was first assumed. They are economical only for a certain number of analyses.

17.9   Radio-immunoassay (RIA)

A radioisotope can be used as a tracer instead of the enzyme used in immuno-chemical tests. This method, introduced in 1960 for the measurement of insulin in serum, is called *radio-immunology*. It is the transposition in immuno-chemistry of the same principle as that used to determine the sulphate ion (cf. 17.5). Radio-immunoassays are similar to ELISA assays in the way in which the analysis is conducted. A known quantity of tracer is added to a sample containing the analyte. If the antibodies are coated on the wall of the tube, measurement of radioactivity will carried out after washing.

However, if the specific antibody is added in sub-stoichiometric proportion, the unbound material must be separated from the bound material by an appropriate means and then the radioactivity is measured.

Radio-immunoassays, which are a popular method for clinical analysis, are not commonly used in chemistry. This is partly due to the desire to avoid unnecessary use of radioisotopes.

17.10 Immunofluorescence assay (IFA)

The use of antibodies labelled with a fluorescent derivative obtained from fluorescein, rhodamine or other fluorophore is well suited for the measurement of human or animal immunoglobulins. However, the same principle is not yet used for simple organic molecules. Discrimination between the labelled and non-labelled species at the isolation stage can occur for small organic molecules. However, there are a few examples of assays in which isolation is followed by fluorescence measurement.

Fluorescein isothiocyanate, a fluorescent label often employed in biochemistry.

■ Many diagnostic techniques are grouped under the name *immunofluorescence.* These methods are not used for quantification but to examine, using a microscope and ultraviolet light, certain preparations obtained by reaction with a fluorescent derivative.

17.11 Neutron activation analysis (NAA)

Some 60 elements can be identified and quantified after having been transformed into radioisotopes by bombardment with particles. This particular labelling technique, whose sensitivity varies with the element (1 000 ppm to ppb), is part of the field of multi-elemental nuclear analysis. Neutron activation is the most widely used of these techniques because it does not involve charged particles.

The addition of one or several neutrons to a target nucleus leads to the formation of an isotope of greater mass. This can be unstable and decompose by β^- emission (see the following equation). The total radioactivity generated in a sample containing many elements – each one having a family of isotopes – leads to a complex emission spectrum.

$$\ce{_Z^A X + _0^1 n \longrightarrow _Z^{A+1}X* ->[\beta^-] _{Z+1}^{A+1}Y* ->[\gamma] _{Z+1}^{A+1}Y}$$

The total activity in the sample decreases with time following a curve that represents the superimposed activities of each radionuclei present. This global activity generally does not allow identification of the elements that have been formed.

Elemental composition cannot be obtained by simple analysis of the radiation since β^- emission is continuous. Thus, γ emission, which accompanies β emission and is characteristic of each nuclide, is analysed. These emissions occur in the same spectral range as X-ray fluorescence.

17.12   Thermal neutrons

The probability of inducing a nuclear transformation depends on the nuclide and the energy of the neutrons. These two factors are combined in a parameter called the *effective cross-section*. One of the simplest ways to bombard a sample with neutrons is to use an intense flux of 10^{14} to 10^{18} neutrons $\cdot$ m$^{-2} \cdot$ s^{-1}, available in nuclear reactors of at least 100 kW. Sufficient activation levels can be obtained in a few minutes for low-concentration samples, even for elements with a long half-life. The process is, however, expensive and imposes constraints on the sample to be analysed. The latter must be thermally stable and preferentially in the solid form. The sample is placed in a tube with an external reference standard and introduced pneumatically into the nuclear reactor for irradiation.

Small neutron sources containing an α emitter (a few µg of ^{241}Am or ^{124}Sb) inserted in a beryllium envelope have been developed to avoid the above limitations. The nuclear reaction generating the neutrons is the following:

$$\tfrac{4}{2}\text{He} + \tfrac{9}{4}\text{Be} \rightarrow \tfrac{12}{6}\text{C} + \tfrac{1}{0}\text{n}$$

Sources of weak intensity have fluxes of a few million neutrons per second. They can be used to measure about 20 elements. A variation of this method consists of using a few µg of ^{252}Cf ($\tau = 2.6$ years) as a rapid source of neutrons (2 MeV) that are slowed by collision with hydrogen atoms (2.4×10^6 neutrons $\cdot$ s$^{-1} \cdot$ µg^{-1}).

17.13   Induced activity – irradiation time

The number of radioactive atoms N^* that can accumulate in sample during irradiation has an upper limit. At a given moment, an increase in the number of nuclei N^* is equal to the difference between the rate of formation, considered to be constant (the number of target nuclei N is large), and the rate of decay:

$$\frac{\mathrm{d}N^*}{\mathrm{d}t} = \varphi\sigma N - \lambda N^* \tag{17.8}$$

where φ represents the neutron flux, λ the radioactive decay constant and σ the effective cross-section of each target atom, of which there are N. This number is related to the mass m of the fraction f of isotopes of the element with atomic mass M. If $\mathcal{N}$ is Avogadro's number, the following relationship is obtained:

$$N = \frac{m}{M} \mathcal{N}/f \tag{17.9}$$

The integrated form of equation (17.8) can be used, allowing evaluation of the number of atoms N^* present after time t, as follows:

$$N^* = \frac{\varphi\sigma N}{\lambda}(1 - \exp[-\lambda t]) \tag{17.10}$$

The induced activity at time t corresponding to $A = \lambda \cdot N^*$ leads to the expression:

$$A = \varphi\sigma N(1 - \exp[-\lambda t]) \tag{17.11}$$

In this equation, the term in parentheses is called the *saturation factor*. This factor tends rapidly to 1 when t is increased. Thus, for $t = 6\tau$, the activity reaches a value of 98%. Experimentally, the time of irradiation never exceeds 4 to 5τ. This method is usually reserved for radioelements with short half-lives.

17.14   Detection by γ-counting – measurement principle

The choice of detection method depends on the nature of the emitter and on the complexity of the spectrum emitted. The most efficient approach consists of recording the γ spectrum that accompanies β^- emission from the radionuclide present in the sample after activation (Fig. 17.8).

Contrary to β^- emission, which is a continuum, the γ spectrum is quantised and thus easier to analyse.

■ The sensor is a crystal of NaI(Tl) that transforms the γ photon into luminescence whose intensity is proportional to the energy of the photon (assuming that the γ photon is entirely absorbed by the crystal). The principle is similar to that of liquid scintillators used to measure ^{14}C. If a Ge(Li) crystal is used, it behaves like the support gas in a Geiger–Müller tube.

17.15   Applications

If the standard and the sample have been irradiated together, under the same conditions, the specific activity of element X will be the same in the sample as it is for the standard. Since the total activity will be proportional to the mass of X, we can write:

$$(\text{Mass of X})_{\text{sample}} = (\text{Mass of X})_{\text{standard}} \frac{(\text{Total Activity})_{\text{sample}}}{(\text{Total Activity})_{\text{standard}}} \tag{17.12}$$

The preceding relationship is applicable if the induced radiation is simple or, alternatively, if the instrument used for counting has filters that allow isolation of the signal due to the element under study. However, the matrix can often be activated and emit radiation that is superimposed on that characteristic of the element. Introducing a time gap between the end of irradiation and the start of the measurement can reduce these interferences. In this way, the signal caused by short-lived emitters will be eliminated.

> ■ For example, suppose that traces of iron in an aluminium sample are to be determined. ^{59}Fe is characterised by γ emission at 1.29 MeV. Aluminium under irradiation will yield ^{24}Na, which is responsible for a γ ray situated at 1.37 MeV via the reaction ^{27}Al$(n, \alpha)^{24}$Na $(\tau = 15\,\text{h})$. Here, a time gap of several days between the end of radiation and measurement permits enough time for the aluminium to disappear.

For certain samples, the radioelement under analysis can be isolated using its stable isotope and a *training* technique. A similar treatment carried out on the reference sample allows the extraction yield to be determined. The result is then normalised to 100%.

Neutron activation is not a widely used method (Fig. 17.8). Some of its applications include characterisation of materials (e.g. high purity metals, semiconductors), the study of the distribution of chemical elements within fossils, ultra-trace analysis in archaeology and geology, and the study of volcanoes.

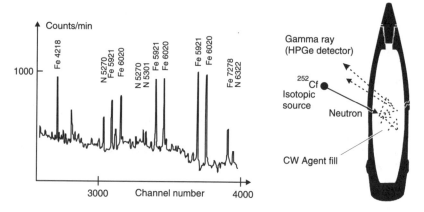

Figure 17.8—*A non-conventional application of neutron activation.* Neutron activation can be used to determine the composition of warheads. The drawing, reproduced by permission of EG&G Ortec, demonstrates the principle of this approach. The graph represents part of a γ spectrum obtained after only a few minutes from a warhead containing a nitrogen compound.

Problems

17.1 To determine the concentration (% mass) of penicillin present in a commercial preparation, an isotopically labelled reference of penicillin is used, whose specific activity is 75 000 Bq/g.

10 mg of labelled penicillin is added to 500 mg of sample. After mixing and equilibration, 1.5 mg of penicillin was recovered whose measured activity was 10 Bq. Calculate the concentration (% mass) of penicillin in the commercial preparation.

17.2 In order to measure, by the methods of isotopic dilution the mass concentration of the orthophosphate ion PO_4^{3-}, in an aqueous solution. 3 mg of labelled phosphate ion $^{32}PO_4^{3-}$ (^{32}P being a β-emitter whose half-life is 14 days), is added to 1 g of the sample solution.
The specific activity of the added $^{32}PO_4^{3-}$ is 3 100 dps/mg.
After mixing, 30 mg of orthophosphate was isolated whose overall activity was 3 000 dps. Find an expression from which the initial quantity of phosphate ion PO_4^{3-} may be calculated, and calculate the % mass of this ion in the aqueous solution.

17.3 Patulin ($C_6H_7O_4$), is a compound, dangerous to health in humans which is found in the juice of damaged apples or grapefruits. The current method of measuring this compound resembles that of a immunoenzymatic ELISA test. A standard solution of pure patulin was prepared in water at a concentration of 1.54 g/l just prior to use.

In the test described, four identical test tubes are used whose inner wall is covered with a suitably adapted antibody. All of the tubes follow the same treatment, differing only in the solution which is introduced in the initial step. The contents are as follows:

— tube 1: a blank, 2 ml of pure water.
— tube 2: 1 ml of pure water plus 1 ml of the standard solution non-diluted.
— tube 3: 1 ml of pure water plus 1 ml of the standard solution diluted twice.
— tube 4: the sample, 2 ml of fruit juice, filtered and diluted twice with water.

(In to each tube the same quantity of conjugated enzyme is introduced, plus all of the other necessary reactants combined).

Following the reaction, the measured absorbance A of the tubes was as follows: tube 1: $A = 1.03$, tube 2: $A = 0.47$, tube 3: $A = 0.58$, tube 4: $A = 0.5$.
1. Calculate the concentration in ppm of the patulin in the standard solution.
2. Calculate the % absorbance (also known as the % inhibition) of the solutions in tubes 2, 3 and 4 with respect to tube 1.
3. Explain why the absorbance of tube 1 is greater than that of the other tubes.
4. Calculate the concentration of patulin in tubes 2 and 3 in µg/l.
5. Plot the calibration curve, % $A = f(\log C)$ ($C = $ concentrations of tubes 2 and 3 in µg/l)
6. Calculate the concentration in mg/l of patulin in the fruit juice also giving the result in ppb.

17.4 To measure elemental chlorine present in a very weak concentration (of the order of a few ppm) in a sample of steel, by the method of neutron activation analysis (NAA), a reactor producing a neutron flux of thermal character $2 \times 10^{16} \, n/m^2/s$ is required.

1. Write clearly, using the symbols provided, the reactions taking place:

$$^{35}Cl(n, \gamma)^{36}Cl(\beta\text{-emitter } \tau = 3.1 \times 10^5 \text{ years})$$

and

$$^{37}Cl(n, \gamma)^{38}Cl(\beta\text{-emitter}, \ \tau = 37.3 \text{ min}).$$

2. Explain why the γ emissions of the isotope ^{38}Cl are preferred for the experiment rather than those of ^{36}Cl.

3. Indicate two other methods which could be used to measure the chlorine in the steel. Give a precise explanation including reasons, advantages and disadvantages of the methods chosen when they are compared to NAA.

   To eliminate the interference afforded by manganese (^{55}Mn) present in steel which leads to ^{56}Mn whose half-life is 2.58 hours and whose radiation is intense, the chlorine is separated by precipitation in the form of silver chloride.

   Simplified operating procedure: For five minutes and under the same conditions in the reactor, a) a sample of the steel to be quantified in Cl, and b) a disc of filter paper onto which 100 µl of a solution of 0.1 g/l of chloride ion had been previously adsorbed, (previously it had been verified that the filter paper contained no Cl), is irradiated. Then the steel sample is dissolved into solution by boiling with 40 ml of 2M HNO_3 to which is added 2.00 g of dry KCl. This resulting solution is then introduced to 50 ml of an aqueous solution of 15% (w/v) $AgNO_3$. The AgCl precipitate is recovered, washed and then dried. The result of the count is displayed in the following table.

	Mass	Mass AgCl	Count γ (1.64 MeV) of ^{38}Cl
Sample	0.51 g of steel	3.726 g	11 203
Standard	10 µg of Cl		48 600

4. Show that the silver nitrate is added in sufficient quantity.

5. Calculate the concentration of elemental chlorine in the steel sample. Recall that $N = 14.007$, $O = 16$, $Cl = 35.453$, $K = 39.098$, and $Ag = 107.863 \, g/mol$.

CHAPTER 18

Potentiometric methods

A high percentage of chemical analyses are based on electrochemistry. Electrochemical methods can be separated into two categories: those that measure voltages and those that measure currents. The first group uses ion selective electrodes (ISE). Measurement of pH, probably the most common and best known electroanalytical measurement, is part of this group. Most measurements include the determination of ions in aqueous solution, but electrodes that employ selective membranes also allow the determination of molecules. Instruments such as low-cost pH meters or sophisticated titrators yield concentration directly. While sensitivity is high for certain ions, the matrix is often responsible for lack of reliability in these methods. When specificity causes a problem, more precise complexometric or titrimetric measurements must replace direct potentiometry. Nonetheless, potentiometry represents an analytical technique with increasing applications.

18.1 General principles

Potentiometric measurements are based on the determination of a voltage difference between two electrodes plunged into a sample solution under null current conditions. Each of these electrodes constitutes a half-cell. The *external reference electrode* (ERE) is the electrochemical reference half-cell, which has a constant potential relative to that of the solution. The other electrode is the *ion selective electrode* (ISE) which is used for measurement (Fig. 18.1). The ISE is composed of an *internal reference electrode* (IRE) bathed in a reference solution that is physically separated from the sample by a *membrane*. The ion selective electrode can be represented in the following way:

Internal reference electrode // internal solution / **membrane**

The potential difference between the internal reference electrode and internal surface of the membrane is constant. Its value is fixed by the design of the electrode (i.e. the nature of the internal reference electrode and internal solution). However, the potential difference generated between the external surface of the membrane and the sample solution depends on the activity of the analyte i to be determined. The measurement string is represented as:

Ion selective electrode / sample solution // external reference electrode

For studies in aqueous solutions, the external reference electrode is often an Ag/AgCl/KCl electrode. Electrical contact with the solution is achieved using a disc-like membrane made of porous fritted glass. Because ions have a tendency to migrate across the membrane, a small potential E_J is generated by this liquid junction. This phenomenon can be minimised by inserting a saturated KCl solution as a salt bridge.

The cell potential is:

$$E_{measured} = E_{ISE} - E_{ERE} + E_J$$

such that

$$E_{measured} = E_{IRE} + E_{membrane} - E_{ERE} + E_J$$

The terms E_{IRE} and E_{ERE} are independent of the concentration c_i of analyte i and, as described previously, the use of saturated KCl minimises the value of the E_J term. Therefore, the potential difference measured is almost solely due to the membrane and is related to the activity of the ionic species under analysis in the sample solution by the following Nernst-type equation:

$$E_{measured} = E' + 2.303 \frac{RT}{zF} \log a_i \qquad (18.1)$$

In equation (18.1), E' is the standard potential and is a constant that includes all other potentials, R is the ideal gas constant, T is the temperature, z is the charge carried by ion i to be measured and whose activity is a_i, F represents Faraday's constant and 2.303 is the logarithmic conversion factor.

■ The activity a_i of an ion i is related to its concentration c_i by the relationship $a_i = \gamma_i c_i$, where γ_i is the activity coefficient which depends on the total ionic strength I. This latter term is a measure of the quantity and charge of all ions present in the medium: $I = 0.5 \sum c_i z_i^2$. Therefore, for ion i of charge z, equation (18.1) becomes:

$$E_{measured} = E' + 2.303 \frac{RT}{zF} \log \gamma_i c_i \qquad (18.2)$$

For dilute solutions, the Debeye–Hückel law ($\log \gamma_i = -0.5 z_i^2 I^{0.5}$) indicates that γ_i will be a constant for a given ionic strength I. Therefore, the same quantity of inert electrolyte, called the *support electrolyte*, must be added to the sample and to the series of standards to increase the concentration of external ions and stabilise the ionic strength. This addition of ISAB (Ionic Strength Adjustment Buffer) is intended to limit variation in γ_i. Under these conditions, the measured difference in potential only depends on the concentration of the ion to be analysed and is given by equation (18.3).

$$E_{measured} = E'' + 2.303 \frac{RT}{zF} \log c_i \qquad (18.3)$$

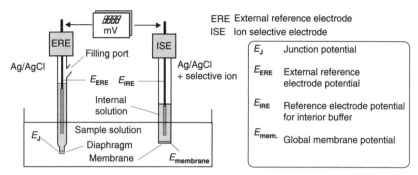

ERE External reference electrode
ISE Ion selective electrode

E_J	Junction potential
E_{ERE}	External reference electrode potential
E_{IRE}	Reference electrode potential for interior buffer
$E_{mem.}$	Global membrane potential

The membrane is permeable to the selective ion

Figure 18.1—*Electrochemical measurement string using an ion selective (or specific) electrode (ISE)*. The membrane potential varies with the concentration of the specific ion in solution. Other potentials are fixed by the construction of the electrode. The junction (diffusion) potential E_J has a low value and is generally constant. Measurement is typically conducted with an ionometer. Manufacturers also supply combined electrodes that include both electrodes (external and ion selective) in the same device. Commonly employed pH electrodes are of this type. The schematics of combined electrodes are much less clear due to the proximity of the membrane to other electrode components.

18.2 A particular ion selective electrode: the pH electrode

This electrode, also called the *glass electrode*, is specific to H^+ ions. Glass in this case does not refer to the material of the electrode body but to the membrane that ensures contact with the solution. The membrane is a thin wall of glass that has a very high sodium content (25%). In the presence of water, hydration occurs and the membrane's surface becomes comparable to a gel while its interior corresponds to a solid electrolyte.

On a microscopic scale, the glass is a network of orthosilicate $Si(OH)_4$ whose open structure contains sodium cations that allow the movement of charges from one side of the membrane to the other (Fig. 18.2). The outside of the membrane is in contact with the sample solution while the inside is in contact with the internal electrolyte, which has a constant acidity (*ca.* pH 7). The membrane is the seat of exchange between Na^+ and H^+ cations as follows:

$$H^+(\text{solution}) + Na^+(\text{glass}) \rightleftharpoons Na^+(\text{solution}) + H^+(\text{glass})$$

When the concentration of H^+ is different on either side of the membrane, a potential difference is generated, which is related to the activity of H^+ ions in solution, i.e. pH. The latter is determined using an electronic millivoltmeter, the pH meter, which monitors the potential difference between the glass electrode and an internal reference electrode of Ag/AgCl (currently preferred to the mercurous chloride (Hg) electrode for environmental purposes). After calibration, the instrument will directly yield the pH of a solution.

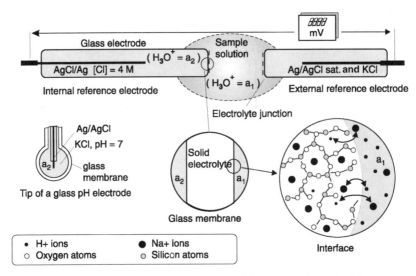

Figure 18.2—*Measurement of pH.* The concentration of H^+ ions can be determined from the potential difference between the reference electrode and the glass electrode. Details of the membrane, which is permeable to the H^+ ion, are shown. When an H^+ ion forms a silanol bond, a sodium ion moves into the solution to preserve electroneutrality. A cross-section of the membrane showing this exchange reaction is presented (IUPAC conventions are not followed to improve clarity in the diagram). Prior to its use, the pH meter is calibrated with a buffer solution of known pH.

18.3   Ion selective electrodes

In the early days of potentiometry, it was realised that other alkaline ions beside Na present in basic solution could interfere with the signal monitored at the electrode depending on the composition of the glass membrane. This is known as the *alkaline error*, which occurs at over 13 pH. The study of this phenomenon led to the development of glasses that could respond to monovalent ions such as Na^+, K^+, Li^+ and Ag^+. Selectivity depends on the material used to ensure contact between the solution and the inside of the electrode. Standard reference half-cells contain KCl as an electrolyte, which normally flows through the diaphragm to ensure contact with the solution. In measurements involving K^+ or Cl^-, an electrolyte gel or a second, non-interfering electrolyte is employed to avoid perturbations (Fig. 18.3).

Presently, about 20 ion selective electrodes (ISE) are commonly used and classified in several groups depending on the membrane they use. These electrodes are used in direct ionometry or as indicator electrodes for many measurements involving titrimetry and complexometry with automatic titrators.

Many types of membrane have been developed and each is adapted to a specific ion or a gas. Technically, it is a matter of developing an ionic material that permits the formation of a concentration equilibrium with specific ions such that if the activity of the ion on both sides of the membrane varies, a potential difference is generated. Measurements involve only ions that are free in solution.

18.3.1 Crystalline mineral membranes

The fluoride electrode is a typical example of an ion selective electrode. Its sensitive element is a crystal of lanthanum trifluoride that allows fluorine atoms to migrate into the network formed by lanthanum atoms (Fig. 18.3). Other electrodes use a mineral membrane obtained as agglomerates of crystalline powders (for measurement of Cl^-, Br^-, I^-, Pb^{++}, Ag^+ and CN^-). Generally, the internal electrolyte can be eliminated (by dry contact). However, it is preferable to insert a polymer layer with a mixed-type conductivity to ensure the passage of electrons from the ionic conductivity membrane to the electronic conductivity electrode (Fig. 18.3).

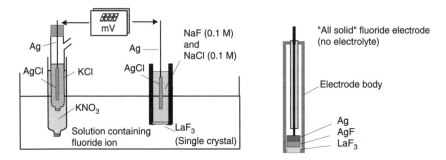

Figure 18.3—*Principle of ISE measurement of fluoride ions in solution using a double junction reference electrode*. The reference electrode is inserted into a separate chamber that contains the auxiliary electrolyte in order to avoid osmosis of KCl into the sample solution. Also, 1 M KNO_3 can be used for F^-, Cl^-, CN^- or Ag^+ determination. The measurement involves the use of a high impedance millivoltmeter (pH meter type). A version of an all-solid fluoride electrode is shown on the right.

18.3.2 Liquid membrane

Separation between the internal and external solutions of the selective electrode is achieved using a porous hydrophobic disc (of 3 mm diameter). The disc is wetted with an organic solvent that is immiscible with water on either side and contains an *ionophore* (Fig. 18.4). Hence, the membrane can be considered as an *immobilised liquid*. The counter ion is a molecule which is soluble in the organic phase.

To measure calcium ions, for example, an aliphatic diester of phosphoric acid $(RO)_2P(O)O^-$ can be used (Fig. 18.5). Concentration equilibrium exists on each side of the wall:

$$[(RO)_2P(O)O]_2Ca \rightleftharpoons [(RO)_2P(O)O]_2^{2-} + Ca^{2+}$$
$$\text{solvent} \qquad\qquad\quad \text{solvent} \qquad\qquad \text{aqueous phase}$$

A measurable difference in potential appears when the concentration of calcium ions is different on each side.

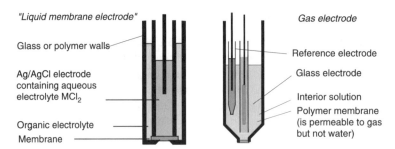

Figure 18.4—*Liquid membrane electrode and gas diffusion electrode.* The gas diffusion electrode is constructed in part from a pH electrode that is plunged into the internal solution.

18.3.3　Polyvinyl chloride membranes

Many membranes are made from a film of polyvinylchloride (PVC) in which one or many ion transporters (ionic or neutral) are inserted. Some 40 chelates are used in about a dozen selective membranes (Fig. 18.5). Electrodes used for ClO_4^- and BF_4^- are of this type.

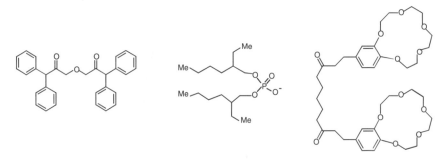

Figure 18.5—*Chemical compounds used for ionophores.* From left to right, three commercially available chelating ionophores: ETH 129 and dioctyl phosphate, which are selective for the ion Ca^{2+}, and K2 ionophore, which is selective for the ion K^+. In the latter compound, a crown ether sub-structure can be seen.

18.3.4　Gas diffusion organic membranes

Several combined electrodes have been developed to measure dissolved gases such as carbon dioxide (CO_2), ammonia (NH_3) and sulphur dioxide (SO_2). These sensor-like electrodes contain an internal solution that is isolated from the external solution by an impermeable membrane. The membrane forbids the passage of water or ions but is transparent to gas molecules dissolved in solution (Fig. 18.4).

For the examples given above, the selective electrode is made of a glass electrode in contact with a solution containing a low concentration (0.01 M of bicarbonate). The bicarbonate solution is separated from the sample solution by a polymer membrane that allows the diffusion of the analyte gas into the bicarbonate solution. This diffusion causes a change in pH in the proximity of the membrane.

For other gases, such as hydrogen cyanide (HCN), fluorhydric acid (HF) or hydrogen suphide (H_2S), the signal is generated by modification of the concentration of anions at the internal wall of the membrane.

> ■ The selectivity of an electrode for other ions can be variable. For each interfering ion, selectivity is defined by a coefficient $K_{specific\ ion/interfering\ ion}$. For example, $K_{Br/Cl} = 2.5 \times 10^{-3}$ indicates that the selectivity of the electrode for the bromide ion will be $1/2.5 \times 10^{-3}$, hence 400 times greater for bromide than for chloride.

18.3.5 Biocatalytic membranes of the future

In principle, the specificity of an electrode is obtained when a membrane can selectively recognise the species to be measured in a sample solution. Biosensors have be developed in which the measuring electrode can detect a particular compound formed through a biochemical enzymatic reaction. For example, if urease is fixed to a membrane sensitive to the ammonium ion, this membrane will be able to detect urea since urease is an enzyme that will decompose urea to form ammonium ions. However, such electrodes do not presently exist in a commercial form.

> ■ These electrodes must be distinguished from biosensors with enzymes at their surface. Electrodes constructed from field effect transistors (see 19.7) lead to voltametric detection.

18.4 Calculations and different methods

There are many methods that allow the determination of the concentration c_i of an ionic species i in a sample. In the presence of an ionic strength adjuster (ISA) or a buffering solution that can fix the pH (TISAB, *Total Ionic Strength Adjustment Buffer*) all of these methods are based on application of equation (18.3).

For a monovalent ion at 298 K, the *slope factor S* that represents the term $2.303RT/F$ in equation (18.3) is equal to 0.0591 V (the theoretical slope). It is also the value of the slope per pH unit for a glass electrode.

$$E_{measured} = E'' + \frac{0.0591}{z} \log c_i \qquad (18.4)$$

In practice, using a set of standard solutions, a calibration plot $E_{measured} = f(c_i)$ allows the determination of the real slope (i.e. the experimental value of the slope factor). This slope can be used as an indicator of the performance of the system.

Table 18.1—Values of ideal slopes at 298 K for some ions.

Valence	Ion	S (mV/decade)
-2	S^{2+}	-29.58
-2	Cl^-, F^-	-59.16
+1	Na^+, K^+	+59.16
+2	Mg^{2+}, Ca^{2+}	+29.58

18.4.1 Direct ionometry

The experimental protocol used to conduct measurements is based on the following principle: a series of standard solutions is prepared by successive dilution of a stock solution and an excess but constant volume of buffer (ISAB or TISAB) is added at each step. Sample solutions are prepared in the same fashion. For each of the standards, the potential across the electrodes is measured and a semi-logarithmic calibration curve $E = f(c_i)$ is obtained (Fig. 18.6). Using this curve and the potential difference obtained for each of the sample solutions, the concentration of species i can be obtained.

Using equation (18.3), the precision of the measurements can be determined as follows:

$$dE = \frac{RT}{zF} \times \frac{dC_i}{C_i} \quad \text{that is} \quad \frac{\Delta C_i}{C_i} = \Delta E \times \frac{zF}{RT} \quad (18.5)$$

It can be noted from equation (18.5) that an uncertainty of 0.2 mV will lead to a lack of precision in the order of 0.8% for a monovalent ion.

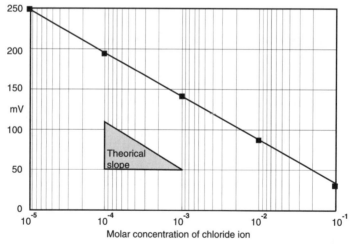

Figure 18.6—*Example of a measurement by direct potentiometry*. The slope for the chloride selective electrode has almost an ideal value. The dynamic range for most selective electrodes spans 4 to 6 orders of magnitude, depending on the ion.

18.4.2   Standard addition method

This technique is useful to eliminate matrix effects and only requires one standard solution and two measurements. However, the values used must be precisely known.

The potential difference E_1 between the two electrodes (ISE and ERE) is measured while they are dipped into a known volume V_X of the sample solution containing concentration C_X of the ion i to be measured (formula (18.4)). After adding a small volume V_R of a reference solution of ion i at concentration C_R, the measurement is redone. Under these conditions, a new potential difference E_2 is obtained. C_X is thus obtained using equation (18.6):

$$C_X = C_R \cdot \frac{V_R}{V_R + V_X} \cdot \left(\frac{1}{10^{\Delta E/S} - \dfrac{V_R}{V_R + V_X}} \right) \qquad (18.6)$$

where S is the actual slope for the electrode used ($0.0591/z$ at 298 K) and $\Delta E = E_2 - E_1$ is the potential difference between the two measurements E_1 and E_2.

■ Instead of adding the reference solution, a volume V_X of the sample solution at concentration C_X can be added to a volume V_R of the reference solution at concentration C_R. This approach is preferable when dealing with relatively concentrated sample solutions. Using ΔE, the measured potential difference, C_X can be obtained using the following equation:

$$C_X = C_R \cdot \frac{V_X + V_R}{V_X} \cdot \left(10^{\Delta E/S} - \frac{V_R}{V_X + V_R} \right) \qquad (18.7)$$

18.4.3   Potentiometric titration

Selective electrodes have a variable specificity. Precision can be increased when they are used as indicating electrodes in potentiometric measurements. The concentration of ions present in solution and the ionic strength will undergo small variations during measurement relative to the concentration of the ion being measured. When two ionic species undergo stoichiometric reaction, this property can be used for their determination. The end point in the measurement is characterised either by the total disappearance of one of the species or by the appearance of an excess of one of the species. The appearance or disappearance of a secondary species can also be used to determine the end point.

■ For example, the aluminium ion, for which there is no selective electrode, can be determined in the following way. Sodium fluoride can be added as a reagent to lead to insoluble AlF_3. The end point of the reaction can be determined by the presence of an excess of fluoride ions in solution, easily measured with a fluoride selective electrode.

A great number of automatic titrators of competing brands allow many similar determinations to be conducted.

18.5 Applications

Potentiometry is often used in routine analysis. The determinations are usually simple and the sensitivity of the electrodes allows sub-ppm concentrations to be measured. However, interferences have to be considered before analysis.

The follow list represents routine applications of the method:

— in agriculture, the analysis of nitrates in soil samples

— in the control of food stuffs, the analysis of ions such as NO_3^-, F^-, Br^- Ca^{2+}, etc. in drinks, milk, meat or fruit juices

— in industry, the analysis of chloride in pulp and paper, cyanide in electrolytic baths, and fluoride and chloride in galvanic processes

— in biomedicine, the analysis of certain ions in serum, biological fluids, saliva and gastric fluids.

■ The *lambda sensor*, which is found in cars with catalytic converters, is an example of an oxygen probe based on the principle of selective electrodes. This sensor, which looks like a spark plug, has a zirconium sleeve (ZrO_2) that behaves as a solid electrolyte. The external wall is in contact with emitted gas while the internal wall (the reference) is in contact with air. Two electrodes measure the potential difference between the two walls, which is indicative of the difference in concentration of oxygen.

Problems

18.1 Consider an aqueous solution of ammonia, 9×10^{-3} M, to which is introduced a quantity of ammonium chloride in order to attain a pH of 9. At this precise moment, what is the concentration of the salt?
(K_B $NH_3 = 1.8 \times 10^{-5}$)

18.2 Explain how the electrode's glass membrane registers the pH of an aqueous solution.

18.3 What is the degree of dissociation and the pH of an aqueous solution of 0.85 M ethanoic acid? ($K_A CH_3CO_2H = 1.8 \times 10^{-5}$ at 20 °C.)

18.4 30 ml of an 10^{-3} M aqueous solution of an Fe^{3+} salt is mixed with 20 ml of a 10^{-3} M solution of a salt of Ti^{3+}. What will be the molar concentration of these salts following reaction in this modified solution? ($Ti^{3+}/Ti^{4+} = 0.2$ V vs SHE; $Fe^{2+}/Fe^{3+} = 0.77$ V vs SHE and SCE/SHE $= +0.25$ V).

18.5 Knowing that the standard potential of an electrode Cd/Cd^{2+} vs SHE, has the value 0.404 V, what happens to this potential if the electrode is plunged into an aqueous solution of 0.01M $CdSO_4$?

18.6 Find the expression used in the standard addition method (using one single addition), recalling that C_X represents the ion concentration to be measured (volume V_X of the sample solution to which is added a small volume, V_R, of a reference solution of concentration C_R of the ion to be measured). ΔE is the difference in potential between the two measurements while S is the slope of real response of the electrode used ($0.0591/z$, at 298 K).

$$C_X = C_R \frac{V_R}{V_R + V_X} \frac{1}{10^{\Delta E/S} - \frac{V_X}{V_R + V_X}}$$

CHAPTER 19

Coulometric and voltammetric methods

Contrary to potentiometric methods that operate under null current conditions, other electrochemical methods impose an external energy source on the sample to induce chemical reactions that would not otherwise spontaneously occur. It is thus possible to measure all sorts of ions and organic compounds that can either be reduced or oxidised electrochemically. Polarography, the best known of voltammetric methods, is still a competitive technique for certain determinations, even though it is outclassed in its present form. It is sometimes an alternative to atomic absorption methods. A second group of methods, such as coulometry, is based on constant current. Electrochemical sensors and their use as chromatographic detectors open new areas of application for this arsenal of techniques.

This chapter is an overview of voltammetric methods used in analytical laboratories and as an example of coulometry the Karl Fischer method is presented.

19.1 General principles

Voltammetric methods consist of applying a variable potential difference between a *reference electrode* (e.g. Ag/AgCl) and a microelectrode, called the *working electrode*, at which an electrochemical reaction is induced (Ox + ne → Red). As the potential at the working electrode reaches a value such that a species present in solution is either reduced or oxidised, the current in the circuit increases. In practice, to avoid flow of current through the reference electrode, a third electrode (called an *auxiliary electrode*) made of inert material or carbon, and a support electrolyte are used to create a conducting medium (Fig. 19.1). Microelectrolysis occurs over a short period during measurements and does not significantly modify the concentration of analytes in solution.

The *voltammogram*, which is a current–voltage curve, $I = f(E)$, corresponds to a voltage scan over a range that induces oxidation or reduction of the analytes. This plot allows the identification and measurement of the concentration of each species. Several metals can be determined by this method.

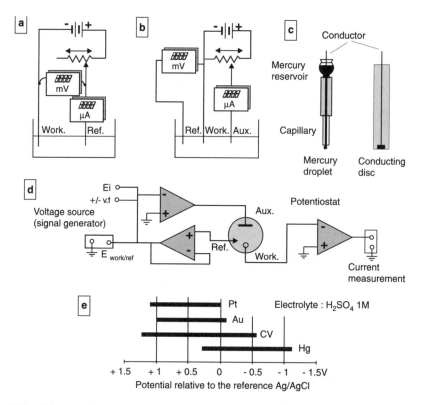

Figure 19.1—*Schematics showing the principle of a voltammetric cell.* a) A set-up based on two electrodes, which is not used in practice because it allows current to flow through the reference electrode and thus should be avoided for several reasons; b) DC set-up in which no current flows through the reference electrode (because of its high impedance); c) two types of electrodes: a disc and a hanging-drop electrode; d) schematic showing several ways to increment voltage; e) the range of use of the four main electrodes (VC stands for vitreous carbon). The mercury electrode can be used over a wide cathodic range with electrolytes such as KCl or NaOH (−2 V). Although overlaps are observed with different electrodes, their sensitivity for a given species can be quite different.

The most widely used working electrodes are vitreous carbon, platinum, gold and mercury (Fig. 19.1). These electrodes are flexible because they can be used between two potential values that depend on the support electrolyte, the pH and the nature of the reference electrode. For example, the limits for the Pt electrode are $+0.65\,V$ relative to the standard calomel reference electrode (SCE) (oxidation of water: $H_2O \rightarrow \frac{1}{2}O_2 + 2H^+ + 2e^-$) and $-0.45\,V$ (reduction of water: $H_2O + 2e^- \rightarrow H_2 + 2OH^-$).

19.2   Dropping-mercury electrode

The most original working electrode is without doubt that which contains micro-droplets of mercury as an active part. This electrode has advantages and limitations that make it different from other electrodes. The voltammetric method what employs this electrode is called polarography, and it was invented by Heyrovsky in the 1920s.

The dropping-mercury electrode is made of a central glass capillary (of 10 to 70 μm diameter) maintained in the vertical position thus allowing the transfer of mercury between a reservoir and the capillary tip where very small droplets are formed. The surface of the droplet increases until it falls (in 3 to 4 s), with the fall often provoked by a small impact on the glass capillary. A new droplet, identical to the previous one, is formed with a refurbished surface.

Mercury is quickly limited at positive potentials (+0.25 V with respect to SCE). Beyond this potential, anodic dissolution of mercury occurs. However, mercury can be used at up to −1.8 or −2.3 V depending on whether the supporting electrolyte is acidic or alkaline. This range offers several possibilities, especially for the determination of heavy metals. The mercury used must be extremely pure (six-time distilled, under nitrogen). Unfortunately, the use of mercury as an electrode is a disadvantage because of its toxicity; the mercury must be recycled after each use.

19.3 Continuous current polarography

In the basic experiment, which is now rarely used, a variable, time-dependent voltage with increments (*scan rate*) in the order of 1 or 2 mV/s with respect to the initial potential E_i is applied to the mercury droplet.

$$E_{\text{working}} - E_{\text{ref.}} = E_i \pm v \cdot t \tag{19.1}$$

In equation (19.1), where E_i represents the initial voltage and E_{working} represents the voltage applied to the working electrode, the ± sign indicates that the scan can be performed on the anodic ($V > 0$) or the cathodic ($V < 0$) side. The resulting *polarogram*, which is an $I = f(E)$ plot, can show one or several steps or waves (Fig. 19.2). The height of each step corresponds to the limiting diffusion current i_D (cf. section 19.4) and is a function of the concentration of the species, which thus allows its quantitative determination. The half-wave potential, measured at $i_D/2$, is characteristic of the analyte. Therefore, many species present in a solution can be studied during the same analysis if their half-wave potentials are sufficiently different. The saw-toothed fine structure observed on the polarogram is due to renewal of the mercury droplets.

■ Voltammetry offers some advantages over atomic absorption. It allows the determination of an element under different oxidation states. For example, Fe^{2+} can be distinguished from Fe^{3+} (cf. Fig. 19.4).

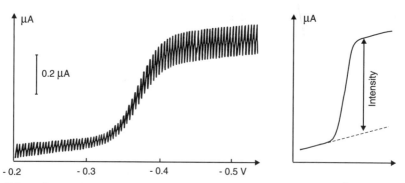

Figure 19.2—*Polarographic wave.* Polarogram of a solution containing 10 ppm Pb^{2+} in 0.1 M KNO_3 obtained with a dropping-mercury electrode. The median position of the wave (-0.35 V) is characteristic of lead and the height of this step is proportional to concentration. To the right, a graph shows the measure of i_D. For a better representation of the curve and measurement of height, the oscillations have been damped.

19.4 Diffusion current

The particular features of a polarographic wave can be explained as follows. Suppose a sample solution contains a very dilute lead salt in a support electrolyte whose concentration is much higher than that of the Pb^{2+} to be measured. If a linearly increasing potential with a starting value lower than that necessary to reduce Pb^{2+} is applied to the working electrode, at a voltage of -0.35 V (vs Ag/AgCl) Pb^{2+} ions will be reduced upon contact with the mercury droplet. At the same time, K^+ ions that cannot be reduced at this potential will form a shield around the droplet and will tend to prevent the normal migration of Pb^{2+} ions in the solution. However, under the effect of a concentration gradient, the Pb^{2+} ions will reach the surface of the electrode – where they will be reduced – by simple diffusion through the layer of K^+ ions. This phenomenon is the basis of the *diffusion current*. The average limiting value of the diffusion current i_D, which is essentially due to the flux of the analyte ion at the instant before the droplet falls, depends on several parameters:

$$\bar{i}_D = 607 \cdot n \cdot D^{1/2} \cdot m^{2/3} \cdot t^{1/6} \cdot C \qquad (19.2)$$

where n represents the number of electrons transferred during electrolysis, D is the diffusion coefficient of the analyte ion (cm^2/s), m is the mass flow of mercury (g/s), t is the lifetime of the droplet (s) and C is the concentration of analyte (mol/ml).

In practice, this pseudo-mathematical expression, known as the Ilkovic equation, which takes into account several factors, is replaced by the simplified formula below where K is a graphical constant that includes several parameters related to the method and the instrument used.

$$i_D = K \cdot C \qquad (19.3)$$

Experimentally, the current passing through the droplet will depend on several phenomena. These are distinguished by:

— the capacitive current, which makes the droplet/solution interface analogous to a capacitor by fixation of electrons that face the support electrolyte (e.g. K^+). This capacitive current decreases rapidly with time because it depends on the rate of change of the droplet's surface area dA/dt, which decreases as the droplet increases.

— the Faraday-type diffusion current, which decreases as $t^{1/2}$ if the surface area of the droplet is constant but which increases because of the increasing applied potential and surface area of the droplet. Expansion of the droplet's surface area more than compensates for the depletion of electroactive substances in the vicinity of the electrode.

The current intensity curve has features shown in Fig. 19.3.

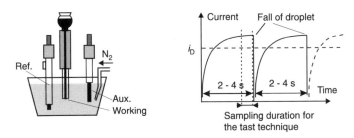

Figure 19.3—*Polarographic cell and diffusion current*. Dissolved oxygen, which leads to an interfering double wave, has to be removed from the sample solution by degassing. On the right features of the diffusion current are shown. These increase with time for every drop of mercury in a static (unstirred) solution. Direct polarography is a slow method of analysis. More than 100 droplets are needed to record the voltammogram.

■ There is a technique (called *Tast*) that consists of measuring the current only for a very short period of time, late in the lifetime of the mercury droplet. This method eliminates the saw-toothed spikes from the current curve. However, it leads to a lower sensitivity because the diffusion current decreases as the analyte flux goes from the bulk of the solution to the surface of the droplet (Fig. 19.3).

19.5   Pulse polarography

Instead of applying a steadily increasing voltage on the mercury droplet, the voltage can be pulsed. The advantage of this approach is increased sensitivity and better distinction between analytes with close half-wave potentials (i.e. which differ by only a few tens of mV). Pulse polarography has two major modes of operation, which are described in the following sections.

19.5.1   Normal pulse polarography (NPP)

In this mode of operation, the standard voltage ramp is replaced by a series of brief pulses (50 to 100 ms) of increasing potential. The pulses are delivered at the same frequency as the droplet is renewed, every 4 or 5 s. Current measurements are made just before the fall of the droplet when the Faraday current has stabilised and the capacitive current is negligible. Using this approach, the current–voltage curve is much smoother. This method also increases sensitivity by two or three decades compared to normal polarography. The threshold voltage is low enough that it does not induce reduction of the analyte: electrolysis is thus blocked between each pulse (Fig. 19.4).

19.5.2   Differential pulse polarography (DPP)

In this mode, each droplet is subject to two measurements. The first measurement is made just before the pulse of 50 mV and the second just before the droplet falls. This way, the two measurements are made on the same droplet but with a 50 mV voltage difference (Fig. 19.4). The graph obtained in DPP is more easily interpreted.

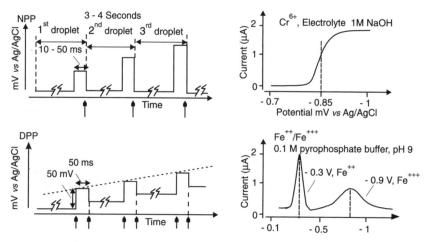

Figure 19.4—*Pulse polarography*. The techniques of NPP and DPP are shown with two examples of measurements. Arrows on the voltage ramp diagram indicate the instant at which the current is measured.

19.6   Amperometric detection in HPLC and CE

Voltammetry has been adapted to HPLC (when the mobile phase is conducting) and capillary electrophoresis (CE) as a detection technique for electroactive compounds. In this usage, the voltammetric cell has to be miniaturised (to about 1 µl) in order not to dilute the analytes after separation. A metal or carbon microelectrode has a defined potential (vs the reference electrode) depending on the substances to be detected (ions or molecules) and the mobile phase flows through the detection cell (Fig. 19.5). This method of amperometric detection in the pulsed mode is very

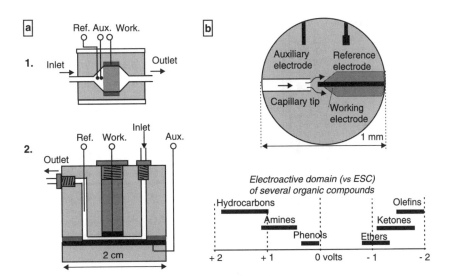

Figure 19.5—*Amperometric detection in HPLC and CE.* a) Two models of amperometric detection are shown. The working electrode, made of porous graphite with a large surface area, operates under coulometric conditions. The flow of the mobile phase at the working electrode ensures renewal of electroactive species; b) expanded diagram of the end of the capillary in CE. Ions exiting the capillary impinge on the working electrode, which is placed on the cathodic side of the instrument. The complete detection cell is not shown.

sensitive. However, few analytically important molecules can be detected by this device besides phenols, aromatic amines and thiols.

19.7 Special sensors

Special electrochemical sensors that operate on the principle of the voltammetric cell have been developed to measure substrates such as oxygen and glucose. In the Clark oxygen sensor, a 1.5 V potential difference is applied between a silver anode and a platinum cathode which are both in contact with a KCl solution separated from the sample by a membrane permeable to oxygen (Fig. 19.6).

After crossing the membrane, the gas comes into contact with the cathode whose potential is sufficiently negative to reduce oxygen (reactions 1 and 2 in Fig. 19.6). The current in the circuit is proportional to the amount of oxygen that crosses the membrane, which is proportional to the oxygen concentration in solution.

Several sensors have been developed to measure substrates in biological fluids. The structure of the membrane in these *biosensors* is quite complex. It generally contains a substrate-sensitive layer that includes a specific enzyme to catalyse a reaction and transform the analyte (or substrate).

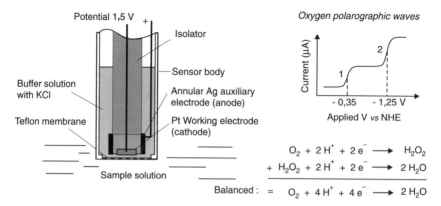

Figure 19.6—*The Clark sensor for oxygen determination.* The Teflon membrane, permeable to gas, must be very close to the cathode so that the double diffusion process through the membrane and the liquid film will lead to a stable signal within a few seconds.

In some systems, for example, the immobilised enzyme is sandwiched between an external cellulose acetate membrane that blocks the passage of heavy molecules and an internal polycarbonate membrane that allows passage of the transformed products to the electrode.

One of the commercial successes in this area is a miniaturised device developed for insulin-dependent diabetes, which uses *glucose oxidase* (GO) to measure glucose in blood (see Fig. 19.7).

■ The mechanism of the glucose sensor's operation is as follows: glucose reacts with the oxidised from of glucose oxidase, an enzyme produced by *Aspergillus niger*, to yield gluconic acid. Two protons and two electrons are transferred to the enzyme, which becomes reduced. The enzyme, by reaction with an oxygen molecule, goes back to its oxidised form and generates a molecule of hydrogen peroxide.

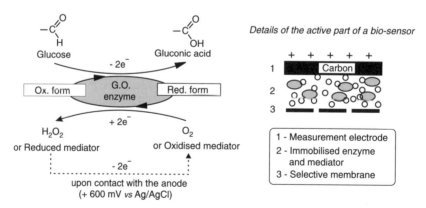

Figure 19.7—*Amperometric detection of glucose.* The reaction cycle is shown on the left. A sandwich-type biosensor involving glucose oxidase co-immobilised with an osmium-based redox mediator in a polyvinyl polymer is shown on the right.

This reaction can be followed by monitoring oxygen consumption or H_2O_2 formation. Possible interferences in the process can be eliminated by replacing oxygen with an oxidant that can be mixed with the enzyme at a known concentration. The oxidant's reduced form can be detected amperometrically since it will be reoxidised at the electrode by yielding two electrons.

A particular type of biosensor can be developed by putting a membrane in contact with the semi-conducting layer of a field effect transistor. If the membrane incorporates an enzyme adapted to transform a particular analyte (Fig. 19.8), reaction of that enzyme will modify the polarity at the surface of the insulating layer. This will in turn modify the conduction between the source and the collector of the field effect transistor. The current flowing through these two electrodes (source and collector) serves as the signal.

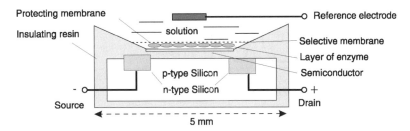

Figure 19.8—*A selective electrode designed from a MOSFET (metal oxide semiconductor field effect transistor)*. A specific reaction can be monitored by putting an enzyme in contact with the electrodes. This schematic shows the three electrodes used for amperometric measurement.

19.8 Stripping voltammetry

Anodic (or cathodic) stripping voltammetric (SV) techniques can be used to measure traces of metal in the environment. These methods are very sensitive and their precision is in the order of 1 to 2%. Stripping voltammetry is carried out in two stages.

A fraction of the electroactive analyte present in the sample is deposited by electrolysis (in 1 to 30 min) on an electrode which is inserted in the stirred solution and held at a constant potential. This electrode, generally a hanging mercury droplet (the drop is not renewed) or vitreous carbon covered with a mercury film, leads to an amalgam M(Hg) when M^{n+} is a metallic cation. Following this preconcentration stage, mixing of the solution is stopped and the potential difference between the mercury and reference electrode drops progressively. Due to the reversibility of redox reactions, the analyte is redissolved. The example given in Fig. 19.9 corresponds to an anodic voltammetric experiment. When combined with DPP, stripping voltammetry generates data that resemble a chromatogram. Obtaining reproducible, quantitative results involves control of the electrochemical parameters as well as the mixing of the solution, the size of the electrode and the duration of electrolysis when analyte is deposited on the electrode.

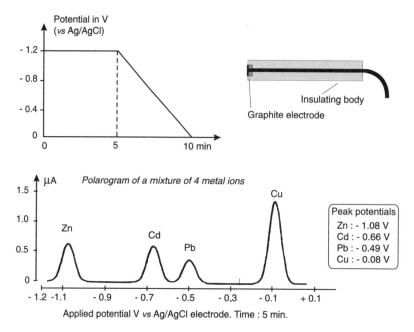

Figure 19.9—*Stripping voltammetry*. The figure shows voltage programming of the working electrode and an example of a determination by differential pulse polarography methods (DPP).

19.9 Coulometric measurements

Previously described methods involved partial electrolysis of the analyte at the working electrode. However, coulometric methods at constant current or constant potential are based on the use of a reagent that can react with or convert the analyte quantitatively.

> ■ For example, the determination of Cl^- anions can be carried out using Ag^+ ions generated at the surface of a silver electrode. The amount of current, in coulombs, allows the precise determination of the quantity of transformed analyte, as long as the current is only used to produce Ag^+ ions. *These determinations do not require the use of standard solutions* because the absolute quantity of ions is measured from the current generated. The amount of current corresponds to $Q(C) = I(A) \cdot t(s)$.

Coulometric determinations are made at a constant current using an amperostat. Coulometric titrators include a cell with a large surface area electrode and an auxiliary electrode isolated from the reaction compartment by a diaphragm. This isolation of the second electrode eliminates the possibility that species formed on its surface may react with the working electrode (see Fig. 19.11).

In this type of determination, it is preferable to generate the titration reagent than to directly transform the species being measured by contact with the electrode. This approach avoids accumulation of the product on the electrode, which may cause its polarisation.

19.10 Karl Fischer coulometric determination of water

Several manufactured products as well as solvents and raw materials are analysed for their water content (or % humidity). Of all the available methods, the Karl Fischer procedure is the most widely used, accounting for almost 500 000 determinations per day world-wide.

This type of analysis is conducted with dedicated instruments. It uses chemical reactions that resemble electrochemical detection. The determination can be carried out in two different ways. The first is conducted in a classical way using titration and the second is a coulometric method using a diaphragm cell. The latter is the more sensitive of the two methods, which makes its use compulsory for the determination of very low levels of water (concentrations in the order of mg/l).

■ Applications of the Karl Fischer method are numerous: food stuffs (butter, margarine, powdered milk, sugar, cheese, processed meats, etc.), solvents, paper, gas, petroleum, etc. Before the determination can be made, solid components that are not soluble must either be ground into powders, extracted with anhydrous solvents, eliminated as azeotropes or heated to eliminate water. Problems are encountered with very acidic or basic media that denature reactants and transform ketones and aldehydes into acetals that interfere with the titration. Special reagents must be used in these instances.

19.10.1 Reactions involved

In the presence of water, iodine reacts with sulphur dioxide through a redox process that is specific to the three compounds present:

$$I_2 + SO_2 + 2H_2O \rightarrow H_2SO_4 + 2HI$$

This reaction can be used to determine water if the transformation is stoichiometric. This is why a base is added to the reaction medium to neutralise the acids formed (pyridine is traditionally used in the Karl Fischer method).

Since iodine is a solid and sulphur dioxide is a gas, a polar solvent must be used for the reaction and as a dilution agent. Methanol is generally used as the solvent; however, methylmonoether glycol or diethylene glycol can also be used. In these conditions, sulphur dioxide is not simply dissolved in the solvent but actually interacts with it. For example, with methanol, SO_2 is transformed into methylhydrogen sulphite that reacts with I_2 in the presence of water. This leads to the following reaction:

$$CH_3OSO_2H + I_2 + H_2O \rightleftharpoons CH_3OSO_3H + 2HI$$

In contrast to the previous reaction, in this case *a single* molecule of water reacts with *two atoms* of iodide. Methylhydrogen sulphite is oxidised to hydrogenosulphate, which is converted in the presence of a base of type RN into an ammonium salt. The Karl Fisher reaction can thus be reformulated as follows:

$$H_2O + I_2 + [RNH]^+CH_3OSO_2^- + 2RN \rightarrow 2[RNH]^+I^- + [RNH]^+CH_3OSO_3^-$$

When pyridine is used as a base, a pyridinium salt $C_5H_5NH^+(CH_3OSO_3)^-$ is formed. Nowadays, pyridine is replaced by imidazole or diethanolamine, which yield more stable and less odorous commercial products.

Because atmospheric humidity must be avoided, the reaction flask is isolated from the atmosphere with drying tubes. Moreover, since the solvent is rarely perfectly anhydrous and will contain traces of water due to its hygroscopic nature, its water content must be measured prior to the determination. The equivalence point of the titration reaction is detected by an electrical method instead of a visual method. The current intensity that passes between two platinum electrodes inserted in the reaction medium is measured (see Fig. 19.10). The reagent, which is a mixture of sulphur dioxide, iodine and a base, is characterised by the *number of mg of water* that can be neutralised by $1\,cm^3$ of this reagent. This is referred to as the equivalent mass concentration of water, or the *titre T* of the reagent.

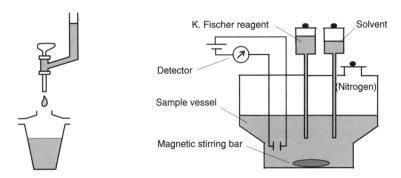

Figure 19.10—*Karl Fischer method for determination of water*. The conventional burette titration with visual detection of the end point leads to imprecise results. Thus, a cell containing two small platinum electrodes is used. As long as no iodide is present in the solution, the current between the electrodes is weak. When excess iodide is present in the solution at the instant the equivalence point is reached, a significant current is registered.

19.10.2 Coulometric adaptation of the Karl Fischer determination

At least $10\,mg$ of water must be present in the sample to obtain an acceptable precision, given the magnitude of the equivalent water titre T of the commercial reagent needed. When the amount of water is lower than this (as low as $10\,\mu g$ of water), the coulometric method must be used to increase sensitivity.

Two types of automated instruments are used for this type of determination. The first type is known as *normal* while the second is *coulometric*.

In the coulometric version of the instrument, the iodine necessary for reaction is generated electrochemically by applying electrical pulses to the electrode. In this case, a modified Karl Fischer reagent is used which contains iodide instead of iodine. An iodide solution is in contact with the anode in one compartment of the electrolytic cell, which has a diaphragm between the anodic and cathodic regions (Fig. 19.11). Thus, the volume of reagent, which is the indicator in the classical titration method, is replaced by a more precise quantity of current, obtained coulometrically.

At the anode, the iodide ion is oxidised to iodine ($2I^- \rightarrow I_2 + 2e^-$). One mole of water necessitates two faradays or alternatively, 1 mg of water corresponds to 11.72 coulombs.

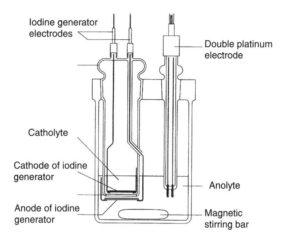

Figure 19.11—*Karl Fischer coulometric cell, model DL-37.* The diaphragm (membrane) is used to avoid oxidation of the reduced ions created at the surface of the cathode. (Reproduced by permission of Mettler Toledo.)

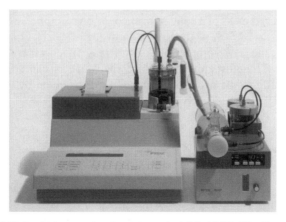

Figure 19.12—*Karl Fischer instrument (coulometric version).* Model DL-37 with an external device that includes an oven to extract water from solid samples. (Reproduced by permission of Mettler Toledo.)

19.11   Determination by the Karl Fischer method

19.11.1   Direct and back titration

Voltammetric determination using the Karl Fischer approach can be carried out using either *direct titration* or *back titration*. In the direct titration procedure, the equivalence point is obtained by adding the necessary amount of Karl Fischer reagent. When back titration is used, an excess of reagent is added and the excess after equivalence is reached is determined with an anhydrous solvent whose humidity content is known.

Two steps are involved in the analysis. First, the water equivalence of the Karl Fischer reagent (equivalent titre T) is measured (cf. 19.11.2). Secondly, the water content of the sample is determined (cf. 19.11.3). Both operations are made in series in a special cell isolated from the atmosphere.

■ The titre of the Karl Fischer reagent varies with time because of the instability of the solutions. For this reason, manufacturers usually sell the Karl Fischer reagent in two different containers; one containing a mixture of sulphur dioxide, methanol and a base, the other containing a solution of either iodine or iodide. The contents are generally mixed a few days before use of the reagent.

19.11.2   Determination of the water equivalence (titre) of the Karl Fischer reagent

— *Direct titration method*

A volume of solvent V_1 is put into the flask (e.g. methanol). Volume V_2 is the volume of reagent necessary to reach the equivalence point of the titration (the solvent blank).

A precise quantity m (in mg) of water is introduced into the flask either using a microsyringe similar to that used in chromatography (graduated to 1/10th of a µl) or as a precisely measured mass of a solid hydrate of known composition, such as oxalic acid · $2H_2O$ (which is 28.57% water) or sodium tartrate · $2H_2O$ (which is 15.66% water). Then V_3 is the new volume (in cm^3) of reagent necessary to return to the equivalence point. Therefore, the titre T of the Karl Fischer reagent (its water equivalence, normally in the order of 5 mg/cm^3) will be given by:

$$T = \frac{m}{V_3} \tag{19.4}$$

— *Back titration method*

A sufficient quantity of the reagent is first added to cover the electrodes in the detection cell (*ca.* 20 cm^3, depending on the concentration). Methanol is then added up to the point of equivalence. After this first dehydration phase (pre-titration) where the volumes introduced are not precisely measured, the following steps take place in this anhydrous medium:

— a volume V_2 of the Karl Fischer reagent is added and neutralised with a volume V_1 of methanol.

— a precise quantity m (in mg) of water is then added, as in the direct method. A volume V_3, that represents an excess of the Karl Fischer reagent, about 2 to 3 cm³, is added to the mixture

— the excess reagent is then neutralised by a volume of methanol, V_4.

The titre T (expressed in mg/cm³) is obtained by the following expression:

$$T = \frac{m}{V_3 - \frac{V_2}{V_1} V_4} \tag{19.5}$$

19.11.3 Determination of the water content of a sample

— *Direct titration method*

The determination is conducted by adding P mg of the sample to the anhydrous solution already in the cell. The sample can be added in its neat form or in solution in a volume V_5 of the same solvent.

If V_6 is the volume of the Karl Fischer reagent needed to reach the equivalence point, the quantity (mass) of water p present in the sample is obtained using the following equation:

$$p = T \left(V_6 - \frac{V_2}{V_1} V_5 \right) \tag{19.6}$$

However, if the sample has been introduced in its neat form without solvent ($V_5 = 0$), the quantity of water p is given by the simplified formula:

$$p = T \times V_6 \tag{19.7}$$

— *Back titration method*

P mg of sample is introduced into the titration flask and V_7 is the *precisely measured* volume of Karl Fischer reagent added, corresponding to an excess. The excess is titrated as previously described by adding a volume V_8 of methanol. Using these values, the amount of water p in P mg of sample is given by:

$$p = T \left(V_7 - \frac{V_2}{V_1} V_8 \right) \tag{19.8}$$

Finally, the water content in the sample, expressed as a percentage, is given by equation (19.9).

$$H_2O\% = 100 \frac{p}{P} \tag{19.9}$$

Problems

19.1   1. Polarography is considered, so far as the sample is concerned, as a non-destructive method of analysis. Is this true?

        2. Why is stirring of the solution avoided in polarography?

        3. The method of standard additions is considered as yielding more reliable results than that of standard solutions. Why?

        4. Why, for the dropping mercury electrode, does the height of the column of Hg have an influence upon the value of the current of diffusion? Upon what is this effect based?

19.2   In a polarography experiment, an aqueous solution containing a concentration of 10^{-3} M $Zn(NO_3)_2$ employs 0.1M KCl as an electrolyte medium.

      Calculate the ratio of the intensities of migration and diffusion for the Zn ion in the proximity of the cathode.

      Ionic mobilities $Zn^{++} = 5.5 \times 10^{-8}$, $NO_3^- = 7.4 \times 10^{-8}$, $K^+ = 7.6 \times 10^{-8}$ and $Cl^- = 7.9 \times 10^{-8}$ m^2 V^{-1} s^{-1}.

19.3   A dropping mercury electrode is regulated such that one drop falls every 4 seconds ($t = 4$ s). The mass of the Hg corresponding to 20 drops is of 0.16 g. Calculate the average mass flow of Hg in mg/s.

      If the flow is proportional to the height of mercury in the reservoir, what would happen to the lifetime of the drops if the height of Hg was increased three-fold?

19.4   Calculate the average current of diffusion i_D for the ion Pb^{2+} present as a concentration of 10^{-3} M in 0.1 M KCl aq. at 0 °C.

      Given: $m = 2.10^{-3}$ g/s, $t = 4$ s, $D = 8.67 \times 10^{-6}$ cm^2 s^{-1}.

19.5   A solution of 1 M KCl, employed as an electrolyte support, is ready to use and contains 0.0001% of Zn. Is this solution richer in zinc than one of 1 M which is prepared by using crystallised KCl for which the content in Zn is of 0.0005%?

19.6   A sample of 25 ml prepared for an electrolysis experiment has a zinc concentration of approximately 2×10^{-8} M which leads to the passage of a current of 1.5 nA. Calculate the time necessary to deposit 3% of the Zn present.

      Show that the technique of stripping is more sensitive.

19.7   In order to measure the quantity of water in powdered milk by the method of Karl Fischer method, the reactant is first standardised as follows:

      15 ml of a solvent comprising methanol and formamide is introduced into a cell. 3 ml of *KF* reactant is added to attain the equivalence point.

      Then, 205 mg of oxalic acid dihydrate is introduced. A further aliquot (13 mol) of *KF* reagent must next be added to attain a new equivalence point. Finally 10 ml of the same solvent containing 1.05 g of dissolved powdered milk is introduced. Again *KF* reagent (12 ml) is added to find a final equivalence point.

      Calculate the % mass of water in the powdered milk.

19.8 Calculate the % mass of water in a portion of anhydrous ether knowing that 1 coulomb is required to attain the equivalence point from 1 ml of ether during Karl Fischer measurements (coulometric version). Recall that two atoms of iodide per molecule of H_2O are needed and that the density of ether = 0.78 g/ml.

19.9 Determine the % impoverishment of a solution of 20 ml, 1×10^{-3} M zinc nitrate during a measurement of five minutes by polarography. The strength of the current in the circuit is 15 μA.

CHAPTER 20

Some sample preparation methods

Application of any analytical method is always easier when the matrix does not contain species that interfere with determination of the analyte. However, when an interference is expected, it is necessary to isolate the component to be measured from the matrix. Therefore, the quality of the result often depends on sample preparation. This preliminary step can have a more important influence on the end result than the measurement itself or the precision of the instrument used. Sample preparation, which follows the so-called sampling procedure, can often be tedious, delicate and time-consuming. Nonetheless, it has become an active area of study that benefits from the recent progress in chemistry and robotics. Currently used instruments that allow fast and selective measurements on very small amounts of sample have encouraged the development of new, rapid sample preparation methods.

20.1 Evolution of methods

The compound to be analysed, the *analyte*, is generally contained in a liquid or solid matrix; it is rarely found in a form that allows direct measurement. Interfering species that may lead to unwanted interactions, particularly during trace analysis in the presence of abundant matrix components, have to be eliminated. As a result, analysts have long acknowledged the need for efficient and reproducible sample preparation methods. The pre-treatment process has to take into account the analyte, matrix and measurement technique chosen. This situation has led to a number of specific sample pre-treatment protocols that describe sample treatment from sampling all the way to recording of the results (Fig. 20.1).

■ The determination of elements by atomic absorption in drinking water at the mg/l level or the recording of a mid-infrared spectrum of a pure organic compound are situations rarely encountered. In these cases, the sample is easily prepared. However, these conditions and similarly those under which analyses made by students in a teaching laboratory are not representative of the difficulties encountered when preparing a real sample for analysis.

Many determinations are conducted in aqueous solutions. To satisfy the demands of analysis, certain classical processes based on traditional methods like adsorption, extraction and precipitation have been improved for sample preparation. The introduction of other methods has principally been to save time, leading in some cases to entirely automated procedures.

The following section reviews some of the sample preparation methods currently used.

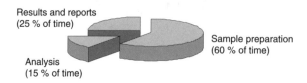

Results and reports
(25 % of time)

Sample preparation
(60 % of time)

Analysis
(15 % of time)

Figure 20.1—*Proportion of the time spent in each stage of a chromatographic analysis.* Sample preparation generally represents a large fraction of the total time required for analysis (*LC-GC Intl.* 1991, **4**(2)).

20.2   Solid-phase extraction using columns or discs

Solid-phase extraction (SPE) has drastically changed the classical approaches of solvent extraction. This process, which is widely used for trace analysis and determination of micropollutants, consists of passing a liquid sample (pure or solution) over a solid sorbent. Figure 20.2 shows a frequently encountered situation. In this example, the analyte is the only compound retained by the sorbent-containing column. Other substances are eliminated by rinsing the column after the analyte has been adsorbed. Following rinsing, the compound of interest is desorbed using an appropriate solvent. This extraction procedure allows not only *isolation* of the analyte but also its *preconcentration*. The enrichment factor can be as high as 100, which is particularly useful in trace analysis. The inverse approach, where the analyte is the only non-retained component, is attractive because it eliminates the elution (desorption) step. On the other hand, this leads to an enrichment factor of 1.

In SPE, the extraction is carried out using a small column (syringe-type or cartridge) containing 0.1 to 1 g of sorbent. The sorbent is typically a modified silica gel or one of many copolymers. This *chemical filter* can only be used once. The low-cost SPE process, which can be readily automated, is generally more useful for hydrophobic or apolar compounds than it is for ionic substances.

■ The stationary phases used in SPE, which are made from modified silica gel and correspond to either apolar, medium polarity or ionic media, closely resemble those used in HPLC. These phases include almost all bonded or non-bonded materials used in normal and reversed phase modes. Because of the nature of the chemical bonding involved, use of these phases is often limited between pH 2 and 8. On the other hand, copolymers such as styrene divinylbenzene that can incorporate functional groups are specific adsorbents that are more stable under severe pH conditions.

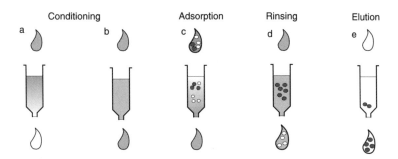

Figure 20.2—*Steps involved in solid-phase extraction (SPE)*. In this example, where the analyte is the only compound retained on the cartridge, the extraction steps are as follows: a) activation of the sorbent; b) rinsing of the column; c) introduction of the sample; d) elimination of interferences by rinsing; e) desorption of the analyte.

This type of extraction offers a number of advantages over the slower, less reproducible and less selective classical procedure of liquid/liquid extraction. Moreover, SPE uses less solvent and avoids the problems of emulsions.

■ In liquid/liquid extraction, the analyte is diluted in a large volume of solvent. When the extracted solution is concentrated, so too are the impurities contained in the solvent. Therefore, this method is not highly efficient unless ultra-pure solvents are used. For example, if 1 mg of analyte is mixed with 1 ml of 99.9% pure solvent, then the equivalent mass of secondary products (i.e. impurities) from the latter is equal to that of the analyte.

Extraction discs (0.5 mm thick, 25 to 90 mm diameter) constitute a variation of column-based SPE. These discs allow rapid extraction of large volumes of sample, which is not possible using a small column. The discs are made of bonded-phase silica particles, a few micrometres in diameter, trapped in a porous Teflon or glass fibre matrix. The discs are operated in a similar way to a paper filter on a vacuum flask. After extraction, the analyte is recovered by percolating a solvent through the filter. The major application of this technique is the isolation of trace amounts of compound dispersed in an aqueous medium.

20.3 Gas-solid extraction using columns or discs

The determination of molecular compounds present at low concentration in a gas sample, for example a polluted atmosphere, is usually carried out by trapping a given volume of the sample in a sorbent-containing column. The column, which typically has dimensions of 4 × 100 mm, is well sealed prior to its use and can only be used for one analysis.

The column (trap) is made of molecular sieves of absorbing materials such as graphite-based carbon black or functionalised organic polymers. In certain

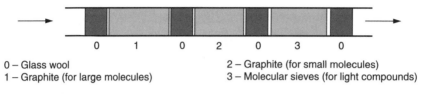

0 – Glass wool
1 – Graphite (for large molecules)

2 – Graphite (for small molecules)
3 – Molecular sieves (for light compounds)

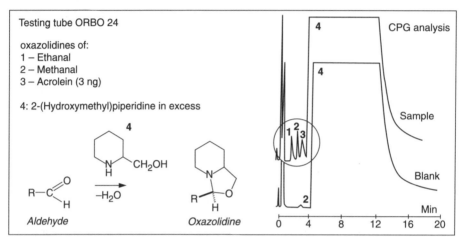

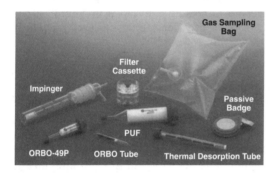

Figure 20.3—*Gas-solid extraction*. The principle of a gas-solid extraction column is shown as well as the chemical reaction used to derivatise an aldehyde in an atmospheric pollution test (adapted from a Supelco Inc. document). Examples of extraction tubes and a passive badge to trap gas samples are shown. (Reproduced by permission of Supelco Inc., USA.)

instances, the trap can be composed of a series of several sorbents with different characteristics (Fig. 20.3). A pump is used to deliver a predetermined volume of gas into the trap, at flow rates between 0.1 and 1 l/min. The adsorbed compounds are recovered by solvent extraction (often using carbon disulphide) or by thermal desorption, which avoids analyte dilution and the introduction of artefacts. Furthermore, the technique is adaptable to gas-phase chromatography where the trap is inserted into a special oven and heated in a few seconds to elevated temperatures (of up to 350 °C). The desorbed compounds are then directly introduced into the GC injector.

An alternative to the trap and special oven method is the use of small-diameter extraction tubes that can be introduced directly into a modified GC injector. Recovery is considered to be satisfactory when it attains 60%, although it is often quantitative with this device. This principle is also applied to badges used in industrial hygiene for monitoring pollution in the workplace or the environment (Fig. 20.3). In the latter application, air flows through the badge in a natural fashion to trap the analyte.

The flexibility of the gas-solid extraction method stems from the choice of many adsorbing phases, each offering a particular selectivity. It is possible to stabilise certain molecules that would normally decompose upon contact with the sorbent by incorporating into the sorbent a reagent that carries out a specific derivatisation. For example, aldehydes can be converted into oxazolidines, which are stable and desorbable compounds, by reaction with 2-hydroxymethylpiperidine.

20.4 Headspace sample analysis

Headspace is a sampling device used in tandem with a gas chromatograph. It is used in the determination of volatile compounds contained in a matrix which does not lend itself to direct analysis by chromatography. The basic principle is very simple, as described in the following examples.

— *Static mode*: the sample (liquid or solid matrix) is placed in a glass phial capped with a septum such that the sample occupies only part of the phial's volume. After thermodynamic equilibrium between the phases has been reached (1/2 to 1 h), a sample of the vapour at equilibrium is taken (Fig. 20.4). Under these conditions, the quantity of each volatile compound present in the headspace above the sample is proportional to its concentration in the matrix. The relationship between the amount of sample injected into the gas chromatograph and its concentration in the matrix can be obtained by calibration (using internal or external standards).

— *Dynamic mode*: instead of working in a closed environment, a carrier gas such as helium is either passed over the surface of the sample or bubbled through it in order to carry the volatile components into a trap where they are adsorbed and concentrated (Fig. 20.5). The sample is then introduced into the chromatograph by thermal desorption. This purge-and-trap technique is semi-quantitative and delivers a sample without residue.

■ This method is very reliable for repetitive analyses involving a stable matrix. However, if the matrix composition fluctuates, it affects the equilibrium and thus precision is lowered. In such cases, cartridge-based extraction is preferred.

■ Solid-phase micro-extraction (SPME), a technique commercialised by Supelco, borrows from column extraction and headspace analysis. In this procedure, a small fibre fixed to the end of a syringe plunger can either exit from or be retracted into

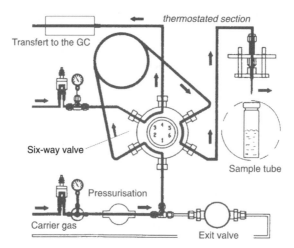

Figure 20.4—*Static mode of headspace sample analysis.* The sampling phial is pressurised with the carrier gas of the chromatograph. After equilibrium, a small volume of the gas containing the volatile compounds is inserted into a sample loop. Rotation of the six-way valve allows introduction of the sample into the injector of the chromatograph. Consequently, this set-up combines sample preparation with sample introduction into the chromatographic column. (Reproduced by permission of Tekmar.)

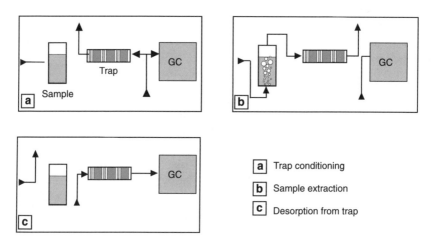

Figure 20.5—*Dynamic mode of headspace sample analysis.* The sample is recovered by thermal desorption ('stripping') from a cartridge appropriate for the compound being measured.

the syringe needle. The porous fused silica fibre, impregnated with a polyacrylate or silicone SE30-type phase, is introduced into a liquid sample or the headspace device. Ensuring its contact with the sample or vapour, the fibre will extract and concentrate volatile compounds from the matrix. Then the fibre is introduced into the injector of a gas chromatograph where the analytes are thermally desorbed and separated. This procedure yields good results for homogeneous samples, provided an adequate calibration is made. The same fibre can be used for approximately 50 analyses.

20.5   Supercritical fluid extraction (SFE)

Extraction using a supercritical fluid (CO_2, N_2O or $CHClF_2$; Freon-22) is a well-known process in the food industry (cf. 6.1). Extractors used for analytical purposes operate on the same principle. They incorporate a highly resistant tubular container into which the sample is placed (in solid or semi-solid form) with the supercritical fluid. Two modes of operation are employed:

— the *off-line* mode consists of depressurising the supercritical fluid. As the fluid reverts to its gaseous state it deposits the analyte, as well as impurities, in a concentrated form on the walls of the extraction container. This is followed by dissolution using classical solvents or solid-phase cartridges to obtain selective extraction.

— the *on-line* mode consists of direct analysis of the extract by introducing the supercritical fluid, while it is still under pressure, into a chromatograph (SFC or HPLC). This process is used for compounds where interferences due to matrix are not likely to occur.

Adding an organic solvent (such as methanol or acetone) to the supercritical fluid can modify its solvating properties. Since the polarity of CO_2 in its supercritical state (at 100 atm and 35 °C) is comparable to that of hexane, it can be altered by introducing a modifier. Nonetheless, isolating analyte from the matrix requires knowledge about the solubility and the transfer rate of solute in the solvent as well as chemical and physical interactions between matrix and solvent (Fig. 20.6).

In summary, there are four major parameters in SFE that can be modified to obtain good selectivity: pressure, temperature, extraction period and choice of modifier. Supercritical fluid extraction is easily automated and can be economically viable if the number of samples is large.

■ An alternative to supercritical fluid extraction is to use a classical solvent combined with microwaves. In a pressurised environment, this can be an efficient and rapid process for the treatment of samples.

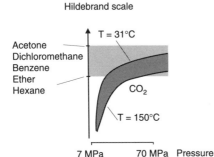

Fluid	Critical Point (°C)	Dipole moment (Debye)
CO_2	32	0
N_2O	37	0.2
$CHClF_2$	96	1.4

Supercritical Fluid Properties

Figure 20.6—*Supercritical fluid extraction.* A comparison of the solvation strength of CO_2 with classical solvents (Hildebrand scale) as a function of temperature and pressure is shown.

20.6   Mineralisation by microwave digestion

Mineralisation consists of the destruction of organic matter. Dry (oven) or humid (acid treatment) methods can be used for this purpose. Due to the absence of a universal method applicable to all mineral elements, it is necessary to adapt mineralisation to the sample being analysed. This stage, which is indispensable for the preparation of many types of samples, particularly those analysed by atomic absorption or emission, can be facilitated by the use of microwave digestion.

■ The classical approach to mineralisation, conducted openly in a fume-hood in the presence of acidic vapours, lends itself to cross-contamination. This operation is long and fastidious and has not changed in over a century. Beyond the practical problems involved, it is impossible to treat some matrices under these conditions (e.g. refractory or volatile materials, certain minerals, carbons and heavy oils).

Microwave digestion devices consist of a Teflon 'bomb' of a few ml in which the sample is placed along with the mineralisation solution (this solution is necessary for carbonisation or oxidation). The heating induced by microwaves is due to the molecular agitation of the dipole of water molecules. The calorimetric-type closed environment avoids losses due to 'bumping' and evaporation of volatile components. Temperatures attained by the acidic mineralisation solution under pressure largely surpass boiling temperatures under normal conditions (known as the pressure cooker effect). Thus, temperature gains up to 90% can be achieved. Such conditions allow sulphuric or perchloric acid digestions, hydrolyses or Kjeldhal mineralisation in an oxidising medium.

CHAPTER 21

Basic statistical parameters

In chemistry, as in many other sciences, statistical methods are unavoidable. Whether it is a calibration curve or the result of a single analysis, interpretation can only be ascertained if the margin of error is known. This section deals with fundamental principles of statistics and describes the treatment of errors involved in commonly used tests in chemistry. When a measurement is repeated, a statistical analysis is compulsory. However, sampling laws and hypothesis tests must be mastered to avoid meaningless conclusions and to ensure the design of meaningful quality assurance tests. Systematic errors (instrumental, user-based, etc.) and gross errors that lead to out-of-limit results will not be considered here.

21.1 Mean value and accuracy

If an experimental measurement is conducted several times (n) on the same sample (in the chemical sense of the term), slightly different individual values are frequently obtained. The measurement is thus considered a random variable. In practice, the correct test result is estimated by replacing individual values by a single one, the *mean value* $\bar{x}$, obtained from the arithmetic average of the measurements:

$$\bar{x} = \frac{x_1 + x_2 + \cdots + x_n}{n} = \frac{\sum x_i}{n} \tag{21.1}$$

If a large number of measurements are involved, the mean is accepted as being better than any individual measurement.

However, it is not simply because a great number of measurements are made or that each individual value is close to the mean that the mean $\bar{x}$ is close to the *exact value* x_0. Systematic errors can occur. A statistical approach to the problem consists of considering each measurement x_i as the real value x_0 plus an *absolute experimental error* ε_i. The error in the ith measurement is thus expressed as:

$$\varepsilon_i = x_i - x_0 \tag{21.2}$$

If the number of measurements is very large (in the case of a statistical population), $\bar{x}$ becomes the *real mean*, μ, which will be identical to the exact value x_0 in the absence of systematic errors. This assumes that the measurements follow a normal, or Gaussian, distribution (Fig. 21.2).

The error (or deviation) of the true mean value corresponds to ε such that:

$$\varepsilon = \mu - x_0 \tag{21.3}$$

If the exact value x_0 is known (e.g. analysis of a standard of known composition), the *total systematic error* ε is characteristic of the *accuracy* of the series of measurements made with the method used (Table 21.1).

The relative error E_R on a measurement (or on the mean value) corresponds to the ratio of the absolute value of the deviation $|\varepsilon_i|$ (or $|\varepsilon|$) over the real value. E_R can be expressed in % or ppm.

$$E_R = \frac{|x - x_0|}{x_0} \qquad \text{where } x \text{ can be } x_i \text{ or } \bar{x} \tag{21.3}$$

■ A value that can sometimes replace the mean, although rarely used in chemistry, is the *median* value. This value is obtained by ranking the experimental data in increasing order and extracting the middle value of the series. If there are an even number of data points, then the average of the two central points is used for the median. The one advantage of using the median value is that it avoids giving weight to an aberrant value. For example, consider the values 12.01, 12.03, 12.05 and 12.68 whose arithmetic average is 12.19 and whose median is 12.04. The latter value appears to better represent the data than the arithmetic mean because it doesn't take into account the apparent outlier, 12.68, chosen in this example. The median plays a role in statistics based on *robust methods* (see Table 21.1 and section 21.8).

21.2   Precision and standard deviation of a group of results

If the real value x_0 is not known in an experiment, which is usually the case in chemical analysis, the experimental error e_i is obtained by replacing x_0 in equation (21.2) by $\bar{x}$.

$$e_i = x_i - \bar{x} \tag{21.5}$$

In equation (21.5), e_i represents the algebraic deviation of the mean for the ith measurement. The easiest way to quantify *precision* is to calculate the *mean* of all the deviations, $\bar{d}$, which represents the average experimental error of n measurements.

$$\bar{d} = \frac{\sum |x_i - \bar{x}|}{n} \tag{21.6}$$

Unless absolute values are taken (equation (21.6)), $\bar{d}$ will tend rapidly towards 0 as n increases.

A problem associated with the use of the mean deviation is that it does not allow statistical interpretation of the data because large and small deviations have the same weight even though they are not equally probable. Thus, it is preferable to sum the square of the deviations. This leads to the most widely used definition of *precision* in statistics, which is based on *variance*, s^2, and is estimated for a set of n measurements by:

$$s^2 = \frac{\sum (x_i - \bar{x})^2}{n - 1} \qquad (21.7)$$

The square root of the variance is called the *standard deviation*, designated either as s when the number of measurements n is small or as σ when a significantly large statistical population is available. The standard deviation s, an indication of the dispersion of a group of measurements, is thus expressed in the same units as x.

Calculators and spreadsheets include the following function to calculate standard deviation (for an infinite number of measurements, s is replaced by σ, $\bar{x}$ is replaced by μ and $n - 1$ is replaced by n):

$$s = \sqrt{s^2} = \sqrt{\frac{\sum (x_i - \bar{x})^2}{n - 1}} \qquad (21.8)$$

Table 21.1—Example of results obtained from an analysis summarising the various parameters presented in sections 21.1 and 21.2 (the real value $x_0 = 20$).

Chemist	1	2	3	4
Measurement 1	20.16	19.76	20.38	20.08
Measurement 2	20.22	20.28	19.58	19.96
Measurement 3	20.18	20.04	19.38	20.04
Measurement 4	20.2	19.6	20.1	19.94
Measurement 5	20.2	20.42	19.56	20.08
Arithmetic average	20.2	20.02	19.8	20.02
Median	20.20	20.04	19.58	20.04
Absolute error	0.2	0.02	0.2	0.02
Relative error	0.01 or 1% or 10^4 ppm	0.001 or 0.1% or 10^3 ppm	0.01 or 1% or 10^4 ppm	0.001 or 0.1% or 10^3 ppm
Variance (s^2)	7.84×10^{-4}	0.1183	0.1772	3.6×10^{-5}
Standard deviation (s)	0.028	0.344	0.421	0.006
Comments on the results	precise not accurate	not precise accurate	not precise not accurate	precise accurate

It is customary when comparing results to express s in a relative fashion. Thus the *relative standard deviation (RSD)* also called the *coefficient of variation (CV)* is expressed as a percentage:

$$CV = 100 \times \frac{s}{\bar{x}} \qquad (21.9)$$

The coefficient of variation is of little interest when the number of measurements is small. On the other hand, the *standard deviation of the mean*, or uncertainty of

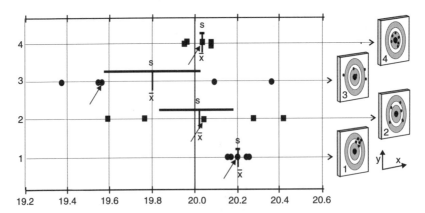

Figure 21.1—*Graphical illustration of the data presented in Table 21.1.* To illustrate *accuracy* and *precision,* the standard deviations are calculated according to equation (21.8). Variations are indicated on the graph for arithmetic averages by a vertical bar and for the corresponding median values by arrows. For the results from chemist 3, the deviation between the mean and median is large. Chemist 1 has probably committed a systematic error. On the right is the classical target illustration of precision and accuracy. This image appears simpler than it is because there are uncertainties in *x* and *y*.

the standard deviation (equation (21.10)), can be useful to specify the scope of μ by using *s*:

$$s_{\bar{x}} = \pm \frac{s}{\sqrt{n}} \tag{21.10}$$

■ When the result of an analysis follows from calculations involving many experimental values, each with its own standard deviation, there will be a propagation of errors. The precision of the result can be obtained using simple equations that are found in most introductory texts on statistics.

21.3   Indeterminate or random errors

In the absence of systematic errors, accidental errors (due to hazards) exist that cannot be controlled because they are indeterminate. The direction and amplitude of these errors varies in a non-reproducible fashion from one measurement to another.

Mathematical analysis of *error curves* leads to the conclusion that the arithmetic mean $\bar{x}$ of individual values is the best estimation of the real mean μ (Fig. 21.2). The features and the symmetry of this curve show that:

— there is an equal number of positive and negative errors relative to the mean

— small errors are more frequent than large errors

— the most encountered value is that of the *mean* μ (without error).

■ If the number of measurements is small, it is important to know their distribution. The distribution curve gives information about the *reliability Re* of the results. The reliability of the mean as an estimate of the true mean value increases with the square root of the number of measurements *n*:

$$Re = K\sqrt{n} \tag{21.11}$$

Thus, when the number of measurements n_1 is increased to n_2, the reliability *Re* increases by the factor k:

$$k = \sqrt{\frac{n_2}{n_1}} \tag{21.12}$$

The normal distribution (bell-shaped Gaussian curve) is the best mathematical model to represent random, or indeterminate, errors (equation (21.13) and Fig. 21.2):

$$f(x) = \frac{1}{\sigma\sqrt{2\pi}} \exp\left[-\frac{(x-\mu)^2}{2\sigma^2}\right] \tag{21.13}$$

To make this expression universal, the origin is taken as *x*, which is the estimate of the true mean μ of a population, with units of standard deviation σ:

$$f(X) = \frac{1}{\sqrt{2\pi}} \exp\left[-\frac{X^2}{2}\right] \quad \text{with} \quad X = \frac{x-\mu}{\sigma} \tag{21.14}$$

The distribution function is the integral of the function f(X). This function indicates that 95.4% of the area lies within an interval of $\pm 2\sigma$ from the mean value. This implies that 95.4% of the measurements of a sample are included in an interval of $\pm 2\sigma$. A high value of the standard deviation σ signifies a broadened distribution

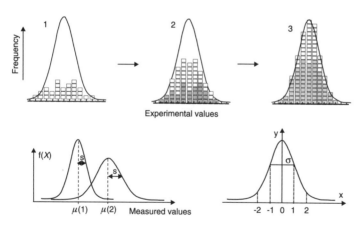

Experimental values

Measured values

Figure 21.2—*Gaussian distribution.* As the number of measurements increases and the width of the interval remains relatively narrow, the envelope of data points (frequency vs measurements) resembles that of a Gaussian curve (normal distribution). The bottom figure represents two series of results with two different means. If the number of measurements is very small, it is not possible to estimate the average distribution. At the bottom right, the reduced form of the Gaussian distribution is shown.

curve. In such cases, if the number of data points is modest (a few values), s is used instead to estimate σ (Fig. 21.2). Concurrence between the real values (μ, σ) which apply to a population, and their estimates $(\bar{x}, s)$ improves as n increases.

■ In practice, neither the true mean μ or the standard deviation σ is known because the number of measurements n is always limited. Thus, s must be used.
■ When a large number of repetitive measurements is available on the same chemical sample, the χ^2 test is used to ascertain whether the frequency distribution significantly differs from that of a normal distribution. This test can be tedious to calculate without the proper software.

21.4 Confidence interval

When the number of measurements is small (for example $4 < n < 15$) with no systematic errors, it is difficult to be certain of the value of $\bar{x}$ knowing that it can be quite different from the real mean μ. A statistical answer to this question is found by imposing a probability (for example, 95%) and then calculating an interval around the estimate of the mean $\bar{x}$ that includes the real value of the mean μ (or x_0 in the absence of all systematic errors):

$$\bar{x} - \frac{t \cdot s}{\sqrt{n}} \leq \mu \leq \bar{x} + \frac{t \cdot s}{\sqrt{n}} \tag{21.15}$$

In equation (21.15), Student's coefficient t is a tabulated statistical factor that depends on the number of measurements n and on the confidence level that has

Table 21.2—Values of Student's coefficient, t.

n	Number of confidence 90%	Number of confidence 95%	Number of confidence 99%
2	6.31	12.706	63.66
3	2.92	4.303	9.93
4	2.35	3.182	5.84
5	2.13	2.776	4.6
6	2.02	2.571	4.03
7	1.94	2.447	3.71
8	1.9	2.365	3.5
9	1.86	2.306	3.36
10	1.83	2.262	3.25
11	1.81	2.228	3.17
12	1.8	2.201	3.11
15	1.76	2.145	2.98
20	1.73	2.093	2.86
30	1.7	2.05	2.76
60	1.67	2.000	2.660
120	1.66	1.980	2.62
9 999	1.65	1.960	2.58

been chosen (see Table 21.2). As n increases, the interval between values of t for a given confidence level decreases. The standard deviation of the series of measurements is estimated by s in equation (21.15). Student's t test permits adjustment of the number of measurements to be conducted as a function of the desired level of confidence.

■ If the real value x_0 is known (or the true mean μ), regardless of $\bar{x}$, it is possible to calculate t (equation (21.16) if $n < 20$, or equation (21.17) for very large n) as a function of the confidence level chosen. A value of t larger than that indicated in the table for the given n value is an indication of systematic error.

$$|t| = \frac{|\bar{x} - \mu|}{s}\sqrt{n} \tag{21.16}$$

$$|t| = \frac{|\bar{x} - x_0|}{\sigma}\sqrt{n} \tag{21.17}$$

21.5 Parametric tests – comparison of results

There are two main families of statistical tests: *parametric tests*, which are based on the hypothesis that data are distributed according to a normal curve (on which the values in Student's table are based), and *non-parametric tests*, for more liberally distributed data (robust statistics). In analytical chemistry, large sets of data are often not available. Therefore, statistical tests must be applied with judgement and must not be abused. In chemistry, acceptable margins of precision are 10, 5 or 1%. Greater values than this can only be endorsed depending on the problem concerned.

Many questions can arise when considering the validity of a statistical result.

21.5.1 Fischer–Snedecor comparison of variances

In order to compare the results between two laboratories for the same sample or, for example, two instruments for the same analysis method, it is essential to know whether the standard deviation s_1 of the first set of results is significantly different from that of the second set, s_2. This is accomplished by using the *variance equality test*. In this test, an F factor is calculated, which is the ratio of the two variances such that $F > 1$:

$$F = \frac{s_1^2}{s_2^2} \tag{21.18}$$

The *null hypothesis* (a term used by statisticians) states that if there are no significant differences in the variances, their ratio approaches 1. In other words, we can refer to the critical values of F, the Fischer–Snedecor parameter, which show that as the number of measurements increases, the ratio of the two variances approaches 1.00 (Table 21.3). If the experimental value obtained for F exceeds that found in the table, then the standard deviations of the two methods are considered to be significantly different. Since s_1^2 must be greater than s_2^2 for the F-test, the second series of measurements is always the more precise one.

Table 21.3—Abridged version of the F threshold values for a confidence level of 95%

Number of measures (denominator)	Number of measures (numerator of the fraction F)						
	3	4	5	6	7	10	100
3	19.00	19.16	19.25	19.30	19.33	19.38	19.50
4	9.55	9.28	9.12	9.01	8.94	8.81	8.53
5	6.94	6.59	6.39	6.26	6.16	6.00	5.63
6	5.79	5.41	5.19	5.05	4.95	4.78	4.36
7	5.14	4.76	4.53	4.39	4.28	4.10	3.67
10	4.26	3.86	3.63	3.48	3.37	3.18	2.71
100	2.99	2.60	2.37	2.21	2.09	1.88	1.00

■ For example, the data in Table 21.1 shows that chemists 2 and 3 have essentially the same precision at a confidence level of 95%. An experimental value of $F = 1.5$ is obtained using equation (21.18), which is less than the tabulated value of $F = 6.39$ for five measurements made by each chemist. This verifies that their standard deviations are not significantly different.

21.5.2   Comparison of two experimental means, $\bar{x}_1$ and $\bar{x}_2$

Sometimes it is desirable to compare results obtained by two different experimental procedures. Since the real value is not known, it must be ascertained whether the two means are similar or significantly different. In such instances, the first step involved is to verify that there are no significant differences between the precision of these means (F-test, cf. 21.5.1). Following this, the global or *pooled* standard deviation, s_p is calculated using equation (21.19) and the corresponding t value is obtained using equation (21.20). The t value is compared to tabulated values for $n = n_1 + n_2 - 2$ at a chosen confidence level. If the t value in the table is greater than the calculated t, it can be concluded that the two means are significantly different.

$$s_p = \sqrt{\frac{(n_1 - 1)s_1^2 + (n_2 - 1)s_2^2}{n_1 + n_2 - 2}} \qquad (21.19)$$

$$t = \frac{|\bar{x}_1 - \bar{x}_2|}{s_p}\sqrt{\frac{n_1 n_2}{n_1 + n_2}} \qquad (21.20)$$

21.5.3 Estimation of the limits of detection for an analyte

Equations (21.19) and (21.20) are useful for estimating the smallest concentration of an analyte that can be detected, but not quantified, with a given confidence level. Consider one of the means, $\bar{x}_2$ for example, to be the result from a set of measurements made on the analytical blank. This mean will be noted as $\bar{x}_b$ with a standard deviation of s_b. If $n_1 = 1$ in equation (21.19), then $s_p = s_b$ and, using a value t taken from Table 21.2 for $n = n_1 + n_b - 2$, we can write:

$$\Delta x = \bar{x}_1 - \bar{x}_b = \pm t \cdot s_b \sqrt{\frac{n_1 + n_b}{n_1 n_b}} \qquad (21.21)$$

■ Given that a set of six measurements on an analytical blank gives a standard deviation of 0.3 µg, a value of 1.31 µg will be obtained for the *limit of detection* if a single measurement is made (with equation (21.21) using a 99% confidence level) and a value of 0.59 µg if the number of measurements is five ($\bar{x}_b = 0$). Empirically, the limit of detection corresponds to a concentration that leads to a signal whose intensity is twice the standard deviation obtained from 10 measurements of the analytical blank (95% confidence level). The *limit of quantification* is always much higher.

21.6 Rejection criteria – Q test (Dixon test)

Sometimes, a value within a set might appear aberrant (this is known as an outlier). Although it might be tempting to reject this data point, it must be remembered that a value can only be aberrant relative to some law of probability. There is a simple statistical criterion on which to base the decision of whether to retain or reject this value. Dixon's test is based on the following ratio (as long as there are at least seven measurements):

$$Q = \frac{|\text{value in question} - \text{its closest neighbour}|}{(\text{largest value} - \text{smallest value})} \qquad (21.22)$$

The calculated value of Q is compared to the tabulated critical Q values as a function of the number of measurements (Table 21.4). If $Q_{\text{calculated}}$ is greater than Q_{critical}, then the value in question can be rejected.

■ Note: The ASTM (American Society for Testing Material) uses a different test for rejection of an outlier, called the reduced central value $z_i = (x_i - \bar{x})/s$, which has its own table of critical values.
   Henry's curve, whose principle can be found in more advanced statistical texts, also constitutes an excellent method for detecting outliers using a visual approach.

Table 21.4—Abridged table of critical values of Q (Dixon's test).

Number of measurements	Confidence level	
n	95%	99%
3	0.94	0.99
4	0.77	0.89
5	0.64	0.78
6	0.56	0.698
7	0.51	0.637

21.7   Linear regression

Linear regression is undoubtedly the most widely used statistical method in quantitative analysis (Fig. 21.3). This approach is used when the signal y as a function of the concentration x is linear. It stems from the principle that if many samples are used (generally dilutions of a stock solution), it becomes possible to perform variance analysis and estimate calibration error or systematic errors.

21.7.1   Simple linear regression

The response of many instruments is linear as a function of the measured variable, if variations due to experimental conditions or the instrument are taken into account. The objective is to determine the parameters of the linear equation that best represents the observations. The primary hypothesis in using the *method of least squares* is that one of the two variables should be without error while the second one is subject to random errors. This is the most frequently applied method. The coefficients a and b of the linear equation $y = ax + b$, as well as the standard deviation on a and on the estimation of y have been obtained in the past using a variety of similar equations. The choice of which formula to use depended on whether calculations were carried out manually, with calculator or using a spreadsheet. However, appropriate computer software is now widely used.

The relationship between the two variables is usually characterised by the dimensionless Pearson *correlation coefficient*, R. A value of $+1$ or -1 for R indicates a strong relationship between the two variables. This approach assumes that the errors in y follow a normal distribution. R^2 is called the *determination coefficient* and indicates what percentage of the variations in x overlap variations in y.

$$a = \frac{n \sum x_i y_i - \sum x_i \sum y_i}{n \sum x_i^2 - (\sum x_i)^2} \tag{21.23}$$

and

$$b = \bar{y} - a\bar{x} \tag{21.24}$$

$$R = \frac{n \sum x_i y_i - \sum x_i \sum y_i}{\sqrt{\left(n \sum x_i^2 - \left(\sum x_i\right)^2\right) \cdot \left(n \sum y_i^2 - \left(\sum y_i\right)^2\right)}} \tag{21.25}$$

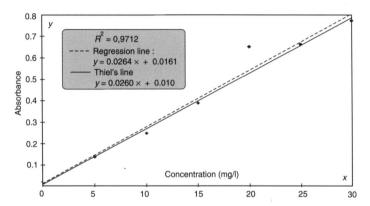

Figure 21.3—*Linear regression line and Thiel's line corresponding to the data in Table 21.5.* Although they appear similar, these two lines lead to different results. For example, for $y = 0.5$, the values $x = 18.85$ (Thiel) and $x = 18.33$ (linear regression) differ by about 3%.

It is often assumed in regression calculations that the experimental error only affects the y value and is independent for the concentration, which is typically placed on the x axis. Should this not be the case, the data points used to estimate the best parameters for a straight line do not have the same quality. In such cases, a coefficient w_i is applied to each data point and a weighted regression is used. A variety of formulae have been proposed for this method.

■ To determine if two methods of analysis are highly correlated, a series of samples can be analysed by both methods and the data compared by simple linear regression. For each sample, the data points are plotted on a graph using the results of the first method as x and the results of the second method as y. The correlation coefficient R is then determined and equation (21.26) is applied. If the t value obtained is higher than that found in the table for $n - 2$ degrees of freedom, the correlation between the two analytical methods is high.

$$t = R\sqrt{\frac{n - 2}{1 - R^2}} \tag{21.26}$$

21.7.2 Multiple linear regression

Multiple linear regression allows the estimation of a result (the *dependent variable*) when it depends on a number of factors (the *independent variables*). Such calculations can be conducted using tables in which the contribution of each factor is determined using standards. If three factors are involved, for example, the result can be estimated using the following general formula:

$$y = a + bx_1 + cx_2 + dx_3 \tag{21.27}$$

■ For example, the protein content in cheese, which is measured as the percentage nitrogen, can be determined using IR absorbance at different wavelengths. This method can also be used to calculate the parameters A, B and C of Van Deemter's equation (cf. 1.11) corresponding to experimental data.

21.8 Robust methods

The statistical tests previously described assume that the data follow a normal distribution. However, the results obtained by several analytical methods follow different distributions. These distributions are either asymmetric or symmetric but not normally distributed. In some approaches, these distributions are considered to be aberrant values superimposed on the normal distribution. In the following approach, the arithmetic mean is replaced by the median (cf. 21.1) and the standard deviation is replaced by the mean deviation, MD.

$$MD = \sqrt{\frac{\pi}{2}} \times \frac{\sum |x_i - \bar{x}|}{n} \tag{21.28}$$

Under such circumstances, the calibration curve can be reconsidered as shown in the following example, which illustrates Thiel's method.

An approach to estimate the best straight line for a series of seven data points (x, y) in a colorimetric determination consists of first ranking the x values in increasing order. An even number of data points is necessary to apply Thiel's method, so the median data set must be rejected in the example given here. The data points and calculation results for this method are shown in Table 21.5.

The slopes of three straight lines are calculated using, for the first slope, the first data point and that immediately following the median, and so on, as shown in Table 21.5. The median of the three slope values is taken as the a term for the linear equation. Then, six values are calculated for the b term where $b_i = y_i - ax_i$. These b_i values are then ranked and the median is taken as the b term of the linear equation (Fig. 21.3).

This method should be more widely used in chemistry to study distributions when the number of measurements is small.

Three advantages of Thiel's method merit note:

— it is not assumed that all errors are in y

— it is not assumed that the errors follow a normal distribution

— the presence of aberrant data do not affect the parameters of the linear equation.

Table 21.5—Determination of the linear equation coefficients using Thiel's method ($y = ax + b$).

Sample $n°$	Conc. (x)	Ab. (y)	b_i $(a_{1.5}\ a_{2.6}\ a_{3.7})$	Term	Sorting
1	0	002		0.02	0.01
2	5	0.14	0.032	0.01	−0.01
3	10	0.25	0.0260	−0.01	0.01
4	15	0.39	(Median eliminated)		
5	20	0.65	0.0260	0.13	0.01
6	25	0.66		0.01	0.02
7	30	0.77		−0.01	0.13

Equation of line: Y= 0.0260 X + 0.010

21.9 Method optimisation using factorial analysis

When a determination relies on a signal (absorbance, fluorescence, etc.) that can be influenced by many factors, it is customary to seek conditions that will lead to the maximum signal.

If the parameters (factors) involved in an analysis are totally independent, which is rarely the situation, it is possible to study the influence of each by varying a single parameter while holding the others constant (Fig. 21.4).

Imagine that a determination is a function of two parameters, x and y. After having fixed the first parameter x to the value x_1, the influence of y on the signal is studied (the signal is measured in the third dimension). It is observed that the signal passes through a maximum for the value Y. Therefore, the parameter y is set

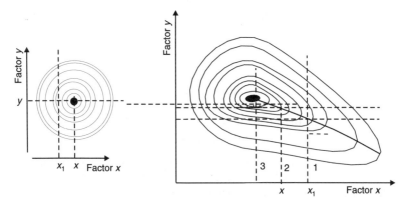

Figure 21.4—*Method of factorial analysis using a single factor.* If response (signal) consists of a continuous surface, the iso-response curves will lead to a variety of optimum situations.

to the value Y and the signal is measured as a function of x in order to find the best value, X. As a result, Y and X become the best values.

In the most general case, the iso-response curves yield an envelope that represents the mathematical interaction between the two factors. In such cases, the method just described will fall upon a false optimum, the maximum of the envelope.

It is always preferable to optimise by varying all the factors simultaneously. The latter approach leads to experimental design and to the simplex method, which is described in a number of specialised monographs.

Problems

21.1 The true value of a particular measurement is 131.9 µg/l. Four chemists (A, B, C and D) each repeat the same procedure five times. The individual values obtained are displayed in the table below. Comment upon these results in terms of accuracy and precision (use an indicator of dispersion or standard deviation, s).

Chemist A	130.7	131.6	133.5	132.3	132.6	129.1
Chemist B	125.0	132.3	136.9	137.9	125.9	131.6
Chemist C	136.7	134.5	134.1	135.4	136.0	137.6
Chemist D	130.7	109.9	131.9	115.6	131.3	132.6

21.2 With a column whose stationary phase is of squalane and under differing conditions of use, the following values for the Kovats factor of benzene (recognised average value = 653) were determined: 650, 652, 648, 651 and 649. For the value of 653 given, is this significantly different from the other values calculated? Consider a confidence level of 95%.

21.3 Supposing that two chemists A and B, whose results figured in question 21.1 above, used two different apparatus. In order to determine if the precision of these apparatus is significantly different apply a test of variance, F.

$$\text{where:} \qquad F = s_A^2 / s_B^2$$

21.4 Once again return to the results of 21.1, and chemists A and C in particular. If the true value is 131.9 µg/l calculate and compare the respective values of parameter t for these chemists. Can the presence of a systematic error be concluded for the data of either A or C?

21.5 The results of experimental measurements repeated four times were the following: 24.24, 24.36, 24.87, 24.20, 24.10. Verify whether the third value, which seems to be high compared to the others, should be considered as a value out of acceptable range.

Consider now two complementary measurements: 24.12 and 24.25.

Consider again the question posed above and conclude by applying the calculation of the standard deviation, s

21.6 The measurements were recorded and then repeated six times for two different methods, for each giving: average 42 ($s = 0.3$), and average 45 ($s = 0.2$).

— Do these two methods yield results that are significantly different?

21.7 A compound is accompanied by a certificate indicating its purity as being 99% with $s = 0.08$, established over five analyses. Four control measurements performed upon the same compound led to the following values: 98.58, 98.62, 98.80 and 98.91. Should the original value be rejected?

21.8 A series of standards used in atomic absorption led to the following results:

Concentration µg/l	1	2.5	5	10	20	30
Absorbance	0.06	0.19	0.36	0.68	1.21	1.58

— Might the method of linear calibration graph be used with these results and, if so, over which range of concentrations?

21.9 A series of measurements must be repeated five times. Eight successive values for the blank used in the analysis yielded the following: 0.3, −0.75, −0.3, 1.5, −0.9, 1.8, 0.6 and 1.2.

— Calculate the limit of detection by applying a 99% confidence level.

Solutions

1 Chromatography, general aspects

1.1 At equilibrium the 40 ml of eluant contains $12 \times 40/10 = 48$ mg of compound. There are therefore $100 - 48 = 52$ mg in the stationary phase. If K represents the ratio of the masses present in 1 ml of each phase in equilibrium, then $K = (52/6)/(48/40) = 7.2$.

1.2 When a molecule of the compound migrates along the column it passes alternately from the mobile phase, where it progresses at the speed of this phase, to the stationary phase where it becomes immobilised. The average velocity for the passage is then reduced in relation to the amount of time spent by the molecule in the stationary phase, increasing by consequence the value for t_S.

Now consider the total mass, m_T, of all of the identical molecules in the sample. Statistically at each moment, there will be a certain number of these molecules within the stationary phase while the rest remain in the mobile phase. The ratio of these molecules fixed in the stationary phase (m_S), and of those present in the mobile phase (m_M), will be the same as the ratio representing the sum of the intervals spent in each phase for a single isolated molecule; that is: $m_S/m_M = t_S/t_M$.

Thus k will correspond to the ratio t_S/t_M and since $t_S = t_R - t_M$ then we can refind the classic expression for the retention factor (or capacity factor), as calculated from the chromatogram:

$$k = (t_R - t_M)/t_M$$

1.3 The separation factor (or selectivity factor), is generally calculated from the retention time t_R, yet in the question it is the retention volumes V_R which are given for the two compounds. However, V_R and t_R are linked by an expression which considers the flow of the mobile phase, D:

$$V_R = D \cdot t_R$$

Therefore if for the expression in α below, each t_R (or t_M) is replaced by V_R (or V_M), we will have:

$$\alpha = (t_{R(2)} - t_M)/(t_{R(1)} - t_M) = (V_{R(2)} - V_M)/(V_{R(1)} - V_M).$$

which leads to $\alpha = 1.2$.

Note: The expression in α is defined from the retention volumes and remains useful with a flow gradient but it is very rare that it is employed since the apparatus does not measure the instantaneous rate of the column. Therefore the calculation of the retention volumes (other than by substituting in the relation, $V_R = D \cdot t_R$) is not permissible.

1.4 The expression giving the retention factor can be written, $t_R = t_M(1 + k)$, where t_M corresponds to the ratio between the length L of a column and the average velocity u of its mobile phase. Equally, $k = KV_S/V_M$, leading upon substitution to

$$t_R = (L/u)(1 + KV_S/V_M).$$

1.5 In order to incorporate k into expression 1, the relation $t_M = t_R/(1 + k)$, is used. Expression 1 then becomes:

$$N_{\text{eff}} = 5.54 \, \frac{\left(t_R - \dfrac{t_R}{1+k}\right)^2}{w_{1/2}^2}$$

simplifying to

$$N_{\text{eff}} = 5.54 \times t_R^2 \, \frac{\left(1 - \dfrac{1}{1+k}\right)^2}{w_{1/2}^2} \qquad \text{and then} \qquad N_{\text{eff}} = N \, \frac{k^2}{(1+k)^2}$$

1.6 The classic formula giving the resolution implies that the peaks are Gaussian. The known relationship between w, $w_{1/2}$ and σ permits the replacement of w by $4 \cdot w_{1/2}/2.35$.

1. The basic formula then becomes,

$$R = 2 \frac{t_{R(2)} - t_{R(1)}}{4(w_{1/2})_1/2.35 + 4(w_{1/2})_2/2.35} = 1.18 \, \frac{t_{R(2)} - t_{R(1)}}{(w_{1/2})_1 + (w_{1/2})_2}$$

The formula, proposed in the question, leads to an error of approximately 20%.

2. If these peaks are adjacent then the widths of their bases will generally be comparable. Allowing therefore $w = w_1 = w_2$ and expressing w as a function of N_2 we will have:

$$R = \frac{t_{R(2)} - t_{R(1)}}{w_2} \qquad \text{with} \qquad \frac{1}{w_2} = \frac{1}{4}\sqrt{N^2} \, \frac{1}{t_{R(2)}}$$

leading, on combination, to:

$$R = \frac{1}{4}\sqrt{N_2} \frac{t_{R(2)} - t_{R(1)}}{t_{R(2)}}$$

k can then be introduced by incorporating the general expression, $t_R = t_M(1 + k)$,

$$R = \frac{1}{4}\sqrt{N_2} \frac{k_2 - k_1}{1 + k_2}$$

then, finally by introducing $\alpha = k_2/k_1$:

$$R = \frac{1}{4}\sqrt{N_2} \frac{k_2 - k_2/\alpha}{1 + k_2} \qquad \text{leading to} \qquad \frac{1}{4}\sqrt{N_2} \frac{k_2}{1 + k_2} \frac{\alpha - 1}{\alpha}$$

1.7 1. In the basic relationship

$$R = 2 \frac{t_{R(2)} - t_{R(1)}}{w_1 + w_2}$$

we can replace w by expressing it as a function of N, $w = 4\dfrac{t_R}{\sqrt{N}}$ which will give

$$w_1 + w_2 = \frac{4}{\sqrt{N}}(t_{R(1)} + t_{R(2)})$$

and then,

$$R = \frac{1}{2}\sqrt{N} \frac{t_{R(2)} - t_{R(1)}}{t_{R(2)} + t_{R(1)}}$$

Since $t_R = t_M(1 + k)$, substitution of this relation leads to,

$$R = \frac{1}{2}\sqrt{N} \frac{k_2 - k_1}{2 + k_1 + k_2}$$

2. In multiplying the second term in the equation above by $k_2 + k_1/k_2 + k_1$ we can write the following intermediate relation:

$$R = \frac{1}{2}\sqrt{N} \frac{k_2 - k_1}{k_2 + k_1} \frac{\bar{k}}{\bar{k} + 1}$$

Then by introducing α and substituting, we arrive at expression 2.

1.8 In supposing that the following equation is true:

$$R = \frac{1}{4}\sqrt{N} \frac{k}{1 + k} \frac{\alpha - 1}{\alpha}$$

then,

$$N = 16R^2 \left(\frac{\alpha}{\alpha - 1}\right)^2 \left(\frac{k + 1}{k}\right)^2 \qquad \text{but} \qquad N_{eff} = N\left(\frac{k}{1 + k}\right)^2$$

which on substitution will lead to the expression given in the question.

1.9 $R_1 = (0.5 \times 60)/(10 + 12) = 1.36$; $R_2 = 1.47$, $(k_2 = 9.5$, $\alpha = 1.056$ and $N_2 = 15\,270)$ $R_3 = 1.5$ $(k_1 = 9)$; $R_4 = 1.5$.

1.10 1. The viscosity of the stationary phase increases with temperature as experienced when a pre-programmed rise in temperature induces a loss of charge across the column. As a result the gas flow vector measured, before the analysis is begun, will be reduced. The average velocity of the gas vector is likely to be different from its optimal value, with an increase in pressure necessary which requires that the chromatograph possesses a flow control.

2. The Van Deemter equation of type $H = f(u)$, has the derivative $d(H)/d(u) = -B/u^2 + C$. This expression cancels to zero when $u = (B/C)^{1/2}$. The value reported in the Van Deemter expression leads to $H = A + 2(BC)^{1/2}$. This calculation becomes important if several experimental points (H_i, u_i) are employed to find the coefficients A, B and C of the equation.

1.11 The solution of this problem returns us to the initial three lines of exercise 1.7, where we considered employing the retention volumes in place of the retention times.

1.12 1. Flow of the mobile phase, k, column temperature.
2. Yes.
3. Yes.

2 Gas phase chromatography

2.1

	Speed of separation	Injection capacity	Retention factor k	Selectivity factor α	Effectiveness
Increase in column length	–	0	0	0	+
Increase in column diam.	0	+	–	0	0
Increase in film thickness	–	+	+	0	+

2.2 1. N_{eff}, k, α.
2. GPC: methane. HPLC: uracil, $D_2O \ldots$
3. Following the experiment three equations can be proposed for the three unknowns: t_M, a and b.

$$\log(271 - t_M) = 6a + b \tag{1}$$

$$\log(311 - t_M) = 7a + b \tag{2}$$

$$\log(399 - t_M) = 8a + b \tag{3}$$

— in multiplying (2) by -1 and adding to (1):

$$\log[(271 - t_M)/(311 - t_M)] = -a \qquad (4)$$

— in subtracting (2) from (3):

$$\log[(399 - t_M)/(311 - t_M)] = a \qquad (5)$$

The resolution of these simultaneous equations yields, $\overset{\cdot}{t}_M = 237.7\,s$.
4. The Kovats graph passes through the points C7 and C8.

$$\log(399 - 237.7) = 2.207 \quad \text{and} \quad \log(311 - 237.7) = 1.863$$

$$\log(246 - 237.7) = 2.033$$

thus $(2.033 - 1.863)/x = (2.207 - 1.863)/1$; so, $x = 0.494$, therefore $I_k = 700 + 49 = 749$. The McReynolds constant of the pyridine on this stationary phase is $749 - 695 = 54$.

2.3 $K = k\beta$ and $\beta = D_i/4e_f$ therefore $e_f = k \cdot D_i/4 \cdot K$. Since $D_i = 200\,\mu m$, $k = (200 - 40)/40 = 4$, and $K = 250$, leading to $e_f = 0.8\,\mu m$.

2.4 The volume V leaving a capillary column per second is equal to the internal volume of the column multiplied by the length u. The internal section of the column is given by: $\pi D_i^2/4$,

$$V = u\pi D_i^2/4$$

The flow in ml/min (if u and D_i are in cm) is given by $D = 60\,V$.
If it is chosen to express D_i in mm, this being a value 10 times larger than that given in cm, we must introduce a factor of 0.1 and the expression would then become: $D = 60u\pi D_i^2\,0.01/4$, leading to $D = 0.47uD_i^2$.

2.5 1. We note that the elution order follows accurately the points of the graph. Since the alkenes are more polar than the corresponding alkanes they migrate faster, as would be expected upon a non-polar column.
2. The selectivity factor is known, $\alpha = k_2/k_1$ and therefore the following values are found for the temperatures indicated: 1.12 (at $-35\,°C$), 1.11 (at $25\,°C$), and 1.09 (at $40\,°C$).
3. Although the column remains the same, the retention times will be decreased as a result of the parallel reduction in the coefficient $K = C_S/C_M$ with the temperature.
4. 138 489 theoretical plates are indicated from the application of the formula recalled from exercise 2.7.
5. $H_{min} = 0.113\,mm$.
6. Coating efficiency: 52%.
7. From these three expressions:

$$\ln 0.408 = a/238 + b$$

$$\ln 0.148 = a/298 + b$$

$$\ln 0.117 = a/313 + b$$

we find that $a = 0.00814$ and $b = 0.00493$.

2.6 Knowing that the retention time for butanol is reduced we can calculate the number of equivalent carbons in an alkane which would have the same retention time. We find 6.45. The Kovats factor of the butanol is therefore 645. The corresponding McReynold's constant will be

$$645 - 590 = 55$$

2.7 1. Knowing the retention time and the capacity factor of the second compound we can use the relation, $t_R = t_M(1 + k)$ to calculate the hold-up time, finding: $t_M = 50\,\text{s}$.
2. The formula proposed enables the effectiveness of the column N_2 to be calculated for the second compound, which is: $N_2 = 22\,861$. Subsequently, knowing equally that:

$$N_2 = 5.54 t_R^2 / w_{1/2}^2$$

it is easy to isolate $w_{1/2}$, and to calculate its value: 4.67 s. Bearing in mind the scale of the chromatogram, this time interval corresponds to:

$$(4.67/60) \times 10 = 0.77 \text{ mm}.$$

3   High performance liquid chromatography

3.1 1. The flow of the mobile phase in a column is proportional to its cross-section and to the linear velocity of progression *for identical contents* ($D = s \cdot v$). The cross-section equally varies with the square of the diameter ($s = k \cdot d^2$). Thus,

$$v_2/v_1 = k\, d_1^2 D_2 / k\, d_2^2 D_1$$
$$v_2/v_1 = (1/0.004) \times (0.3/4.6)^2 = 1.06$$

2. The hold-up volume is $V_M = t_M D = 4 \times 4 = 16\,\mu\text{l}$.
3. For the final peak: $t_R = 48\,\text{min}$, thus $V_R = 48 \times 4 = 192\,\mu\text{l}$ or 0.192 ml (around 4 drops!). The first advantage of a narrow column is its economy in the use of solvent.
4. The efficiency of the column, as calculated by the classical formula, gives 67 034 plates.
5. $k = (48 - 4)/4 = 11$.
6. The stationary phase is of a functionalised silica-type, as indicated by different annotations on the base of the chromatogram.
7. The hold-up volumes are in the same ratio as the two cross-sections of the columns

$$v_{0.3}/v_{4.6} = (0.3/4.6)^2 = 0.0043$$

Elsewhere the coefficient K (or the values of k) of the compounds does not change from one column to another, therefore the retention volumes, $V_R = V_M(1 + k)$, remain the same as the hold-up volumes. The retention volume of the narrow column (0.3 mm) is therefore equal to that of the broader column (4.6 mm) × 0.0043 (a factor of 235 times less).

8. Since the volume injected was the same, it will be 235 times more concentrated in the mobile phase of the narrow column and therefore in theory the signal should be 235 times more intense at the maximum of the peak.

The second advantage of narrow columns is that, for detection, they are more sensitive.

3.2 At pH 9, the acids are in the form of their corresponding carboxylate ions. These ions have the tendency to be strongly attracted by the polar aqueous phase. The hydrophobic part (the carbon chain), guards their contact with the stationary phase. The polarity of this chain will decrease in the compound order 1, 3 and 2, which is the order of elution for the compounds.

3.3 1. c – Reverse phase
2. a – Gel permeation
3. d – Ionic chromatography
4. b – Normal phase

3.4 These compounds are very polar and are therefore little retained by a phase of type RP-18. At pH 6, the ATP is the first to be eluted from the column since it is the most polar. ADP will follow and then finally AMP. If, however, an ammonium salt of type phase transfer is added to the mobile phase then dipole–dipole interactions will be created with the three nucleotides. The complex made with the ATP is the strongest and is therefore retained the longest, hence the order of elution is reversed.

3.5 1. If the selectivity factor $\alpha = 1$, it must be that the peaks are superimposed and that the retention times are the same which implies that the respective retention factors k, and therefore $\log k$, also have the same value. The % MeCN is then verified by solving the following expression:

$$-0.0107[\%\text{MeCN}] + 1.5235 = -6.075 \times 10^{-3} [\% \text{ MeCN}] + 1.3283$$

leading to MeCN = 42.2%.

2. The values of $\log k$ can be found from the equations given in the text:

$$\text{Compound } A: k_{30} = 14; k_{70} = 8$$

$$\text{Compound } B: k_{30} = 16; k_{70} = 6$$

It is now possible to calculate values of α for the two compositions. When using 30% of MeCN: $\alpha = 16/14 = 1.1428$ while for 70% of MeCN: $\alpha = 8/6 = 1.333$.

In applying the formula yielding the resolution as a function of k and α: $R = 1/4N^{1/2} (\alpha - 1)/\alpha \cdot k/(k + 1)$, we find that substituting the terms in α and k will yield 0.222 for 70% and 0.1176 for 30%. Therefore it is preferable to utilise the MeCN in a proportion of 70%, which also conveys the advantage of leading to a more rapid separation.

4 Ion chromatography and quantitative analysis

4.1 The stationary phase is of a cationic type. In the dry state it remains in a non-dissociated form but when water is added and in the presence of a high concentration of sodium ions, the following equilibrium is attained in which there is a liberation of H^+:

$$RSO_3H + Na^+ \rightleftharpoons R-SO_3Na + H^+$$

The calculation of the molar capacity of the stationary phase is as follows:

$$25.5 \times 0.105/1\,000 = 2.6775 \times 10^{-3} \text{ mol/g}$$

The equilibrium above is long in establishing itself such that the addition of the sodium hydroxide solution removes all of the protons liberated. A certain time is required for them to reappear in solution.

Note: The *helianthin* $(Me)_2N(C_6H_4)N = N(C_6H_4)-SO_3Na$ is another name for *methyl orange*, which colours red for $pH < 3.2$ and yellow for $pH > 4.4$ ($\lambda_{max} = 505$ nm).

4.2 1. Such a phase corresponds to a polymer ($250\,000 < M < 900\,000$) comprising acid groups, $-CH_2COOH$, and is therefore weakly cationic. The pH of the separation is inferior to the pI of proteins, thus their amino functionality will be fully protonated ($-NH_3^+$).

2. If the pH is increased then the ionic character of the column is reduced leading to a faster elution of the proteins. Here $pI_1 < pI_2 < pI_3$.

4.3 1. The concentration factor is 5 at the beginning since we have 1 ml of plasma and, following treatment of this extract, there will be 200 μl remaining of the solution containing the same amount of cyclosporin as the 1 ml at the beginning.

2. No, because we cannot weigh the collected cyclosporins A or D.

3. a) *Method of internal standard.*

Following convention we will give the letter A to cyclosporin A and D to cyclosporin D as used in the internal standard:

$$k_{A/D} = m_A A_D/m_D A_A = C_A A_D/C_D A_A$$

To calculate the relative response coefficient we will take for example the solution containing 400 ng/ml of cyclosporin A:

$$k_{A/D} = (400/250) \times (1/2.04) = 0.784$$

$$C'_A = C'_D k_{A/D} A'_A/A'_D$$

The chromatogram of the sample for which $A'_A/A'_D = 9/14$ (heights of the peaks measured from the original chromatogram), leads to

$$C'_A = 250 \times 0.784 \times 9/14 = 126 \text{ ng/ml}$$

b) *Method of calibration curve*. Plot the graph of the ratios of the peak heights against concentration C of cyclosporin A.

By the method of least squares we find:

$$R_h = 0.005092C - 0.00411$$

For $R_h = 9/14$, $C = 127.1$ ng/ml.

Note: If we evaluate R_h incorrectly then we commit an error of some importance which is of course translated into the result. However, frequently, for very weak concentrations, we can estimate with respect to a limit below which the error holds less importance, e.g. in taking $R_h = 8/15$ rather than 9/14, the difference will be 20% and we will find that $C = 104$ ng/ml.

4.4 Through the application of the general expression we will have a total of four formulae giving the % mass of the various esters. Below is an example for the methyl butanoate ester:

$$\%_{ME} = 100 \frac{A_{ME} \times 0.919}{A_{ME} \times 0.919 + A_{EE} \times 0.913 + A_{PE} \times 1.06 + A_{BE}}$$

We will solve this by taking and substituting the areas given in the question:

$$\%_{ME} = 16.6$$

$$\%_{EE} = 16.6$$

$$\%_{PE} = 33.4$$

$$\%_{BE} = 33.4$$

Note: The scale in mV corresponds to UV detection.

4.5 1. The addition of N-methylserotonin before the extraction is made removes the necessity of counting any eventual loss of product due to the different intermediate manipulations. We can suppose that the yield of the extraction is the same for the two compounds, which are structurally very similar.
2. To determine the relative response factor $k_{S/NMS}$ of the serotonin (S), with respect to its N-methyl derivative (NMS), the following relation is used:

$$k_{S/NMS} = \frac{m_S}{m_{NMS}} \times \frac{A_{NMS}}{A_S} = \frac{5}{5} \times \frac{30\,956\,727}{30\,885\,982} = 1.002$$

3. Measuring the sample:

$$m_S = m_{NMS} k_{S/NMS} \frac{A'_{NMS}}{A'_S} = 30 \times 1.002 \frac{25\,738\,22}{17\,198\,18} = 45 \text{ ng/ml}$$

which gives a concentration of approximately 45 ppb, for an aqueous solution.

5 Planar chromatography

5.1 1. $R_{F(A)} = 27/60 = 0.45$; $R_{F(B)} = 33/60 = 0.55$

$$N_A = 16 \times 27^2/2^2 = 2\,916; N_B = 16 \times 33^2/2.5^2 = 2\,788$$

$$H_A = x_A/N_A = 9.26 \times 10^{-4}\,\text{cm}; H_B = x_B/N_B = 1.18 \times 10^{-3}\,\text{cm}$$

2. $R = 2(33 - 27)/(2 + 2.5) = 2.67$
3. From the definition for the selectivity factor and from the relation linking R_F and k, we arrive at:

$$\alpha = (R_{F(B)}/R_{F(A)}) \cdot (1 - R_{F(A)})/(1 - R_{F(B)}) = 1.49$$

5.2 1. Since it arises from a normal phase, the more polar compound will be retained the longest. In order of increasing migration distances we will have C, then B and finally A.
2. The order of elution will be A, B then C, reflecting the migration times.
3. Reverse order.
4. Approximate value for $R_{F(A)} = 22.5/50 = 0.45$. HETP $= L/N$ with $N = 5.54x^2/d^2 = 5.54 \times 22.5^2/2^2 = 701$, leading to HETP $= 50/701 = 0.07\,\text{mm}$.

7 Size exclusion chromatography

7.1 In size exclusion chromatography $K <$ or $= 1$, excepting cases where interactions develop between the solute and stationary phase since this creates a partition phenomenon which superimposes itself upon the diffusion in the pores.

7.2 By placing the following two columns end to end, first column C will separate the masses of 3×10^6 and 1.1×10^6 Da from the remaining two. These two masses will subsequently be separated by the column A. On the corresponding chromatogram we will obtain four distinct peaks.

7.3 1. See graph.
2. The total exclusion or interstitial volume is approximately 4.2 ml. The intra-particle volume of the pore is close to: $7.9 - 4.4 = 3.5\,\text{ml}$.
3. For the mass of 3 250 Da, $K = (5.4 - 4.4)/(7.9 - 4.4) = 0.29$

8 Capillary electrophoresis

8.1 1. Since it is known that $\mu_{app} = \mu_{EP} + \mu_{EOS}$ therefore $\mu_{EP} = \mu_{app} - \mu_{EOS}$ and since the experiment allows access to the velocities: v_{app} and v_{EOS}. Then by substitution

$$v_{app} = 24.5/(60 \times 2.5) = 0.163\,\text{cm/s}$$

and

$$v_{EOS} = 24.5/(60 \times 3) = 0.136\,\text{cm/s}.$$

As a result: $v_{EP} = 0.163 - 0.136 = 0.027\,\text{cm/s}.$

$$\mu_{EP} = v_{EP}/E = 0.027 \times 32/30\,000 = 2.88 \times 10^{-5}\,\text{cm}^2\,\text{s}^{-1}\,\text{V}^{-1}.$$

2. $D = l^2/(2Nt_m) = 24.5^2/(2 \times 80\,000 \times 150) = 2.5 \times 10^{-5}\,\text{cm}^2\,\text{s}^{-1}.$

8.2 1. See the scheme in the book.

2. No, since it arises from a non-treated internal surface, which at the pH considered acts as a polyanion. This results in the creation of an electro-osmotic flow. The fact that the compound migrates towards the cathode does not necessarily mean that it is carrying a net positive charge. The possibility remains that it is being trained in that direction, even if carrying a negative charge. Following convention:

3. $l = v_{app}t$ and $\mu_{app} = v_{app}/E = v_{app}L/V$ therefore $\mu_{app} = lL/Vt.$

$$\mu_{app} = 7.5 \times 10^{-4}\,\text{cm}^2\,\text{s}^{-1}\,\text{V}^{-1}$$

4. Following the same reasoning, this leads to $\mu_{EOS} = lL/Vt_m.$

$$\mu_{EOS} = 1.5 \times 10^{-3}\,\text{cm}^2\,\text{s}^{-1}\,\text{V}^{-1}$$

5. $\mu_{EP} = \mu_{app} - \mu_{EOS}$ thus $\mu_{EP} = -7.5 \times 10^{-4}\,\text{cm}^2\,\text{s}^{-1}\,\text{V}^{-1}$

6. The negative character of μ_{app} implies that it originates from a species carrying a net negative charge. The compound will migrate more slowly than a neutral marker.

7. If the internal lining is rendered neutral there would be no more electro-osmotic flux and as a result the compound will no longer migrate towards the cathode but will reappear in the anode compartment.

8. If the pI is 4 for all pH less than 4, then the compound will be in the form of a cation. In this case, the migration time will normally be shorter than for a neutral marker.

9. In using the formula recalled in the question, we will find 337 500 theoretical plates.

10. A small molecule diffuses faster than a larger one. Therefore the effectiveness is greater for molecules of greater mass.

8.3 The two control compounds enable the following simultaneous equations to be written:

$$\log 45\,000 = 1.5a + b \tag{1}$$

$$\log 17\,200 = 5.5a + b \tag{2}$$

The resolution of these two equations lead to $a = -0.1$ and $b = 4.8$. Therefore, $\log M = -0.1v + 4.8$ and by substituting $v = 3.25$ we find

$$\log M = 4.48 \quad \text{and so} \quad M = 29\,854\,\text{Da}.$$

Note that the stationary phase does not behave like an SEC gel since it creates obstacles for the larger molecules which are forced to migrate more slowly than their smaller counterparts.

8.4 If the isoelectric pH of B is superior to that of A then this is because the pH at B oversteps with respect to the position of A. Therefore the pH increases from the right to the left of the capillary.

   If B is displaced further towards the right it will become positively charged and will therefore migrate towards the left and its original position which is stable. The positive pole is therefore to the right and the negative one to the left. The electric field therefore applies itself from right to left on the corresponding diagrams.

8.5

Protein	pI	pH = 3	pH = 7.4	pH = 10
Insulin	5.4	+	−	−
Pepsin	1	−	−	−
Cytochrome C	10	+	+	0
Haemoglobin	7.1	+	−	−
Albumin serum	4.8	+	−	−

9 Nuclear magnetic resonance spectroscopy

9.1 $\gamma = \mu_z/m$ with $m = h/4\pi$, thus $\gamma = 4\pi\mu_z/h = 1.41. \quad 10^{-26} \times 4 \times \pi/(6.6262 \times 10^{-34})$. $\gamma = 2.674 \times 10^8$ rad $T^{-1} s^{-1}$

9.2 The ratio of the populations is: $N_{E(1)}/N_{E(2)} = e^{\Delta E/KT}$ with $\Delta E = \gamma h B_0/2\pi$ from the calculation 1.0000095. If $B_0 = 7\,T$, this ratio becomes 1.0000477.

9.3 1. If on the scale 4 cm represents 200 Hz (then 40 mm = 200 Hz), 7 H will correspond to

$$7/200 \times 40 = 1.4\,mm.$$

2. If the apparatus functions at 200 MHz for ^{1}H, it is $n_C = n_H \cdot \gamma_C/\gamma_H$ for ^{13}C. $\gamma_C = 200 \times 1/3.98 = 50$ MHz and 7 Hz will now correspond to $(7/50) \times 40 = 5.6\,mm.$

9.4 1. The frequency of the apparatus is

$$2.6752 \times 10^8 \times 1.879/(2 \times 3.1416) = 80\,MHz.$$

The chemical shift is therefore: $220/80 = 2.75$ ppm.
2. The shift in Hz will be: $90 \times 200/60 = 300$ Hz.
3. The chemical shifts do not change when they are expressed in ppm.

9.5 The proton spectrum of A presents a doublet of doublets. Two different values for the coupling constants can be noted; one large, the other small ($A = 2$).

   The proton spectrum of B presents a triplet, therefore the proton is located at the same distance from two atoms of fluorine since the value of the coupling constant is weak and can be considered to originate from compound 3 ($B = 3$).

The third isomer gives a triplet whose coupling constant will be of the order of the distance separating the doublets of B, $1.2 \times 60 = 72\,\text{Hz}$. The structure corresponding to this spectral description is isomer 1.

$$1: CHF_2CCl_3 \qquad 2: CHFClCFCl_2 \qquad 3: CHCl_2CF_2Cl$$

9.6 When the nucleus of spin $I = 1/2$ is introduced into the magnetic field B_o, it self-projects into an orientation which follows the rule, $m = (\tfrac{1}{2})/(\tfrac{h}{2}\pi)$. Knowing the value for the spin vector we can write:

$$(\sqrt{3}/2)(h/2\pi)\cos\theta = 1/2(h/2\pi)$$

which is

$$\cos\theta = 1/\sqrt{3} \qquad \text{or} \qquad \theta = 54.7°$$

9.7 In NMR the intensity of the signals reflects the molar concentrations. Vanillin has a molar mass of $152\,\text{g}$, thus for 25 mg of this compound:

$$25 \times 10^{-3}/152 = 1.645 \times 10^{-4}\,\text{mole}$$

Since 1 proton of the vanillin corresponds to a value of integration which is half as strong as that of the proton of the unknown compound, we can deduce that the number of moles of the latter is twice as great, being $1.645 \times 10^{-4} \times 2 = 3.29 \times 10^{-4}\,\text{mole}$ in 0.1 g of sample. The molar mass of the compound is therefore $0.1/(3.29 \times 10^{-4}) = 304\,\text{g/mol}$.

9.8 From the gyromagnetic ratio we are able to calculate that the resonance frequency of the chlorine is $188.255\,\text{MHz}$. Therefore between the two nucleii there are $200 - 188.255 = 11.745\,\text{MHz}$.

As $20\,\text{cm} = 10 \times 200\,\text{Hz} = 2\,000\,\text{Hz}$ and $1\,\text{cm}$ corresponds to $100\,\text{Hz}$, then the distance between the signals will be: $11.745 \times 10^6/100 = 117\,450\,\text{cm}$ or $1\,174.5\,\text{m}!$

9.9 Due to the reduction not going to completion the reaction mixture comprises both acetone and isopropanol. 1 proton of acetone corresponds to $24/6$ while a proton of isopropanol is $8/1 = 8$. The yield will be $R = 100 \times 8/(4 \times 8) = 66.7\%$.

9.10 All of the compounds present in the mixture are submitted to the same conditions. The spectrum recorded corresponds to a superimposition of the spectra of the individual constituents (the dilution of the sample in a solvent does not modify its initial composition). Each constituent leads to either a single or several signals: at 1.6 and 3.3 ppm for C_2H_5I, at 5.2 ppm for CH_2Cl_2 and at 7–7.5 ppm for C_6H_5Br. To find the composition of the mixture we will follow a reasoning close to that used in internal normalisation. An important difference concerns the relative response factors which can be established without undertaking an analysis of a standard reference. A molecule of dichloromethane (2H) will give a response which will be 2/5 that of a molecule of bromobenzene (5H), or iodoethane (5H). If we divide the respective areas indicated for each (84, 22.5, 31.5) by the corresponding number of protons (5, 2, 5), we will have the value (16.8, 11.25, 6.3) which will be proportional to the molar concentrations. Then, knowing the molar masses of the three constituents (156, 84 and 157 g/mol), we can establish the % mass with the aid of three similar formulas, as follows: take for example CH_2Cl_2:

$$\%CH_2Cl = 100 \, [(\text{area } CH_2Cl_2/2) \times 84]/[(\text{area } CH_2Cl_2/2) \times 84$$
$$+ (\text{area } C_2H_5I/5) \times 156 + (\text{area } C_6H_5Br/5) \times 157]$$

$$\%C_2H_5I = 100 \times [16.8 \times 156]/[(16.8 \times 156) + (11.25 \times 84)$$
$$+ (6.3 \times 157)] = 51\%$$

$$\%CH_2Cl_2 = 100 \times [11.25 \times 84]/[(16.8 \times 156) + (11.25 \times 84)$$
$$+ (6.3 \times 157)] = 18.4\%$$

$$\%C_6H_5Br = 100 \times [6.3 \times 157]/[(16.8 \times 156) + (11.25 \times 84)$$
$$+ (6.3 \times 157)] = 30.6\%$$

10   Infrared spectrometry

10.1 1. A wavenumber of $1\,000\,cm^{-1}$ corresponds to a wavelength of $10\,\mu m$ as $E = h\nu = hc/\lambda$. $E = 6.62 \times 10^{-34} \times 3 \times 10^8/(10 \times 10^{-6}) = 1.99 \times 10^{-20}$ J. For one mole we would have 12 000 J (that is 2.87 kcal/mol).

2.
$$\bar{\nu} = \frac{1}{15 \times 10^{-4}} = 666.67 \, cm^{-1}$$

3. The absorbance A is such that: $A = \log 1/T$. If $T = 0.05$, $A = \log 1/0.05 = 1.3$.

10.2 We can accept that the force constant k remains the same for the two species since they are isotopes of the same element. Thus $\mu_H = (12 \times 1)/(12 + 1)$ amu and $\mu_D = (12 \times 2)/(12 + 2)$ amu.

$$\frac{\bar{\nu}_H}{\bar{\nu}_D} = \sqrt{\frac{\mu_D}{\mu_H}}$$

We will find $\bar{\nu}_D = 2\,215\,cm^{-1}$ which represents a difference of less than 2% from the experimental value.

10.3 The approximate mass of the carbonyl group will be:

$$\mu_{CO} = \frac{12 \times 16}{12 + 16} \, 1.66 \times 10^{-27} = 1.138 \times 10^{-26} \, kg.$$

then substituting,

$$k = 4\pi^2 \overline{\nu}^2 c^2 \mu_{CO} = 4\pi^2 \times 1\,710^2 \times (3 \times 10^{10})^2 \times 1.138 \times 10^{-26} = 1\,183\,\text{N/m}$$

(Note the speed of light must be expressed in cm/s here).

10.4 A photon corresponding to $2\,000\,\text{cm}^{-1}$ will transport an energy of

$$E = hc/\lambda = 3.972 \times 10^{-20}\,\text{J}.$$

This energy is transformed into mechanical energy: $E_{TOT} = E_{KIN} + E_{POT}$. At the maximum of the elongation Δx, $E_{KIN} = 0$ because the velocity is zero. The energy of the photon E is therefore entirely in the form of potential energy: $\Delta E_{POT} = 1/2k(\Delta x)^2$. $\Delta E_{POT} = E$ carried by the photon.

$$\Delta x = \sqrt{\frac{2\Delta E}{k}} = \sqrt{\frac{2 \times 3.972 \times 10^{-20}}{1000}} = 8.91 \times 10^{-12}\,\text{m}$$

which is about 6% of a bond whose length is 0.15 nm.

10.5 The two sets of vibrational frequencies have for their origin the two isotopes of chlorine in the sample of HCl (75% of ^{35}Cl and 25% of ^{37}Cl). When the same transition is considered for the two molecules, reunited in the sample, only the reduced mass differs, leading to the separation. The calculation leads to: $\mu_{(35)Cl}/\mu_{(37)Cl} = 0.9985$.

$$\frac{\overline{\nu}_{37}}{\overline{\nu}_{35}} = \sqrt{0.9985} = 0.9992$$

Thus towards $3\,000\,\text{cm}^{-1}$ the difference is $2.3\,\text{cm}^{-1}$.

10.6 A similar calculation as exercise 10.3 leads to $k = 1\,846\,\text{N/m}$.

10.7 1. By linear regression we find: $A/d = 0.0009x + 0.0003$ (1), where x is the VA% in film.
2. By the method of least squares we find: $A_{1030}/A_{720} = 0.0531x + 0.0047$ (2).
3. From equation (1), for the unknown film VA% $= 8.44$; from equation (2) VA% $= 8.46$.

10.8 1. From the application of the Felggett advantage: in calling S/N the signal to noise ratio:

$$\text{S/N} = 10 \times (1/16)^{1/2} = 2.5$$

2. $\Delta = 2x$ therefore $R = 1/2x$ and so $x = 1/2R = 0.5\,\text{cm}$.
3. If $\nu = 15\,800\,\text{cm}^{-1}$, $\lambda = 0.633\,\mu\text{m}$. In passing from one minimum to the next the mirror must be displaced by a distance corresponding to a half-wavelength which is $0.316\,\mu\text{m}$.

11 Ultraviolet and visible spectroscopy

11.1 The application of the expression $E = h.\nu$ for one mole is written $E = N.h.\nu = N.h.c/\lambda$ leading to:

$$E = 6.022 \times 10^{23} \times 6.6262 \times 10^{-34} \times 3 \times 10^{8}/(300 \times 10^{-9})$$

$$= 399\,030\,\text{J},\ 399\,\text{kJ or},\ 399/4.18 = 95.5\,\text{kcal}.$$

11.2 $A = \varepsilon.l.C = 650 \times 7 \times 10^{-4} \times 2 = 0.91$. If the optical path length is doubled the absorbance will equally be doubled: $A = 2 \times 0.91 = 1.82$.

11.3 1. If $T = 0.5$, $A = \log l/0.5 = 0.3$.
  Since, $A = \varepsilon.l.C$, $\varepsilon = 0.3/1.28 \times 10^{-4} = 2\,344\,\text{l}\,\text{mol}^{-1}\,\text{cm}^{-1}$.
  2. If the concentration is doubled, $A = 0.6$, therefore $\log l/T = 0.6$ thus $T = 0.25$ and the percentage transmittance is 25%.

11.4 The concentration of a solution of 0.1 ppm is $0.1 \times 10^{-3}\,\text{g/l}$. The molar concentration is therefore $0.1 \times 10^{-3}/52 = 1.92 \times 10^{-6}\,\text{mol/l}$. We can thus calculate:

$$l = A/(\varepsilon.C) = 0.4/(41\,700 \times 1.92 \times 10^{-6}) = 4.98\,\text{cm}$$

Therefore a cell of 5 cm pathlength is well adapted.

11.5 If 90% of the radiation is absorbed, $T = 0.1$. Thus $A = 1$ and $C = 1/(0.03 \times 15\,000) = 2.22 \times 10^{-3}\,\text{mol/l}$. For $M = 500\,\text{g/mol}$, we find, that $m = 1.11\,\text{g/l}$.

11.6 1. $207 + 12 = 219\,\text{nm}$ (exp $\lambda_{max} = 218$ in EtOH);
  2. $215 + 12 + 12 = 239\,\text{nm}$.
  3. $215 + 60 + 5 + 12 + 18 + 39 = 349\,\text{nm}$ (exp $\lambda_{max} = 348$ in EtOH);
  4. $215 + 5 + 10 + 12 = 242\,\text{nm}$.

11.7 To resolve this problem we use the fact that absorbances are additive:

$$A_T = C_A.l.\varepsilon_A + C_B.l.\varepsilon_B$$

From these two reference solutions we can calculate ε_A and ε_B at 510 nm:

$$\varepsilon_A = A_A/(C_A.l) = 0.714/(1.5 \times 10^{-1}) = 4.760$$

$$\varepsilon_B = A_B/(C_B.l) = 0.298/(6 \times 10^{-2}) = 4.967$$

and calculate the same for ε'_A and ε'_B at 280 nm:

$$\varepsilon'_A = A'_A/(C_A.l) = 0.097/(1.5 \times 10^{-1}) = 0.647$$

$$\varepsilon'_B = A'_B/(C_B.l) = 0.757/(6 \times 10^{-2}) = 12.617$$

By virtue of the addition of absorbances we can now write for the mixture:

$$0.671 = 4.76C_A + 4.967C_B$$

$$0.330 = 0.647C_A + 12.617C_B$$

The resolution of this two equation system leads to:

$$C_A = 1.2 \times 10^{-1}\, \text{mol/l} \quad \text{and} \quad C_B = 2.0 \times 10^{-2}\, \text{mol/l}.$$

11.8 1. See graph.
2. The equation of the graph (absorbance against concentration) is:

$$A = 1.3771C + 0.0024 \qquad (R^2 = 0.9987).$$

3. For an absorbance of 0.45, $C = 0.325\, \text{mol/l}$.

11.9 1. The equation of the calibration graph is: $Y = 0.232C + 0.162$ $(R^2 = 0.9834)$.
2. For a reading of $Y = 3.67$, $C = 15\, \text{mg/l}$ in the sulphate ion.

12   Fluorimetry

12.1 The wavenumber corresponding to a wavelength of 400 nm is $25\,000\, \text{cm}^{-1}$ $(1/400 \times 10^{-7} = 25\,000)$. The Raman peak is displaced towards longer wavelengths and will be located at $25\,000 - 2880 = 22\,120\, \text{cm}^{-1}$. This value corresponds to a wavelength of 452 nm $(10^7/22\,120)$.

12.2 A wavelength of 250 nm corresponds to a wavenumber of $40\,000\, \text{cm}^{-1}$. The Raman peak of water will therefore be situated at $40\,000 - 3380 = 36\,620\, \text{cm}^{-1}$, corresponding to a wavelength of 273.1 nm.

12.3 1. The use of the same control source of fluorescence to compare the measurements made under the same conditions leads to the following simple calculations: the concentration of the unknown sample solution is $(60/40) \times 0.1 = 0.15\, \text{ppm}$ or 150 ppb.
2. We can calculate the difference in cm^{-1} between the light of the source and 456 nm (measurement), to see if the displacement is $3\,380\, \text{cm}^{-1}$. By the introduction of a fluorospectrometer we could equally judge, by studying the spectra, the effects upon the fluorescence of a displacement of the wavelength of excitation of several nanometers.

12.4 1. The benzopyrene has a rigid polycondensed aromatic structure, typical of fluorescent compounds. Forming part of the PAH, it is currently measured by HPLC in drinking water by using a method of detection by fluorescence.
2. The fluorescences being additive, the analysis begins with the correction of the values read for the three solutions by subtracting the value read for the blank. These new values become 21, 35.1 and 29.8. To calculate the concentration of the sample solution one can write: $(1.25 - 0.75)/(35.1 - 21) = (x - 0.75)/(29.8 - 21)$, finding $x = 1.062\, \mu\text{g/ml}$.
3. This value corresponds to the mass of benzopyrene which is found in 1 litre of air $(\rho = 1.3\, \text{g/l})$. The mass concentration is therefore

$$1.06 \times 10^{-6}/1.3 = 0.815\, \text{ppm} \quad \text{or} \quad 815\, \text{ppb}.$$

12.5 The method followed is that of addition by measurement: we will resolve this problem easily. In 1 ml of the solution of iron 5.15×10^{-5} M there are 5.15×10^{-8} moles of Fe. We will therefore have

$$29.6/(5.15 \times 10^{-8} + x) = 16.1/x,$$

which leads to $x = 6.142 \times 10^{-8}$ moles for 2 ml. From this the concentration of the solution of Fe II is 3.07×10^{-5} M.

12.6 1. Hydroxyquinoline and zinc form a chelate (one atom of the metal is inserted between two molecules of the complexant), which can be extracted with tetrachloromethane and which yields fluorescence. The four solutions treated being each in the same manner suggests that we might suppose the yields of the extraction to be identical.
   2. By application of the least squares calculation the following equation can be derived:

$$I_f = 1.5V + 7.74 \qquad \text{(Vml)}$$

   3. For an intensity of fluorescence $I_f = 0$, we will find $V = 5.16$ ml. In 1 litre of the solution B there is 0.0136 g of zinc chloride, which is 6.538 µg/ml of Zn. Therefore 5 ml of unknown solution contains $6.538 \times 5.16 = 33.736$ µg of Zn. Per litre there will be 200 times more, i.e. 6747.2 µg, representing a concentration of 6.75 ppm.

12.7 1. The five standards (1 to 5) correspond to the following concentrations C (in g/l), $(1) = 0.0001$; $(2) = 0.00008$; $(3) = 0.00006$; $(4) = 0.00004$; $(5) = 0.00002$. With these values derived from the method of least squares the equation of the graph is:

$$I_f = 1\,773\,285C + 2.22$$

   2. The sample of drink whose fluorescence is 113 following a dilution by a factor of 1 000 corresponds to a calculated concentration of 6.25×10^{-5} g/l. The drink, non-diluted, yields a concentration of 6.25×10^{-2} g/l which is approximately 62.5 ppm.
   3. When we follow a protocol we make measurements of fluorescence at a fixed wavelength which renders the trace of the spectrum redundant.

13   X-ray fluorescence spectroscopy

13.1 1. The solution is an equilibrated mixture of the elements constituting the sample. In 8 g of KI (166 g/mole), there are:

8 × 39/166 = 1.98 g of potassium and 8 × 127/166 = 6.12 g of iodine.

In 92 g water (18 g/mole), there are:

92 × 2/18 = 10.22 g of hydrogen and 92 × 16/18 = 81.78 g of oxygen.

The balance of the mixture by gram is:

$$\mu_m = (1.98 \times 16.2 + 6.12 \times 36.3 + 1.2 \times 81.78 + 10.22 \times 0.4)/100$$
$$= 3.56 \, cm^2/g.$$

2. $\mu = 3.56 \times 1.05 = 3.738 \, cm^{-1}$ since $P = P_0 \exp[-\mu x]$, thus finding $P/P_0 = 0.024$ which is 2.4%.

13.2 $E = h.c/\lambda$ therefore $\lambda = h.c/E$

$$\lambda_{nm} = 10^9 \times (6.626 \times 10^{-34} \times 2.998 \times 10^8)/(E_{eV} \times 1.602 \times 10^{-19})$$
$$= 1\,240/E_{eV}$$
$$\lambda_{\overset{\circ}{A}} = 1\,240 \times 10/(E_{keV} \times 1\,000) = 12.4/E_{keV}$$

13.3 1. $2d \sin \theta = k.\lambda$ (here $k = 1$)

$$\sin \theta = 8.126/(2 \times 4.404) = 0.9226 \, therefore \, \theta = 67.31°$$

The deviation $2\theta = 134.62°$.
2. $E = h.c/\lambda$ substituting $(6.626 \times 10^{-34} \times 3 \times 10^8)/(0.209 \times 10^{-10}) = 9.517 \times 10^{-15} \, J$ or 59 480 eV.

13.4 1. The crystal sweeps through the angle θ between two extreme values. By the application of $2d \sin \theta = k.\lambda$, if $\theta = 10°$, $\lambda = 0.047 \, nm$ and if $\theta = 75°$, $\lambda = 0.262 \, nm$.
2. The formula for the conversion $\lambda_{\overset{\circ}{A}} = 12.4/E_{keV}$ leads, for the two wavelengths above, to $E = 26.38 \, keV$ and $E = 4.73 \, keV$, respectively.

13.5 1. $\Delta E/E = \Delta\lambda/\lambda$, therefore $\Delta E = E\Delta\lambda/\lambda = (h.c/\lambda).(\Delta\lambda/\lambda) = (h.c/\lambda^2).\Delta\lambda$
$\Delta E = 1.9284 \times 10^{-19} \, J$ or $1.9284 \times 10^{-19}/(1.6 \times 10^{-19}) = 1.2 \, eV$.
2. We will not be able to distinguish these two transitions arising from the sulphur atom.
3. A variation of $2 \times 10^{-4} \, nm$ corresponds to a difference in energy of less than 1 eV, which will be invisible on the spectrum.

13.6 The table of mass attenuation coefficients indicates that for aluminium and the $K\alpha$ of copper, $\mu_m = 264 \, cm^2/g$. Knowing the density of this metal leads us to the linear coefficient:

$$\mu = 264 \times 2.66 = 702 \, cm^{-1}.$$

The fraction transmitted: $P/P_0 = e^{-702 \times 12.10^{-4}} = 0.43$. By consequence 57% of the radiation is absorbed by the aluminium film.

For the emission $K\alpha$ of silver, $\mu_m = 2.54 \, cm^2/g$. By application of the same calculations we will find that $P/P_0 = 0.99$. In this case only 1% of this more energetic radiation will be absorbed.

In initial calculation of the coefficient for the latex ($M = 68 \, g/mole$), we are aided by the values given in the question:

$$\mu_m = (60 \times 25.6 + 8 \times 0.43)/68 = 22.64 \, cm^2/g.$$

The same sequence of calculations will lead to $P/P_0 = 0.89$.

13.7   1. At least one electron is required in the shell 'L' i.e. $n = 2$.
2. This is because the constituents of air absorb weakly energetic radiation of fluorescence X. The helium is almost completely transparent.

13.8   For the X-ray beam to reach the detector with sufficient intensity the incident radiation will be required to hit the crystal over a surface large enough such that thousands of atoms reflect the light in an identical fashion. If the angle of refraction is different from the angle of incidence then successive atoms from the same horizontal reticular plane will return the radiation with an optical displacement which will appear as interference.

To avoid this we observe the reflection through an angle equal to the angle of incidence. The differences of pathway only cause interference due to successive horizontal reticular planes.

13.9   We begin by establishing the ratio Mn/Ba for the two solid standard solutions. Therefore sol $1 = 0.813$ and sol $2 = 0.9625$ we can write:

$$(x - 0.25)/(0.886 - 0.811) = (0.35 - 0.25)/(0.9625 - 0.811)$$

leading to: $x = 0.30\%$.

13.10 The coefficient of mass attenuation for water for the emission $k\alpha$ of Ni is:

$$\mu_m = (16 \times 13.8 + 2 \times 0.43)/18 = 12.31 \, \text{cm}^2/\text{g}.$$

Since $\rho = 1$, $\mu = 12.31 \, \text{cm}^{-1}$. We will find that $P/P_0 = e^{-12.31} = 4.5 \times 10^{-6}$. Therefore 1 cm of water is sufficient to stop almost all radiation of this particular strength.

13.11 The mass attenuation coefficient of air for a radiation whose energy is $2 \, \text{keV}$ is $\mu_m = (0.8 \times 494 + 0.2 \times 706) = 536.4 \, \text{cm}^2/\text{g}$. Since $\rho = 0.0013 \, \text{g/ml}$, $\mu = 0.697 \, \text{cm}^{-1}$ and $P/P_0 = 0.03$. Attenuation attains 97%.

14   Atomic absorption and flame emission

14.1 For a given temperature, $R = N_e/N_o = \gamma \cdot \exp[-\Delta E/kT]$. By combining this equation for the tube values of R such that R_2/R_1 the ratio of the two values of R to $2\,000 \, \text{K}$ and $2\,500 \, \text{K}$ leads to

$$\frac{R_2}{R_1} = \exp\left[\frac{\Delta E}{k}\left(\frac{1}{T_1} - \frac{1}{T_2}\right)\right]$$

For the resonance emission of sodium the value of $\Delta E = h.c/\lambda$ is

$$\Delta E = 3.37 \times 10^{-19} \, \text{J (or 2.1 eV)}.$$

Knowing that $k = 1.38 \times 10^{-23} \, \text{J K}^{-1}$, $R_2/R_1 = 11.5$. The measurement will therefore be almost 12 times more sensitive at 2500 than at 2000 K.

14.2 EDTA yields complexes with a ratio of 1:1 with many metals. Better known among these complexes are those of bivalent cations which lead to hexacoordinated ligands bound through the four acid functions and the two nitrogen atoms. The chelated ions pass more easily to the atomic state in the flame since their volatility is increased.

14.3 Only a small proportion of sodium atoms are excited to the atomic state in the flame, which leads to an underestimate of the quantity of this element in a sample. If one adds potassium salt in a sufficient quantity with respect to the atoms of sodium, then a large number of potassium atoms will be present in the flame. These atoms lead to an oxydo-reduction reaction,

$$Na^+ + K \rightarrow Na + K^+$$

In fact, according to the two ionisation potentials given, less energy is required to remove the least held electron within the potassium atom than for the corresponding electron of sodium.

14.4 The quantity of potassium ions introduced and expressed in moles is:

$$0.2 \times 10 \times 10^{-6} = 2 \times 10^{-6} \text{ mole}$$

If the total volume of the solution has not changed and we call x the number of moles of potassium in the sample: $32.1/x = 58.6/(x + 2 \times 10^{-6})$ from where it can be deduced that $x = 2.42 \times 10^{-6}$ mole. This quantity is present in 0.5 ml of serum, which is:

$$2.42 \times 10^{-6} \times 1\,000/0.5 = 4.84 \times 10^{-3} \text{ mol/l.}$$

14.5 If we call the signal of absorbance of the sample extracted from paprika A_X and the sample solution A_R, m_X and m_R are the corresponding masses of lead and we will have:

$$A_X/A_R = m_X/m_R \quad \text{therefore} \quad m_X = m_R A_X/A_R$$
$$A_X = 1\,220, \; A_R = 1\,000; \quad m_R = 10 \times 0.01/1\,000 = 1 \times 10^{-4} \text{ g}$$

The sample introduced into the graphite furnace contains:

$$m_X = 1 \times 10^{-4} \times 1\,220/1\,000 = 1.22 \times 10^{-4} \text{ g.}$$

The % mass is therefore $1.22 \times 10^{-4}/10^{-2} = 1.22\%$.

14.6 Calcium chloride dihydrate has a molar mass of 147.1 g. The concentration of the parent solution is $1.834/147.1 = 0.01247$ mol/l. Solution diluted 10: 1.247×10^{-3} mol/l. The different solutions made up as standards have the following molar concentrations in Ca^{++}:

Standard 1/20:	0.623×10^{-4} mol/l	signal 10.6
Standard 1/10:	1.247×10^{-4} mol/l	20.1
Standard 1/5:	2.494×10^{-4} mol/l	38.5
Analytical blank:	0 mol/l	1.5

The equation of the calibration curve is: $[\text{signal}] = 148\,900C + 1.407$. We will find that the unknown solution is such that $C = 1.89 \times 10^{-4}$. The parent solution has a concentration 25 times greater, being $4.73 \times 10^{-3}\,\text{mol/l}$, or $0.19\,\text{g/l}$ in Ca^{++}.

14.7 1. This parameter arises from the spectral band which is selected by the exit slit and will reach the detector. This is not the physical width of the exit slit which cannot be less than several micrometres.

2. EDTA has the molecular formula $C_{10}H_{16}N_2O_8$ while the mixed salt of zinc and sodium is $C_{10}H_{12}N_2O_8Na_2Zn$.

   The mass concentration of the parent solution is $35.7 \times 10 = 357\,\text{mg/l}$. The diluted solution has the concentration $C = 2 \times 357/100 = 7.14\,\text{mg/l}$. The apparatus indicates that this solution corresponds to a concentration of $0.99\,\text{mg/l}$ in zinc. As a result the molar mass of the salt is $7.14 \times 65.39/0.99 = 471.6\,\text{g/mol}$.

   The salt in the anhydrous state has a molar mass of $399.6\,\text{g/mol}$. We can therefore deduce that the difference $(471.6 - 399.6) = 72\,\text{g}$ represents the mass of water, per mole, of this salt, which is $72/18 = 4$ moles. We conclude therefore that the hydrate comprises four molecules of water.

3. $A = 0.321C + 0.03$, for $A = 0.369$, $C = 1.14\,\text{mg/l}$, a value too high, which shows that for this measurement the calculation of least squares is not well adapted.

4. The general answer to this question is no, since the experimental points are not aligned such that the range of concentrations extends more than $1\,\text{mg/l}$. The apparatus proposes curves of preference.

15 Atomic emission

15.1 The elements which are measured are often in much smaller concentrations than all those constituting the matrix. Often it is necessary to identify characteristic emissions of trace elements (in concentrations of the order of ppm or less), mixed with elements which equally will yield emissions but whose concentrations can only attain 10 or 20%.

   We observe ionic emissions because the temperatures are very high and the plasma is a medium rich in argon ions and in free electrons which provoke ionisation due to collisions with non-ionised atoms.

15.2 From $E = h.\nu$, and if $\Delta E.\Delta t > h/2\pi$ then $\Delta\nu > 1/2\pi\,\Delta t$. Then $\nu = c/\lambda$, so $\Delta\nu = (c/\lambda^2)\,\Delta\lambda$.

   Therefore,

$$\Delta\lambda \geq \frac{1}{2\pi\Delta t}\,\frac{\lambda^2}{c}$$

If $\lambda = 589\,\text{nm}$ and $\Delta t = 10^{-9}\,\text{s}$, we will find $\Delta\lambda > 1.84 \times 10^{-13}\,\text{m}$ $(1.84 \times 10^{-4}\,\text{nm})$.

Note: The imperfections of spectrometers and the Doppler and Stark effects contribute to an important enlargement of the image of the entrance slit of the apparatus.

15.3 From the data below we can construct the calibration graph of the ratio of the signal emission Pb/Mg against concentration.

Equation of the graph: [ratio Pb/Mg] $= 8.219C + 0.345$

For the two solutions: $A = 0.118$ mg/l and $B = 0.376$ mg/l.

15.4 The expression declares that the transition in energy which corresponds to the emission of resonance is $16\,960\,\text{cm}^{-1}$. This unit is employed to measure the energies $(E = h.c/\lambda)$ with $\lambda = 1/\nu$ and $\lambda = 589.62$ nm. It is the first component of the doublet called the resonance line $(E = 2.102\,\text{eV})$. The second is not a resonance emission.

15.5 Measuring radioactivity using a Geiger counter requires a sufficient number of disintegrations to be accumulated in order to obtain a reliable result. If the half-life of the radionucleus is very long and the element not very abundant then the rhythm of the decompositions will not be significant enough to be distinguished from the background noise. An isolated counting can differ enormously from the average since radioactivity results from a series of random events which do not follow Gaussian behaviour.

Alternately, the method of atomic emission works upon the total atomic population of the isotope.

Example: 1 pCi/l of ^{237}Np (half-life 2.2×10^6 years) corresponds to a population of $N = A/\lambda = 3.7 \times 10^{12}$ atoms, which can be represented as 1.5×10^{-9} g/l (or 1.5 ppt).

Such a concentration is at the limit of the method based upon counting yet nonetheless is reliable for atomic emission spectroscopy (AES).

15.6 The linear dispersion represents the distance which separates, in mm, two wavelengths which differ by 1 nm. If the exit slit is 20 µm $(2 \times 10^{-2}$ mm), the bandwidth reaching the detector will be:

$$1 \times (2 \times 10^{-2})/2 = 1 \times 10^{-2} \text{ nm (10 pm).}$$

16   Mass spectrometry

16.1 If x_n and x_s are fractions of natural and synthetic vanillin, and $\delta_n = -20$; $\delta_s = -30$; $\delta_m = -23.5\%$ the ratio of the isotopes of these two in natural, synthetic and a mixture of these variants, then:

$$x_n + x_s = 1$$

$$\delta_n x_n + \delta_s x_s = \delta_m$$

then $$x_n = (\delta_m - \delta_s)/(\delta_n - \delta_s)$$

Substituting, $x_n = 6.5/10 = 0.65$. The composition found is therefore 65% natural vanillin and 35% of the synthetic variant.

16.2 1. $X = 2 \times (174.97)/(389.42) = 0.899$ g of the element lutetium.

2. The extracted volume of 1 litre contains: $(175.94)/(390.4) \times 20 = 9.013$ µg of ^{176}Lu.

3. Coupled method ICP/MS.

4. The ^{175}Lu is the only isotope whose mass spectrum displays a peak at 175 mass units, while a peak at 176 could be due to either Yb or Hf. These elements constitute isotopic families. We could verify beforehand the absence of peaks for the masses for Yb at 171, 172, 173 and 174, and for Hf at 177, 178 and 179 mass units.

5. Let x be the mass (in µg) of Lu in the extraction of 1 l. This mass, before addition of ^{176}Lu, comprises $0.974x$ of ^{175}Lu and $0.026x$ of ^{176}Lu. Following the addition of ^{176}Lu the mass of ^{176}Lu (in µg) is now $0.026x + 9.013$. Since the analysis indicates that the ratio of the intensities of the peaks is such that ^{175}Lu/^{176}Lu $= 90/10$, then:

$$(0.974x)/(0.026x + 9.013) = 9 \times 174.97/175.90 = 8.95.$$

Thus $x = 108.85$ µg, which leads to a volume of $0.899/108.85 \times 10^{-6} = 8\,260$ l.

16.3 1. $m/e = R^2 B^2/2U$, therefore $B = (2U)^{1/2}(m/e)^{1/2}/R$. For $M = 20$ mass units, $B_{20} = 0.182$ T. For $M = 200$ mass units, $B_{200} = 0.566$ T. The ratio between these two values limiting the field is $B_{200}/B_{20} = 10 = 3.16$.

2. If we sweep by a variation in the electric field then there will be, for higher masses, a voltage ten times weaker resulting in a ten-fold decrease in kinetic energy, lowering the resolving power.

16.4 Intensity of the peak: 720: $I_{720} = 0.989^{60} = 0.515$.
Intensity of the peak: 721: $I_{721} = 0.989^{59} \times 0.011 \times 60 = 0.344$.
The ratio of the two intensities: $I_{721}/I_{720} = 100 \times 0.344/0.515 = 66.7\%$.

16.5 1. Approximate answer: If, following mixing, the peaks 50 and 51 have the same area and if we accept that the intensity of the signal is proportional to the mass of the element (and that the intensity reflects the number of ions formed), we would be able to accept that there is 1 µg of ^{51}V in 2 g of steel, namely 0.5 µg/1 g. The mass concentration is therefore 0.5 ppm.

2. More accurate answer:

$$\left(^{51}V/^{50}V\right)_{mass} = (50.944/49.947)\left(^{51}V/^{50}V\right)_{int} = 1.02(1/1) = 1.02$$

If we call x the quantity of V to be found in 2 g of steel:

$$\left(^{51}V/^{50}V\right)_{mass} = 1.02 = (0.9975x)/(0.0025x + 1)$$

we find $x = 1.025$ µg, or 0.513 ppm.

3. If Ti or Cr were present in the steel, there would be a peak of nominal mass 51 due to Cr while the intensity indicating a mass of 50 would be disturbed by the presence of the Ti.

16.6 1. Calculation of the % mass of each of the two isotopes of copper:

$$\% \,^{63}\text{Cu} = \frac{82\,908 \times 62.9296}{82\,908 \times 62.9296 + 37\,092 \times 64.9278} \times 100 = 68.42$$

$$\% \,^{65}\text{Cu} = \frac{37\,092 \times 64.9278}{82\,908 \times 62.9296 + 37\,092 \times 64.9278} \times 100 = 31.58$$

Note: On the recording of the mass spectra the areas of the peaks are proportional to the population of the corresponding ions. For the isotopes of an element which do not have the same mass, the % masses will be a little different.

2. We have added 250 µl of a solution of ^{65}Cu at 16 mg/l, therefore

$$0.25 \times 16/1\,000 = 4 \times 10^{-3} \text{ mg or 4 mg of } ^{65}\text{Cu}.$$

Following the addition we must calculate, as above, the new % masses of the two isotopes of copper:

$$\% \,^{63}\text{Cu} = \frac{31\,775 \times 62.9296}{31\,775 \times 62.9296 + 79\,325 \times 64.9278} \times 100 = 27.97$$

$$\% \,^{65}\text{Cu} = \frac{79\,325 \times 64.9278}{31\,775 \times 62.9296 + 79\,325 \times 64.9278} \times 100 = 72.03$$

If we call x the mass of the copper in the sample (x comprises ^{63}Cu and ^{65}Cu), we have (in µg):

$$72.03/27.97 = (0.3158x + 4)/0.6842x$$

We find $x = 2.7658\,\mu g$ (in the 250 mg of the original unknown solution to measure). Thus for 1 g, there will be $4 \times 2.7658 = 11.06\,\mu g$. This solution contains 11.06 ppm of copper.

Note: If in place of the two areas we considered the ratio $^{63}\text{Cu}/^{65}\text{Cu} = 2.2352$ before adding and 0.4006 following the addition, the calculations would lead us to the values above.

For example: If y is the % mass in ^{63}Cu:

$$y/(100 - y) = 2.2352\,(62.9296/64.9278) = 2.16641$$

leading to $y = 68.42\%$ and 31.58% of ^{65}Cu.

16.7 1. The precise mass corresponding to the molecular formula $C_{15}H_{12}O$ calculated with the more abundant isotopic masses present is as follows:

$$15 \times 12.0000 + 12 \times 1.007825 + 1 \times 15.994915 = 208.088815 \text{ amu}.$$

The peak $M + 1$ comprises the sum of the three least abundant isotopic species:

a) $^{12}C_{14}\,^{13}C\,^1H_{12}\,^{16}O$
b) $^{12}C_{15}\,^1H_{11}\,^2H\,^{16}O$
c) $^{12}C_{15}\,^1H_{12}\,^{17}O$

2. The two modes of decomposition correspond to a loss of 28 amu. The ions formed ($m/z = 180$), follow the general rule for radical cations of type CHO.

3. By the loss of CO: $m/z = 208.08822 - (12.000 + 15.99492) = 180.0939$. By the loss of C_2H_4, $m/z = 208.08822 - (24.0000 + 4.0313) = 180.05732$.

   Following examination of the position of the peaks we can conclude that the more upright peak corresponds to the ions formed when the parent ion loses CO. The molecular formula of this ion is therefore $C_{14}H_{12}$, while the second peak has the molecular formula $C_{13}H_8O$.

4. The resolving power is defined by $R = M/\Delta M$. In considering the scale of the spectrum, ΔM at the mid-height of the largest peak (FWHM) corresponds to approximately 0.012 amu, leading to $R = 15\,000$.

   In calculating the exact masses of the three isotopomers of the peak $M + 1$ (molecular formulae given in Part One), $a = 209.09217$; $b = 209.095054$ and $c = 209.09321$ amu. The differences between these values are very much smaller than the value of ΔM calculated from the spectrum. Under the conditions of recording the spectrum these three types of molecules would appear superimposed.

16.8 1. For macromolecules which cannot be vaporised, particular modes of ionisation are used which are similar to methods of impact (FAB or MALDI). Equally ionisation is attained in solution at atmospheric pressure before the sample penetrates the MS (processes of electrospray, thermospray or ionspray). The number of elementary charges carried vary from one macromolecule to another. Since the apparatus records the ratio m/z it will appear as a series of molecular peaks for the same compound. Following the resolution of the MS we will be able to determine M by calculation from the two peaks (in species carrying different charges), or from the isotopic distribution molecular weight (in species of the same charge state). The recording allows both types of calculation to be made.

2. a) The ratios m/z of the two principal peaks have values close to 1 224 and 1 429. We can express two relations:

$$m/z_1 = 1\,224 \quad \text{and} \quad m/z_2 = 1\,429.$$

   Furthermore, we might suppose that the number of charges, greater for the peak to the left, differ by one unit from the peak on the right. Therefore, $z_1 = z_2 + 1$. This will lead to three equations for the three unknowns. We begin by calculating z_1, finding $z_2 = 5.97$. Since z must be a whole number we will choose $z_2 = 6$. Following on,

$m = 6 \times 1\,429 = 8\,574$. The ubiquitin has therefore a molecular mass of about 8 574 Da.

b) In this method we will consider an enlargement of the isotopic mass $1\,429.2 < m/z < 1\,430$. Between two successive peaks, m varies by a single unit of mass. We have for example $m/z = 1\,428$ and $(m + 7)/z = 1\,429.2$. So, $7/z = 1.2$ and $z = 5.83$. If we take $z = 6$ and the most intense peak located at $m/z = 1\,428.55$, we find $M = 8\,571$ Da.

17 Isotopic and related methods

17.1 Calling the specific activity of the marker per g A_S, the mass of this marker used in g, m_S, the activity per g after recuperation A_X and the unknown mass (in g) of penicillin in the sample extracted m_X.

$m_S = 1 \times 10^{-2}$ g, $A_S = 75\,000$ Bq/g and $A_X = 10 \times 1\,000/1.5 = 6\,666.7$ Bq/g.

$m_X = 1 \times 10^{-2} \times [(75\,000/6\,666.7 - 1)] = 0.102$ g

In 1 g there is twice as much penicillin, i.e. 0.204 g or 20.4%.

17.2 Following the convention of the previous exercise:

$$m_S = 3 \text{ mg}, \qquad A_S = 3\,100 \text{ Bq/mg},$$
$$A_X = 3\,000/30 = 100 \text{ Bq/mg}$$

where

$$m_X = 3 \times (3\,100 - 100)/100 = 90 \text{ mg or } 9\%.$$

17.3 1. The aqueous sample solution of patulin $(M = 154 \text{ g/mol})$, contains 1.54×10^{-3} g/l of this compound. The concentration is therefore $154 \times 10^{-3}/154 = 1 \times 10^{-5}$ M (or 1.54 ppm).

2. The % absorbance (or inhibition) with respect to tube 1:
   — tube 2: $0.47/1.03 = 45.63\%$
   — tube 3: $0.58/1.03 = 56/63\%$
   — tube 4 (sample): $0.50/1.03 = 48.54\%$

3. The absorbance of tube 1 is greater because there is no analyte introduced into the tube. All of the antibody sites are occupied by the conjugated enzyme and the absorbance is, as a result, more intense.

4. The quantity of patulin in tube 2 (2 ml) is $1 \times 10^{-5}/1\,000 = 1 \times 10^{-8}$ mole, equally 1.54 μg or 770 μg/l (with $\log C = 2.8865$). In tube 3 the molar quantity of patulin is only half, being 0.5×10^{-8} mole, which is 0.77 μg or again 385 μg/l (with $\log C = 2.5854$).

5. Along the x-axis, $x = \log C$ (C being in μg/l), and along the y-axis, the % of inhibition, which will lead to the following expression:

$$(48.54 - 45.63)/(56.63 - 45.63) = (x - 2.5854)/(2.8865 - 2.5854)$$

leading to $x = 2.6651$. The value of C is thus $462.4\,\mu g/l$. The initial solution is twice the concentration and contains $0.93\,mg/l$ ($930\,ppb$) patulin.

17.4 1. $^{35}Cl + {}^{1}n \rightarrow {}^{36}Cl^* \rightarrow {}^{36}Ar^* \rightarrow {}^{36}Ar$ ($\tau_{\beta-} = 3.1 \times 10^5$ years)
       $^{37}Cl + {}^{1}n \rightarrow {}^{38}Cl^* \rightarrow {}^{38}Ar^* \rightarrow {}^{38}Ar$ ($\tau_{\beta-} = 37.3$ min)

2. The γ emissions of ^{38}Cl are preferred since their half-life is short. Counting over a period of time (several hours) and with the aid of software we can identify the fraction of this isotope with respect to the radiation constant of the emitter of long half-life (less intense).

3. Atomic absorption, but possible loss of volatile elements during the treatment of the sample. Fluorescence X would be, essentially, analysis of the surface only.

4. KCl and AgCl have $74.551\,g$ and $143.321\,g$ for their respective molar masses. $2\,g$ of KCl corresponds to $2/74.551 = 0.0268$ mole. If AgCl was recovered totally, its mass would be $143.321 \times 0.0268 = 3.845\,g$. However, we recover $3.726\,g$, representing a mass yield of 97%.

     $50\,ml$ of a solution of $AgNO_3$ ($M = 203.868\,g/mol$, at 15% corresponds to a mass of $7.5\,g$ of this salt, being $7.5/203.868 = 0.037$ mole. We calculate that the quantity of the silver ion is $0.037/0.0268 = 1.38$ times the stoichiometric quantity.

5. The numerical value of the γ count for a total recovery of the isotope ^{35}Cl is $11\,203/0.97 = 11\,549$. The quantity of chlorine in the sample is $11\,203/48\,600 \times 10 = 2.38\,\mu g$, which is $2.38/0.51 = 4.66\,\mu g/g$ (or $4.7\,ppm$).

18   Potentiometric methods

18.1 If we call x the molar concentration of the added NH_4Cl, at $pH = 9$, $[H^+] = 10^{-9}$ and $[OH^-] = 10^{-5}$
As a result: $K_B = [NH_4][OH]/[NH_3] = (x + 10^{-5})(10^{-5})/(9 \times 10^{-3} - 10^{-5})$.
    Then $x = 1.6 \times 10^{-5}\,M$

18.2 Two solutions whose concentrations in H^+ ions are different, being C_{int} and C_{ext}, are found in part of the special glass membrane which constitutes the inner wall.

    — Potential across the internal face: $E_{int} = RT/F \ln(\gamma C_{int})$

    — Potential across the external face: $E_{ext} = RT/F \ln(\gamma C_{ext})$

Thus in the chain of potentials between the two limits the contribution of the membrane:

$$\Delta E_{memb} = 0.059 \log(\gamma C_{ext}) - 0.059 \log(\gamma C_{int})$$

Since C_{int} is constant, and if there are no modifications of other characteristics of the electrodes, then γC_{ext} alone will interfere with the potential of the external face of the membrane. The signal depends upon the pH: 0.059 pH in mV.

18.3 If x is the concentration in H^+ ions:

$$CH_3COOH \rightleftharpoons CH_3COO^- + H^+ \qquad K_a = 1.8 \times 10^{-5}$$

$K_a = x^2/(0.85 - x)$ therefore $x = 0.0039$ and so $H^+ = 0.0039$ M with the pH $= 2.4$. The degree of dissociation will be $(0.0039/0.85) \times 100 = 0.46\%$.

18.4 The reaction is the following:

$$Ti^{+++} + Fe^{+++} \rightarrow Ti^{4+} + Fe^{++}$$

Since we have added 1.5 times the stoichiometric quantity of Fe^{+++} there will be no more Ti^{+++}. Thus $[Ti^{+++}] = 0$.

$$[Ti^{4+}] = (20/1\,000) \times 10^{-3} \times 1\,000/50 = (20/50) \times 10^{-3} = 4 \times 10^{-4} \text{ M}$$

$$[Fe^{++}] = 4 \times 10^{-4} \text{ M}$$

$$[Fe^{+++}] = 2 \times 10^{-4} \text{ M (half of the preceding value)}$$

18.5 $[Cd^{++}] = 0.01$. $E = E_0 - RT/nF \ln[\text{Red}]/[\text{Ox}]$ or $E = E_0 - 0.059/n. \log[\text{Red}]/[\text{Ox}]$

$$E = -0.403 - (0.059/2) \times \log(1/0.01) = -0.462 \text{ V}.$$

18.6 The first measurement made with the sample solution was of the following type:

$$E_1 = E'' + S \log C_x$$

The second was of the following type:

$$E_2 = E'' + S \log(C_x V_x + C_R V_R)/(V_x + V_R)$$

The term $(C_x V_x + C_R V_R)/(V_x + V_R)$ represents the new concentration of the compound measured when we add the volume V_R of concentration C_R to the volume V_x of concentration C_x. Therefore:

$$\Delta E = E_2 - E_1 = S \log(C_x V_x + C_R V_R)/[(V_x + V_R)C_x]$$

being:

$$10^{\Delta E/S} = (C_x V_x + C_R V_R)/[(V_x + V_R)C_x]$$

If we isolate C_x from the second member of the expression above then we refind the formula proposed in the question.

19 Voltametric and coulometric methods

19.1 1. Not quite since approximately 0.5% of the sample is electrolysed.
2. The reproducibility of the phenomenon measured to the level of the drops requires that the diffusion establishes itself in the solution at rest. Stirring

renders the system unstable. However, an electrode which turns regularly can be used.

3. Often the sample is complex and the method of standards is not represen-tative of the effect of the matrix for the metallic species or the organic molecules which can react with the electrodes (reductions for example), producing interference.

19.2 We calculate the ratio of the number of transport of zinc with respect to the number of migrating or diffusing ions present in the solution:

$$t_{Zn^{++}} = (2 \times 5.5 \times 10^{-8} \times 10^{-3})/DC$$
$$DC = (2 \times 5.5 \times 10^{-8} \times 10^{-3}) + (1 \times 7.4 \times 10^{-8} \times 2 \times 10^{-3})$$
$$+ (1 \times 7.9 \times 10^{-8} \times 0.1) + (1 \times 7.6 \times 10^{-8} \times 0.1)$$
$$t_{Zn^{++}} = 6.98 \times 10^{-3}$$

Thus for each coulomb exchanged, the Zn^{++} ion will transport 6.98×10^{-3} C and the rest through the solution $(1 - 6.98 \times 10^{-3}) = 0.993$ C. As a result $i_m/i_D = 7.03 \times 10^{-3}$ or 0.007. The transport of zinc towards the electrode is therefore controlled by the diffusion.

19.3 20 drops (approximately 0.16 g) fall in 80 s. The flow of mercury is therefore:

$$0.16/80 = 2 \times 10^{-3} \text{ g/s}.$$

For a distance three times as high, the flow will be three times as much, i.e. 6×10^{-3} g/s. The drops always have the same mass and follow a rhythm of:

$$(0.16/20)/(2 \times 10^{-3}) = 1.33 \text{ s}.$$

19.4 By application of the equation of Ilkovic (i in μA):

$$i_D = 607 \times 2 \times (8.67 \times 10^{-6})^{1/2} \times 2^{2/3} \times 4^{1/6} \times 1 = 7.15 \ \mu A.$$

Note: D is in cm^2/s; m in mg/s and C in mmol/l.

Different units can be used, such as D in m^2/s, m in kg/s and C in mol/m^3, leading to i_D expressed in amperes.

19.5 The 1 M solution of KCl ready to use is such that $[Zn] = 1$ ppm. If we prepare a 1 M solution with crystallised KCl, then we will have $74.6 \times 5/10^6 = 3.73 \times 10^{-4}$ g of zinc in 74.6 g of this salt found in 1 litre (which is approxi-mately 1 kg), of solution 0.37 ppm of zinc. The ready-to-use solution is there-fore three times more concentrated in zinc than that prepared with crystallised KCl.

19.6 The number of moles of Zn in the experiment: $25/1\,000 \times 2 \times 10^{-8} = 5 \times 10^{-10}$ mole. We would like to deposit 3% i.e: $0.03 \times 5 \times 10^{-10} = 1.5 \times 10^{-11}$. It requires therefore 3×10^{-11} moles of electrons, representing a charge of:

$$Q = 96\,500 \times 3 \times 10^{-11} = 2.895 \times 10^{-6} \text{ C}$$

Since $Q = it$, $t = 2.895 \times 10^{-6}/1.5 \times 10^{-9} = 1.93 \times 10^3$ s which is 32.16 min.

Note: A current of 1.5 nA is practically undetectable. However, if we employ stripping voltametry, we will redissolve this quantity of zinc in approximately 1 s, which will then produce a much easier signal to detect:

$$i = 2.895 \times 10^{-6}/1 = 2.895 \ \mu A.$$

19.7 Standardising the reactant: following the question, 1 ml of this solvent has a water content of 3/15 ml of KF reagent. In oxalic acid dihydrate ($M = 126$ g/mol), the mass concentration of water is $36/126 = 28.57\%$. The information given permits the following calculation to find the titre of the reactant:

$$T = 0.2857 \times 205 \times 1/13 = 4.51 \ mg/ml$$

Measurement: 10 ml of this solvent neutralises $10 \times 3/15 = 2$ ml of reactant, thus 1.05 g of powdered milk will react with $12 - 2 = 10$ ml of this same reagent. There are therefore $4.51 \times 10 = 45.1$ mg of water in the sample, which is a concentration of

$$(45.1/1\,050) \times 100 = 4.3\% \ mass.$$

19.8 In the coulometric version of KF titration 1 molecule of water requires 2 atoms of iodine and 2 electrons. Therefore $2 \times 96\,500$ coulombs (C), are required for one mole, which is 1.8×10^4 mg of water. 1 C corresponds to $1.8 \times 10^4/(2 \times 96\,500) = 0.0933$ mg of water. This then corresponds to the quantity of water of 1 ml of ether. There are therefore 93 mg of water per litre. Considering the density of ether (0.78 g/ml), the mass concentration will be:

$$93 \times 1/0.78 = 120 \ mg/kg \ (120 \ ppm)$$

19.9 Two electrons are required to reduce a zinc ion. The quantity of current used in the experiment is $15 \times 10^{-6} \times 5 \times 60 = 4.5 \times 10^{-3}$ C, which will effectively reduce:

$$4.5 \times 10^{-3}/(2 \times 96\,500) = 2.33 \times 10^{-8} \ mole \ of \ zinc$$

knowing that at the beginning of the experiment there were $(20/1\,000) \times 1 \times 10^{-3} = 2 \times 10^{-5}$ mole of zinc. The impoverishment is therefore:

$$(2.33 \times 10^{-8})/(2 \times 10^{-5}) \times 100 = 0.12\%$$

21 Basic statistical parameters

21.1

Chemist	Average value (x)	Standard deviation (s)	ε	RSD %	Conclusion
Chemist A	131.6	1.56	0.3	1.18	Correct and precise
Chemist B	131.6	5.37	0.3	4.08	Correct yet imprecise
Chemist C	135.7	1.33	3.8	0.98	Incorrect yet precise
Chemist D	125.3	9.93	6.6	7.93	Incorrect and imprecise

21.2 The average value is 650 with $s = 1.581$. For the level of confidence indicated, the value of $t = 2.776$. We can then calculate:

$$t.s/(n)^{1/2} = 1.963$$

The results determine a range of 650 ± 1.963 in which we have a 95% chance of finding the true average. There is probably a systematic error in these experiments. However, if we fixed a level of confidence of 99% ($t = 4.6$), we would have $s/(n)^{1/2} = 3.25$ and therefore a range of 650 ± 3.25. The value of 653 would be included in this interval and would thus be considered as a viable result.

21.3 $F = (s_1/s_2)^2 = 11.88$. For two series of measurements, the frontier value is 5.05. Therefore the precision of the two apparatus is significantly different.

21.4 The value of t calculated (with $n = 6$) for the chemist A ($s = 1.559$) is 0.471. This value is smaller than those presented in the table of t values: 2.57 (for 95%), and 4.03 (for 99%). There is probably not a systematic error. However, for chemist C ($s = 1.325$), we will find $t = 4.05$, a value which indicates a high likelihood of a systematic error.

21.5 The value of $Q = (24.8 - 24.36)/(24.8 - 24.10) = 0.63$ is inferior to that which is found in the table for five measurements and a level of confidence of 95% (value 0.64). We should therefore not reject the value 24.8. However, if we recalculate following the addition of the two new values indicated, Q remains unchanged but in the table of seven values we have 0.51. The value 24.8 is now rejected.

Note: The values of s and of the average in including ($s = 0.239$ and average 24.29), or by rejecting ($s = 0.095$ and average 24.21) these measurements are very different. If we take the middle values we have 24.24 (with), and 24.22 (without). In this case the middle seems preferable to the average.

21.6 The pooled standard deviation s_p should be calculated, as follows:

$$s_p^2 = \frac{6 \times 0.3^2 + 6 \times 0.2^2}{14 - 2} = 0.2549^2 \qquad t = \frac{3}{0.2549}\sqrt{\frac{7^2}{14}} = 22$$

In the table $t = 2.2$, therefore the two methods do not produce at the same result.

21.7 The problem is centred around the comparison of two averages. That for the first percentage purity is 99%, while the average value for the four analyses reported is 98.73 with $s_{n-1} = 0.155$. The 'pooled standard deviation' of the two series of values is:

$$s_p = \{(4 \times 0.08^2 + 3 \times 0.155^2)/7\}^{1/2} = 0.118$$

The value of t based upon $(5 + 4 - 2)$ degrees of freedom leads, according to the table, to 2.365 for a level of confidence of 95%.

We must next calculate $2.365 \times 0.118 \times \{(4+5)/(4 \times 5)\}^{1/2} = 0.187$ and compare this value with the difference in the averages of the two series, $99 - 98.73 = 0.23$. The result of this comparison appears rather large and therefore leads to the conclusion that the two averages can be considered to be incompatible. Thus the original value will not be retained.

21.8 If we consider that the law of variation of the absorbance with the concentration is a straight line then this will have the equation: $A = 0.05 \, [\text{conc}] + 0.08$. The differences are relatively large. A quadratic adjustment would be preferred and across a narrower range of concentrations.

21.9 For the blank, $s_{n-1} = 0.82$. The value of t calculated for $5 + 8$ measurements is 3.17. Therefore $\Delta x = 3.17 \times 0.82 \times [(5+8)/(5 \times 8)]^{1/2} = 1.48$. The limit of the detection is around 1.5 mg.

APPENDIX

List of acronyms

AAS	Atomic absorption spectroscopy
AC	Alternative current
ADC	Analogic-digital converter
AED	Atomic emission detector
AES	Atomic emission spectroscopy
API	Atmospheric pressure ionisation
ASTM	American society for testing material
ATR	Attenuated total reflectance
BSA	Bovine serum albumin
CCD	Charge-coupled device
CE	Capillary electophoresis
CEC	Capillary electrochromatography
CGE	Capillary gel electrophoresis
CI	Chemical ionisation
CID	Collision-induced dissociation
CIEF	Capillary isoelectric focusing
CZE	Capillary zone electrophoresis
CW	Continuous wave
DAD	Diode array detector (or detection)
DPP	Differential pulse polarography
DTGS	Deuterated triglycine sulfate
ECD	Electron capture detector
EDTA	Ethylenediamine-tetracetic acid
EI	Electron impact (or ionisation)
ELISA	Enzyme linked immuno sorbent assay
EOF	Electro-osmotic flow
ESCA	Electron spectroscopy for the chemical analysis
FAB	Fast atom bombardment
FES	Flame emission spectroscopy
FID	Flame ionisation detector
FID	Free induction decay
FPD	Flame photometry detector
FT	Fourier transform
FTIR	Fourier transform infrared
FTMS	Fourier transform mass spectrum
FWHM	Full width at half maximum
GC	Gas chromatography
GC-ICP	Gas chromatography-inductively coupled plasma
GLP	Good laboratory practice
HPCE	High performance capillary electrophoresis
H_2S	Hydrogen sulfide
HCL	Hollow cathod lamp
HClO	perchloric acid

HETP	Height equivalent to a theorical plate
HF	Fluorhydric acid
HOMO	Highest occupied molecular orbital
HPLC	High performance liquid chromatography
HPTLC	High performance thin layer chromatography
IC	Ion chromatography
ICP	Inductively coupled plasma
ICP-MS	Inductively coupled plasma-mass spectrometry
IEA	Immunoenzymological assay
IEC	Internal electron capture
IFA	Immunofluorescence assay
IR	Infrared
ISE	Ionic selective electrode
KRS-5	Thallium-bromide-iodide
LIDAR	Light detection and ranging
LOMO	Lowest occupied molecular orbital
MALDI	Matrix-assisted Laser desorption/ionisation
MCA	Multi-component analysis
MCT	Mercury cadmium tellure
MEKC	Micellar electrokinetic capillary chromatography
MIKE	Mass ion kinetic energy
MS	Mass spectrometry
MS-MS	Tandem mass spectrometry
MSD	Mass spectrometer detector
NAA	Neutron activation analysis
NADH	Nicotinamide adenine dinucleotide, reduced form
NMR	Nuclear magnetic resonance
NPD	Thermoionic detector
NPP	Normal pulse polarography
OES	Optical emission spectrophotometry
PAH	Polynuclear aromatic hydrocarbons
PMT	Photo multiplier tube
POPOP	1,4-Bis(5-phenyloxazol-2-yl)benzene
ppb	parts per billion
ppm	parts per million
PPO	2,5-diphenyloxazole
ppt	parts per trillion
RIA	Radio-immunoassay
RSD	Relative standard deviation
SCE	Standard calomel electrode
SEC	Size exclusion chromatography
SFC	Supercritical fluid chromatography
SFE	Supercritical fluid extraction
SEM	Scanning electron microscope
SPE	Solid-phase extraction
SPME	Solid-phase micro-extraction
TIC	Total ion chromatogram
TISAB	Total ionic strength adjustment buffer
TLC	Thin layer chromatography
TMS	Tetramethylsilane
TOF	Time of flight
UV	Ultraviolet
VOC	Volatile organic compound
WCOT	Wall coated open tubular

Bibliography

Baker, D.R. (1995) Capillary Electrophoresis, John Wiley, ISBN 0-471-11763-3.

Christian, G. (1994) Analytical Chemistry (5th edn), John Wiley, ISBN 0-471-59761-9.

Christian, G.D. and O'Reilly, J.E. (1986) Instrumental Analysis (11th edn), Allyn & Bacon Intl, ISBN 0-205-08685-3.

Crosby, N.T., Davy, J.A., Hardcastle, W.A., Holcombe, D.G. and Treble, R.D. (1995) Quality in the Analytical Chemistry Laboratory, ACOL series, ISBN 0-471-95470-5.

Day, R.A. and Underwood, A.L. (1991) Quantitative Analysis, Prentice Hall, ISBN 0-13-747361-3.

De Graeve, J., Berthou, F. and Prost, M. (1986) Methodes Chromatographiques Couplees à la Spectrometrie de Masse, Masson, ISBN 2-225-80627-6.

De Hoffmann, E., Charette, J. and Stroobant, V. (1996) Mass Spectrometry: Principles and Applications, John Wiley, ISBN 0-471-96697-5.

Deportes, C. (1994) Electrochimie des Solides, Presses Universitaires de Grenoble, ISBN 2-7061-0585-2.

Fried, B. and Sherma, J. (1996) Practical Thin-Layer Chromatography, Springer-Verlag, ISBN 0-8493-2660-5.

George, W.O. and McIntyre, P.S. (1987) Infrared Spectroscopy, John Wiley, ISBN 0-471-91389-9.

Gunther, H., Suffert, J.-J. and Ourisson, G. (1994) La Spectroscopie de RMN, Masson, ISBN 2–225-84029-6.

Kellner, R., Mermet, J.M., Otto, M. and Widmer, H.M. (1998) Analytical Chemistry Wiley-VCH, ISBN 3-527-28881-3.

Maurice, J. (1993) Jugement Statistique sur Echantillons en Chimie, Polytechnica, ISBN 2-84054-013-4.

Miller, J.C. and Miller, J.N. (1993) Statistics for Analytical Chemistry (3rd edn), Ellis Horwood/Prentice Hall, ISBN 0-13-030990-7.

Murray, R. (1992) Instrumentation in Analytical Chemistry, Louise Voress, ISBN 0-8412-2202-9.

Rosset, R., Caude, M. and Jardy, A. (1991) Chromatographie en Phase Liquide et Supercritique, Masson, ISBN 2-225-82308-1.

Sandra, P. (1989) High Resolution Gaz Chromatography (3rd edn), in (ed.) Hyver, K.J., Hewlett-Packard Co, ISSN 5950-3562.

Skoog, D.A., Holler, F.J. and Nieman, T.A. (1998) Principles of Instrumental Analysis, Saunders College Publishing, ISBN 0-03-002078-6.

Skoog, D.A., West, D.M. and Holler, F.J. (1996) Fundamentals of Analytical Chemistry (7th edn), Saunders College Publishing, ISBN 0-03-005938-0.

Smith, M. and Busch, K.L. (1999) Understanding Mass Spectra: A Basic Approach, John Wiley, ISBN 0-471-29704-6.

Techniques de l'ingenieur (various years) Analyse Chimique et Caracterisation, Vol. P1, P2, P3, P4, Istra, ISSN 0245-9639.

Tranchant, J. (1994) Manuel Pratique de Chromatographie en Phase Gazeuse, Masson, ISBN 2-225-84681-2.

Table of physico-chemical constants

Physical constants

Quantity	Symbol	Value	Units
Speed of light in vacuo	$c = (\varepsilon_0 \mu_0)^{-1/2}$	2.9979×10^8	m/s
Vacuum permittivity	ε_0	8.8542×10^{-12}	F/m
Vacuum permeability	μ_0	$4\pi \times 10^{-7}$	H/m
Proton charge	e	1.6022×10^{-19}	C
Electron rest mass	m_e	9.1095×10^{-31}	kg
Proton rest mass	m_p	1.6726×10^{-27}	kg
Neutron rest mass	m_n	1.6750×10^{-27}	kg
Atomic mass unit	amu	1.6605×10^{-27}	kg
Planck's constant	h	6.6262×10^{-34}	J s
Planck's constant	$\hbar = h/2\pi$	1.0546×10^{-34}	J s
Avogadro number	N_A	6.0221×10^{23}	/mol
Faraday constant	$F = N_A e$	9.6485×10^4	C/mol
Boltzmann constant	k_B	1.3807×10^{-23}	J/K
Gas constant	$R = N_A k_B$	8.3145	J/K mol
Gravitational constant	G	6.6720×10^{-11}	N m^2/kg^2

Unit conversion
$1\,\text{Å} = 10^{-10}$ m
$1\,\text{eV} = 1.6022 \times 10^{-19}$ J

$1\,\text{cm}^{-1} = 2.9979 \times 10^{10}$ Hz
$p^{\ominus} = 1\,\text{atm.} = 1.0133 \times 10^5$ Pa: $1\,\text{bar} = 10^5$ Pa

Prefixes

Symbol	f	p	n	μ	m	c	d	k	M	G	T
Name	femto	pico	nano	micro	milli	centi	deci	kilo	mega	giga	tera
Factor	10^{-15}	10^{-12}	10^{-9}	10^{-6}	10^{-3}	10^{-2}	10^{-1}	10^3	10^6	10^9	10^{12}

Index